Birkhäuser

Control Engineering

Series Editor
William S. Levine
Department of Electrical and Computer Engineering
University of Maryland
College Park, MD
USA

For further volumes:
http://www.springer.com/series/4988

Yuri Shtessel • Christopher Edwards
Leonid Fridman • Arie Levant

Sliding Mode Control
and Observation

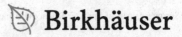 Birkhäuser

Y. Shtessel
Department of Electrical and Computer
  Engineering
University of Alabama in Huntsville
Huntsville, AL, USA

C. Edwards
College of Engineering, Mathematics
  and Physical Science
University of Exeter
Exeter, UK

L. Fridman
Department of Control
Division of Electrical Engineering
Faculty of Engineering
National Autonomous University of Mexico
Mexico

A. Levant
Department of Applied Mathematics
School of Mathematical Sciences
Tel-Aviv University
Israel

ISBN 978-1-4899-9122-5   ISBN 978-0-8176-4893-0 (eBook)
DOI 10.1007/978-0-8176-4893-0
Springer New York Heidelberg Dordrecht London

Mathematics Subject Classifications (2010): 93B12, 93C10, 93B05, 93B07, 93B51, 93B52, 93D25

Printed on acid-free paper

Springer is part of Springer Science+Business Media (www.birkhauser-science.com)

We dedicate this book with love and gratitude to
  Yuri's wife Nina,
  Chris' parents Shirley and Cyril,
  Leonid's wife Millie,
  Arie's wife Irena.

Preface

Control in the presence of uncertainty is one of the main topics of modern control theory. In the formulation of any control problem there is always a discrepancy between the actual plant dynamics and its mathematical model used for the controller design. These discrepancies (or mismatches) mostly come from external disturbances, unknown plant parameters, and parasitic dynamics. Designing control laws that provide the desired closed-loop system performance in the presence of these disturbances/uncertainties is a very challenging task for a control engineer. This has led to intense interest in the development of the so-called robust control methods, which are supposed to solve this problem. In spite of the extensive and successful development of robust adaptive control [159], $\mathcal{H}_\infty$ control [48], and backstepping [121] techniques, sliding mode control (SMC) remains, probably, the most successful approach in handling bounded uncertainties/disturbances and parasitic dynamics [67, 182, 186].

Historically sliding modes were discovered as a special mode in variable structure systems (VSS). These systems comprise a variety of structures, with rules for switching between structures in real time to achieve suitable system performance, whereas using a single fixed structure could be unstable. The result is VSS, which may be regarded as a combination of subsystems where each subsystem has a fixed control structure and is valid for specified regions of system behavior. It appeared that the closed-loop system may be designed to possess new properties not present in any of the constituent substructures alone. Furthermore, in a special mode, named a sliding mode, these properties include insensitivity to certain (so-called matched) external disturbances and model uncertainties as well as robustness to parasitic dynamics. Achieving reduced-order dynamics of the compensated system in a sliding mode (termed partial dynamical collapse) is also a very important useful property of sliding modes. One of the first books in English to be published on this subject is [85]. The development of these novel ideas began in the Soviet Union in the late 1950s.

The idea of SMC is based on the introduction of a "custom-designed" function, named the sliding variable. As soon as the properly designed sliding variable becomes equal to zero, it defines the sliding manifold (or the sliding surface). The proper design of the sliding variable yields suitable closed-loop system performance while the system trajectories belong to the sliding manifold. The idea of SMC is to steer the trajectory of the system to the properly chosen sliding manifold and then maintain motion on the manifold thereafter by means of control, thus exploiting the main features of the sliding mode: its insensitivity to external and internal disturbances matched by the control, ultimate accuracy, and finite-time convergence of the sliding variables to zero.

The first well-cited text in English on SMC was by Itkis and published in 1976 [113]. By 1980, the main contributions in SMC theory had been completed and subsequently reported in Utkin's 1981 monograph (in Russian) and its subsequent English version [182]. A comprehensive review was published by DeCarlo et al. in [56]. In these publications (see also the advanced results presented in the later works [186] and [67]), the two-step procedure for SMC design was clearly stated.

The first step involves the design of a switching function so that the system motion on the sliding manifold (termed the sliding motion) satisfies the design specifications. The second step is concerned with the selection of a control law, which will make the sliding manifold attractive to the system state in the presence of external and internal disturbances/uncertainties. Note that this control law is not necessarily discontinuous.

SMC-based observers allow estimation of the system states in the presence of unknown external disturbances, which can also be explicitly reconstructed online by an observer.

Control chattering still remained a problem impeding SMC implementation. Addressing control chattering was the main motivation for the emerging so-called second-order sliding. Thus, the already matured conventional SMC theory received a significant boost in the middle of the 1980s: when new "second-order" ideas appeared [132] and then, in the beginning of 2000s, when "higher-order" [124] concepts were introduced. The introduction of these new paradigms was dictated by the following reasons:

1. The conventional sliding mode design approach requires the system relative degree to be equal to one with respect to the sliding variable. This can seriously constrain the choice of the sliding variable.
2. Also, very often, a sliding mode controller yields high-frequency switching control action that leads to the so-called chattering effect, which is difficult to avoid or attenuate.

These intrinsic difficulties of conventional SMC are mitigated by higher-order sliding mode (HOSM) controllers that are able to drive to zero not only the sliding variable but also its $k - 1$ successive derivatives (kth-order sliding mode). The novel approach is effective for arbitrary relative degrees, and the well-known chattering effect is significantly reduced, since the high-frequency control switching is "hidden" in the higher derivative of the sliding variable.

When implemented in discrete time, HOSM provides sliding accuracy proportional to the kth power of the sampling time, which makes HOSM an enhanced-accuracy robust control technique. Since only the kth derivative of the sliding manifold is proportional to the high-frequency switching control signal, the switching amplitude is well attenuated at the sliding manifold level, which significantly reduces chattering.

The unique power of the approach is revealed by the development of practical arbitrary-order real-time robust exact differentiators, whose performance is proved to be asymptotically optimal in the presence of Lebesgue-measurable input noises. The HOSM differentiators are used in advanced HOSM-based observers for the estimation of the system state in the presence of unknown external disturbances, which are also reconstructed online by the observers. In addition HOSM-based parameter observers have been developed as well.

The combination of a HOSM controller with the above-mentioned HOSM-based differentiator produces a robust and exact output-feedback controller. No detailed mathematical models of the plant are needed. SMC of arbitrary smoothness can be achieved by artificially increasing the relative degree of the system, significantly attenuating the chattering effect. For instance, the continuous control function can be obtained if virtual control in terms of the control derivative is designed in terms of SMC. In this case, the control function will be continuous, since it is equal to the integral of the high-frequency switching function. In the case of parasitic/unmodeled dynamics the SMC function will switch with lower frequency (the control chattering). Designing the SMC in terms of the derivative of the control function yields chattering attenuation.

The practicality of conventional SMC and HOSM control and observation techniques is demonstrated by a large variety of applications that include DC/DC and AC/DC power converters, control of AC and DC motors and generators, aircraft and missile guidance and control, and robot control.

SMC is a mature theory. This textbook is mostly based on the class notes for the graduate-level courses on SMC and Nonlinear Control that have been taught at the Department of Electrical and Computer Engineering, the University of Alabama in Huntsville; at the Department of Engineering, the University of Leicester; at the Department of Control Engineering and Robotics, the Engineering Faculty, the National Autonomous University of Mexico and at the Department of Applied Mathematics, the Tel Aviv University for the last 10–15 years. The course notes have been constantly updated during these years to include newly developed HOSM control and observation techniques.

This textbook provides the reader with a broad range of material from first principles up to the current state of the art in the area of SMC and observation presented in a pedagogical fashion. As such it is appropriate for graduate students with a basic knowledge of classical control theory and some knowledge of state-space methods and nonlinear systems. The resulting design procedures are emphasized using Matlab/Simulink software.

Fully worked out design examples are an additional feature. Practical case studies, which present the results of real sliding mode controller implementations, are used to illustrate the successful practical application of the theory. Each chapter is equipped with exercises for homework assignments.

The textbook is structured as follows.

In Chap. 1 we "intuitively" introduce the main concepts of SMC for regulation and tracking problems, as well as state and input observation using only basic control system theory. The sliding variable and SMC design techniques are demonstrated on tutorial examples and graphical expositions. The reaching and sliding phases of the compensated system dynamics are identified. Advanced concepts associated with conventional sliding modes, including sliding mode observers/differentiators and second-order sliding mode controllers, are studied on a tutorial level. Robust output tracking controller design based on a relative degree approach is studied. The design framework comprises conventional and second-order sliding modes as well as sliding mode observers. The main advantages of SMC and HOSM control, including robustness, finite-time convergence, and reduced-order compensated dynamics, are demonstrated through numerous examples and simulation plots.

In Chap. 2 we formulate and rigorously study the conventional multivariable SMC problem using linear algebra and Lyapunov function techniques. The interpretation of the sliding surface design problem as a straightforward linear state-feedback problem for a particular subsystem is emphasized. A variety of methods for sliding surface design, including linear quadratic minimization and eigenvalue placement algorithms, are presented. Possible control design strategies to enforce a sliding motion, including the unit-vector control structure, are described, and the problem of smoothing undesirable discontinuous signals is addressed. The output-feedback SMC techniques that do not require measurement of the system states are presented. Integral sliding modes (ISM) that are a special type of conventional SMC are discussed in detail. The ability of ISM to be initiated without a reaching phase is emphasized. The specific property of ISM that consists of retaining the order of the compensated system is studied. The use of ISM for disturbance compensation is discussed together with a linear quadratic regulation (LQR) problem, "robustified" via ISM. Several examples illustrate the ISM concept.

In Chap. 3 a detailed coverage of conventional sliding mode observers (CSMOs) for state estimation and unknown input reconstruction in dynamic systems is presented. The design techniques for a variety of CSMOs are rigorously studied using linear algebra and Lyapunov function techniques. The robustness properties of CSMO are discussed. Several examples illustrate the CSMO design and demonstrate their performance via simulations. The chapter ends with a list of exercises for homework assignments.

In Chap. 4 second-order sliding mode (2-sliding mode or 2-SM) control is studied as a new generation of conventional SMC. The main definitions, properties and design frameworks for 2-SM control, and associated observers/differentiators are rigorously presented. The essential properties of 2-SM control, including finite-time convergence to zero of the sliding variable and its derivative in the presence of disturbances/uncertainties as well as the ability of computer-implemented 2-SM control to provide enhanced stabilization accuracy that is proportional to the square of the time increment, are emphasized. Several particular types of 2-SM control algorithms, including twisting and super-twisting controllers, the suboptimal control algorithm, the control algorithm with prescribed convergence law, and the quasi-continuous control algorithm, are introduced. A special case of 2-SM, super-twisting SMC with variable gains, is also studied analytically and experimentally. An output regulation problem solution is described in terms of the above-mentioned 2-SM controllers. In particular the chattering attenuation capabilities of 2-SM controllers are emphasized. Numerous examples illustrate the advantages in terms of performance of 2-SM controllers. The chapter culminates with a list of exercises for homework assignments.

In Chap. 5 we study a very important robustness property of conventional SMC and 2-SM-based controllers to parasitic dynamics using frequency-domain techniques. The describing function technique is used to estimate both amplitude and frequency of the switching control oscillation as soon as the transient response is over. The robustness of conventional SMC to first- and second-order parasitic dynamics is described. The analysis of oscillations with finite amplitude and frequency in 2-SM controllers, including the twisting and super-twisting controllers and the quasi-optimal controller, in the presence of first- and second-order parasitic dynamics, is performed. Numerous examples illustrate the performances of conventional SMC and 2-SM-based controllers. Exercises for homework assignment are presented at the end of the chapter.

In Chap. 6 the concept of 2-SM control is generalized by introducing HOSM control that is a new generation of SMC. The ability of HOSM control to drive the sliding variable and its $k - 1$ successive derivatives (a so-called kth-order sliding mode) to zero in finite time is rigorously derived and discussed. Two families of HOSM control algorithms, a nested SMC algorithm and a quasi-continuous control algorithm, are introduced. Homogeneity and contractivity-based techniques that are used for HOSM control analysis and design are described. The efficacy of HOSM control for systems with arbitrary relative degree with respect to the sliding variable is identified. Significant attenuation of the well-known chattering effect via HOSM control is described. The HOSM-based arbitrary-order online robust exact differentiator is introduced and discussed. Several examples are presented to illustrate the performance of HOSM controllers and differentiators. The application of the HOSM controllers and differentiators to blood glucose regulation, using an insulin pump in feedback, illustrates the HOSM algorithms. A list of exercises for homework assignments completes the chapter.

In Chap. 7 we revisit the state observation and identification problem, previously studied in Chap. 3. In this chapter state observation, identification, and input reconstruction are discussed using algorithms based on HOSM exact differentiators. HOSM observers for nonlinear systems are described. Parameter identification algorithms using HOSM techniques are presented and discussed. Several examples, including pendulum and satellite dynamics estimation and identification, are presented to illustrate the performance of the HOSM observation and identification algorithms. The exercises for homework assignments are presented at the end of the chapter.

In Chap. 8 we describe output regulation and tracking problems addressed by conventional SMC and HOSM controllers driven by sliding mode disturbance observers (SMC–SMDO). The particular features of the application of SMC/HOSM observers to the above-mentioned output regulation/tracking problems, including the necessity to differentiate the measured output in order to implement the SMC or HOSM controller, as well as the possibility of reconstructing unknown external disturbances via SMC/HOSM observers with the possibility to compensate for them within a traditional continuous controller, are emphasized. The continuous SMC–SMDO design techniques are illustrated with two case studies: launch vehicle attitude control and satellite formation control. A variety of exercises are presented at the end of this chapter to facilitate homework assignments.

The contribution of the authors to this textbook is as follows: Dr. Shtessel has written the Preface and Chaps. 1 and 8 and has contributed to Chap. 5 by writing Sect. 5.2 and Chap. 6 by writing Sect. 6.11. Dr. Edwards has written Chaps. 2 and 3. He has also carried out the bulk of the editorial work. Dr. Fridman has written Chaps. 5 and 7 and has contributed to Chap. 2 by writing Sect. 2.7 and to Chaps. 4 and 6 by writing Sects. 4.7, 4.8, and 6.11. Dr. Levant has written Chaps. 4 and 6.

The authors would also like to acknowledge the graduate and postdoctoral students of the Department of Automatic Control, the National Autonomous University of Mexico, Francisco Bejarano, Jorge Davila, Lizet Fraguela, Ana Gabriela Gallardo, Tenoch Gonzalez, Antonio Rosales, and Carlos Vazquez, for their invaluable help in preparing examples and exercises. We would also like to thank our colleagues, Professors Igor Boiko, Leonid Freidovich, Elio Usai, and Vadim Utkin for their careful reading of early drafts of the manuscript and for their constructive criticisms and suggestions for improvement.

Huntsville, USA   Y. Shtessel
Exeter, UK   C. Edwards
Mexico, Mexico   L. Fridman
Tel Aviv, Israel   A. Levant

Contents

Chapter 1
Introduction: Intuitive Theory of Sliding Mode Control

In the formulation of any practical control problem, there will always be a discrepancy between the actual plant and its mathematical model used for the controller design. These discrepancies (or mismatches) arise from unknown external disturbances, plant parameters, and parasitic/unmodeled dynamics. Designing control laws that provide the desired performance to the closed-loop system in the presence of these disturbances/uncertainties is a very challenging task for a control engineer. This has led to intense interest in the development of the so-called robust control methods which are supposed to solve this problem. One particular approach to robust controller design is the so-called *sliding mode control* technique.

In Chap. 1, the main concepts of sliding mode control will be introduced in an intuitive fashion, requiring only a basic knowledge of control systems. The sliding mode control design techniques are demonstrated on tutorial examples and via graphical exposition. Advanced sliding mode concepts, including sliding mode observers/differentiators and second-order sliding mode control, are studied at a tutorial level. The main advantages of sliding mode control, including robustness, finite-time convergence, and reduced-order compensated dynamics, are demonstrated on numerous examples and simulation plots.

For illustration purposes, the single-dimensional motion of a unit mass (Fig. 1.1) is considered. A state-variable description is easily obtained by introducing variables for the position and the velocity $x_1 = x$, $x_2 = \dot{x}_1$ so that

$$\begin{cases} \dot{x}_1 = x_2 & x_1(0) = x_{10} \\ \dot{x}_2 = u + f(x_1, x_2, t) & x_2(0) = x_{20}, \end{cases} \tag{1.1}$$

where u is the control force, and the disturbance term $f(x_1, x_2, t)$, which may comprise dry and viscous friction as well as any other unknown resistance forces, is assumed to be bounded, i.e., $|f(x_1, x_2, t)| \leq L > 0$. The problem is to design a feedback control law $u = u(x_1, x_2)$ that drives the mass to the origin asymptotically. In other words, the control $u = u(x_1, x_2)$ is supposed to drive the state variables to

Y. Shtessel et al., *Sliding Mode Control and Observation*, Control Engineering, DOI 10.1007/978-0-8176-4893-0_1, © Springer Science+Business Media New York 2014

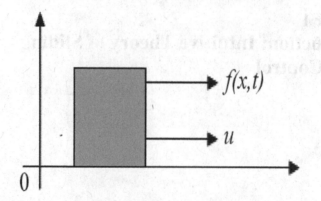

Fig. 1.1 Single-dimensional motion of a unit mass

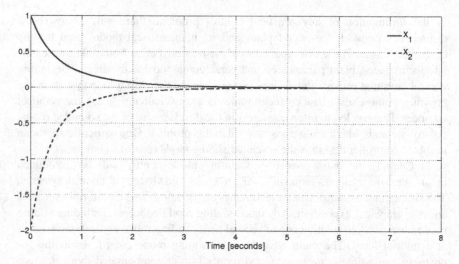

Fig. 1.2 Asymptotic convergence for $f(x_1, x_2, t) \equiv 0$

zero: i.e., $\lim\limits_{t \to \infty} x_1, x_2 = 0$. This apparently simple control problem is a challenging one, since asymptotic convergence is to be achieved in the presence of the unknown bounded disturbance $f(x_1, x_2, t)$. For instance, a linear state-feedback control law

$$u = -k_1 x_1 - k_2 x_2, \quad k_1 > 0, \quad k_2 > 0 \tag{1.2}$$

provides asymptotic stability of the origin only for $f(x_1, x_2, t) \equiv 0$ and typically only drives the states to a bounded domain $\Omega(k_1, k_2, L)$ for $|f(x_1, x_2, t)| \leq L > 0$.

Example 1.1. The results of the simulation of the system in Eqs. (1.1), (1.2) with $x_1(0) = 1$, $x_2(0) = -2$, $k_1 = 3$, $k_2 = 4$, and $f(x_1, x_2, t) = \sin(2t)$, which illustrate this statement, are presented in Figs. 1.2 and 1.3.

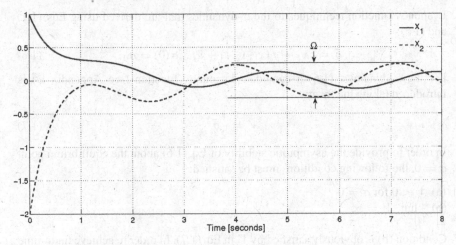

Fig. 1.3 Convergence to the domain Ω for $f(x_1, x_2, t) = \sin(2t)$

The question is whether the formulated control problem can be addressed using only knowledge of the bounds on the unknown disturbance.

1.1 Main Concepts of Sliding Mode Control

Let us introduce desired compensated dynamics for system (1.1). A good candidate for these dynamics is the homogeneous linear time-invariant differential equation:

$$\dot{x}_1 + cx_1 = 0, \quad c > 0 \tag{1.3}$$

Since $x_2(t) = \dot{x}_1(t)$, a general solution of Eq. (1.3) and its derivative is given by

$$
\begin{aligned}
x_1(t) &= x_1(0) \exp(-ct) \\
x_2(t) &= \dot{x}_1(t) = -cx_1(0) \exp(-ct)
\end{aligned}
\tag{1.4}
$$

both $x_1(t)$ and $x_2(t)$ converge to zero asymptotically. Note, no effect of the disturbance $f(x_1, x_2, t)$ on the state compensated dynamics is observed. How could these compensated dynamics be achieved? First, we introduce a new variable in the state space of the system in Eq. (1.1):

$$\sigma = \sigma(x_1, x_2) = x_2 + cx_1, \quad c > 0 \tag{1.5}$$

In order to achieve asymptotic convergence of the state variables x_1, x_2 to zero, i.e., $\lim_{t \to \infty} x_1, x_2 = 0$, with a given convergence rate as in Eq. (1.4), in the presence of the bounded disturbance $f(x_1, x_2, t)$, we have to drive the variable σ in Eq. (1.5) to zero in finite time by means of the control u. This task can be achieved by applying

Lyapunov function techniques to the σ-dynamics that are derived using Eqs. (1.1) and (1.5):

$$\dot{\sigma} = cx_2 + f(x_1, x_2, t) + u, \quad \sigma(0) = \sigma_0 \tag{1.6}$$

For the σ-dynamics (1.6) a candidate Lyapunov function (see Appendix D) is introduced taking the form

$$V = \frac{1}{2}\sigma^2 \tag{1.7}$$

In order to provide the asymptotic stability of Eq. (1.6) about the equilibrium point $\sigma = 0$, the following conditions must be satisfied:

(a) $\dot{V} < 0$ for $\sigma \neq 0$
(b) $\lim\limits_{|\sigma| \to \infty} V = \infty$

Condition (b) is obviously satisfied by V in Eq. (1.7). In order to achieve finite-time convergence (global finite-time stability), condition (a) can be modified to be

$$\dot{V} \leq -\alpha V^{1/2}, \quad \alpha > 0 \tag{1.8}$$

Indeed, separating variables and integrating inequality (1.8) over the time interval $0 \leq \tau \leq t$, we obtain

$$V^{1/2}(t) \leq -\frac{1}{2}\alpha t + V^{1/2}(0) \tag{1.9}$$

Consequently, $V(t)$ reaches zero in a finite time t_r that is bounded by

$$t_r \leq \frac{2V^{1/2}(0)}{\alpha}. \tag{1.10}$$

Therefore, a control u that is computed to satisfy Eq. (1.8) will drive the variable σ to zero in finite time and will keep it at zero thereafter.
The derivative of V is computed as

$$\dot{V} = \sigma\dot{\sigma} = \sigma\left(cx_2 + f(x_1, x_2, t) + u\right) \tag{1.11}$$

Assuming $u = -cx_2 + v$ and substituting it into Eq. (1.11) we obtain

$$\dot{V} = \sigma\left(f(x_1, x_2, t) + v\right) = \sigma f(x_1, x_2, t) + \sigma v \leq |\sigma| L + \sigma v \tag{1.12}$$

Selecting $v = -\rho\, \text{sign}(\sigma)$ where

$$\text{sign}(x) = \begin{cases} 1 & \text{if } x > 0 \\ -1 & \text{if } x < 0 \end{cases} \tag{1.13}$$

and

$$\text{sign}(0) \in \begin{bmatrix} -1, & 1 \end{bmatrix} \tag{1.14}$$

with $\rho > 0$ and substituting it into Eq. (1.12) we obtain

$$\dot{V} \leq |\sigma| L - |\sigma| \rho = -|\sigma| (\rho - L) \qquad (1.15)$$

Taking into account Eq. (1.7), condition (1.8) can be rewritten as

$$\dot{V} \leq -\alpha V^{1/2} = -\frac{\alpha}{\sqrt{2}} |\sigma|, \quad \alpha > 0 \qquad (1.16)$$

Combining Eqs. (1.15) and (1.16) we obtain

$$\dot{V} \leq -|\sigma| (\rho - L) = -\frac{\alpha}{\sqrt{2}} |\sigma| \qquad (1.17)$$

Finally, the control gain ρ is computed as

$$\rho = L + \frac{\alpha}{\sqrt{2}} \qquad (1.18)$$

Consequently a control law u that drives σ to zero in finite time (1.10) is

$$u = -cx_2 - \rho \operatorname{sign}(\sigma) \qquad (1.19)$$

Remark 1.1. It is obvious that $\dot{\sigma}$ must be a function of control u in order to successfully design the controller in Eq. (1.8) or (1.19). This observation must be taken into account while designing the variable given in Eq. (1.5).

Remark 1.2. The first component of the control gain Eq. (1.18) is designed to compensate for the bounded disturbance $f(x_1, x_2, t)$ while the second term $\frac{\alpha}{\sqrt{2}}$ is responsible for determining the sliding surface reaching time given by Eq. (1.10). The larger α, the shorter reaching time.

Now it is time to make definitions that interpret the variable (1.5), the desired compensated dynamics (1.3), and the control function (1.19) in a new paradigm.

Definition 1.1. The variable (1.5) is called *a sliding variable*

Definition 1.2. Equations (1.3) and (1.5) rewritten in a form

$$\sigma = x_2 + cx_1 = 0, \quad c > 0 \qquad (1.20)$$

correspond to a straight line in the state space of the system (1.1) and are referred to as *a sliding surface*.

Condition (1.8) is equivalent to

$$\sigma \dot{\sigma} \leq -\frac{\alpha}{\sqrt{2}} |\sigma| \qquad (1.21)$$

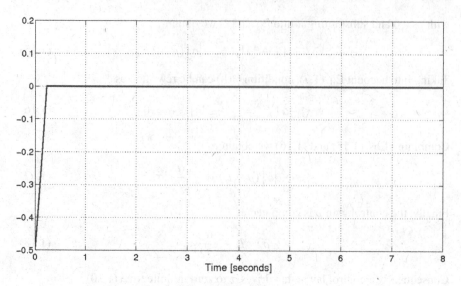

Fig. 1.4 Sliding variable

and is often termed the *reachability condition*. Meeting the *reachability* or *existence* condition (1.21) means that the trajectory of the system in Eq. (1.1) is driven towards the sliding surface (1.20) and remains on it thereafter.

Definition 1.3. The control $u = u(x_1, x_2)$ in Eq. (1.19) that drives the state variables x_1, x_2 to the sliding surface (1.20) in finite time t_r, and keeps them on the surface thereafter in the presence of the bounded disturbance $f(x_1, x_2, t)$, is called a *sliding mode controller* and *an ideal sliding mode* is said to be taking place in the system (1.1) for all $t > t_r$.

Example 1.2. The results of the simulation of system (1.1) with the sliding mode control law (1.5), (1.19), the initial conditions $x_1(0) = 1$, $x_2(0) = -2$, the control gain $\rho = 2$, the parameter $c = 1.5$, and the disturbance $f(x_1, x_2, t) = \sin(2t)$ (which is used for simulation purposes only) are presented in Figs. 1.4–1.9.

Figure 1.4 illustrates finite-time convergence of the sliding variable to zero. Asymptotic convergence of the state variables x_1, x_2 to zero in the presence of the external bounded disturbance $f(x_1, x_2, t) = \sin(2t)$ is shown in Fig. 1.5. The phase portrait, which is given in Fig. 1.6, demonstrates such phenomena as a *reaching phase* (when the state trajectory is driven towards the sliding surface) and a *sliding phase* (when the state trajectory is moving towards the origin along the sliding surface).

A zoomed portion of the phase portrait (Fig. 1.7) illustrates the "zigzag" motion of small amplitude and high frequency that the state variables exhibit while in the *sliding mode*. Sliding mode control, which is presented in Figs. 1.8 and 1.9, is a

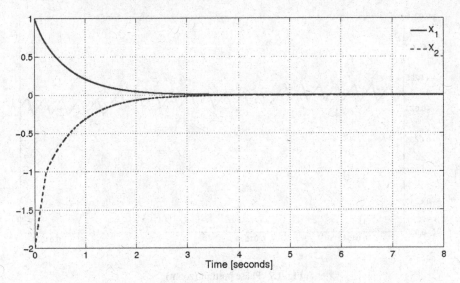

Fig. 1.5 Asymptotic convergence for $f(x_1, x_2, t) = \sin(2t)$

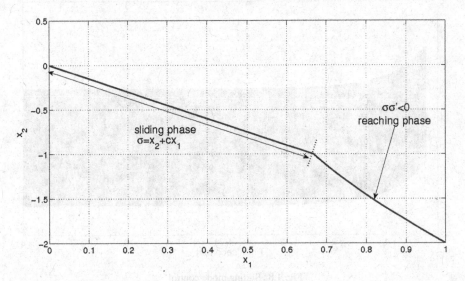

Fig. 1.6 Phase portrait

high frequency switching control with a switching frequency inversely proportional to the time increment 10^{-4} s used in the simulation. Apparently, this high-frequency switching control causes the "Zigzag" motion in the sliding mode (Fig. 1.7). In an ideal sliding mode the switching frequency is supposed to approach infinity and the amplitude of the "zigzag" motion tends to zero.

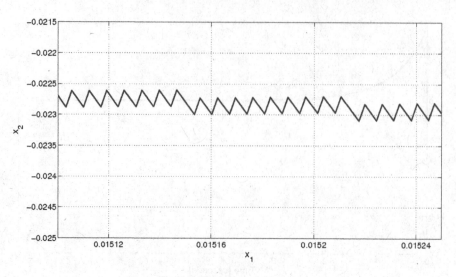

Fig. 1.7 Phase portrait (zoom)

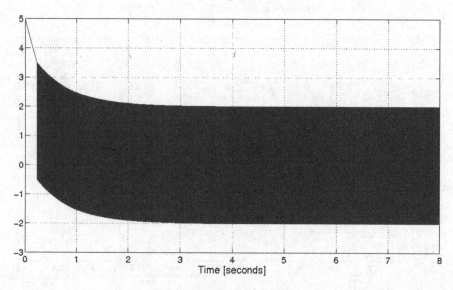

Fig. 1.8 Sliding mode control

As we see in Figs. 1.7 and 1.9, the imperfection in the sign-function implementation yields a finite amplitude and finite frequency "zigzag" motion in the sliding mode due to the discrete-time nature of the computer simulation. This effect is called *chattering*.

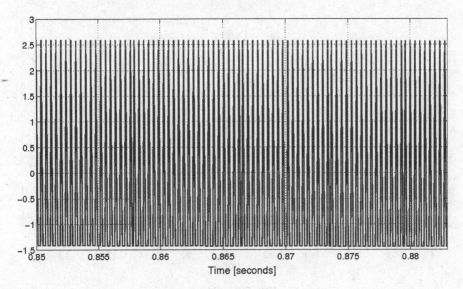

Fig. 1.9 Sliding mode control (zoom)

1.2   Chattering Avoidance: Attenuation and Elimination

In many practical control systems, including DC motors and aircraft control, it is important to avoid control chattering by providing continuous/smooth control signals: for instance, aircraft aerodynamic surfaces cannot move back and forth with high frequency, but at the same time it is desirable to retain the robustness/insensitivity of the control system to bounded model uncertainties and external disturbances.

1.2.1   Chattering Elimination: Quasi-Sliding Mode

One obvious solution to make the control function (1.19) continuous/smooth is to approximate the discontinuous function $v(\sigma) = -\rho \, \mathrm{sign}(\sigma)$ by some continuous/smooth function. For instance, it could be replaced by a "sigmoid function"

$$\mathrm{sign}(\sigma) \approx \frac{\sigma}{|\sigma| + \varepsilon} \tag{1.22}$$

where ε is a small positive scalar. It can be observed that point-wise

$$\lim_{\varepsilon \to 0} \frac{\sigma}{|\sigma| + \varepsilon} = \mathrm{sign}(\sigma)$$

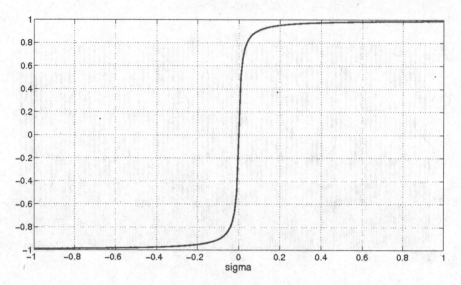

Fig. 1.10 Sigmoid function

for $\sigma \neq 0$. The value of ε should be selected to trade off the requirement to maintain an ideal performance with that of ensuring a smooth control action. The sigmoid function (1.22) is shown in Fig. 1.10.

Example 1.3. The results of the simulation of system (1.1) with the pseudo-sliding mode control

$$u = -cx_2 - \rho \frac{\sigma}{|\sigma| + \varepsilon} \tag{1.23}$$

using the same parameters as those in Example 1.2 are presented in Figs. 1.11–1.14.

The smooth control function (Fig. 1.11) cannot provide finite-time convergence of the sliding variable to zero in the presence of the external disturbance $f(x_1, x_2, t)$ (see Fig. 1.12). Furthermore, the sliding variable and the state variables do not converge to zero at all, but instead converge to domains in a vicinity of the origin (Figs. 1.12–1.14) due to the effect of the disturbance $f(x_1, x_2, t) = \sin(2t)$. The price we pay for obtaining a smooth control function is a loss of robustness and, as a result, a loss of accuracy. The designed smooth control (1.23) is technically not a sliding mode control and there is no ideal sliding mode in the system (1.1), since the sliding variable has not been driven to zero in a finite time. However, the system's performance under the smooth control law in Eq. (1.23) is close to the system's performance under the discontinuous sliding mode control (1.19). This gives us grounds for calling the smooth control law in Eq. (1.23) a *quasi-sliding mode control* and the system's motion, when the sliding surface converges to a close vicinity of the origin, a *quasi-sliding mode*.

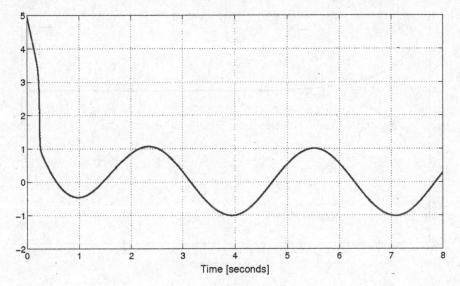

Fig. 1.11 Smooth control

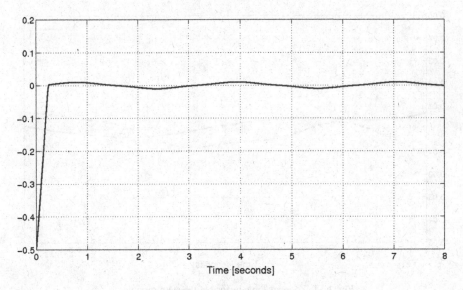

Fig. 1.12 Sliding variable

1.2.2   Chattering Attenuation: Asymptotic Sliding Mode

In this section we consider another approach to designing continuous control that is robust to bounded disturbances. The idea is to design an SMC in terms of the control function derivative. In this case the actual control, which is the integral of the

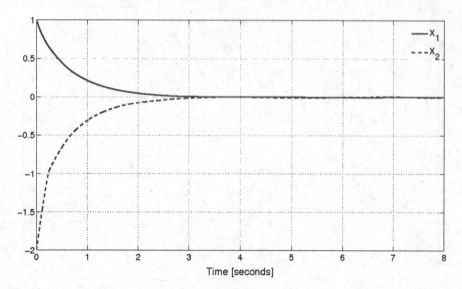

Fig. 1.13 Time history of the state variables

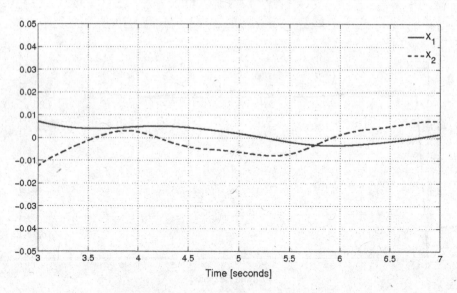

Fig. 1.14 Zoomed Time history of the state variables

high-frequency switching function, is continuous. This approach is called *chattering attenuation*, since some periodic residual is observed in the sliding mode control after the integration of the high-frequency switching function.

To proceed, the system in Eq. (1.1) is rewritten as

$$\begin{cases} \dot{x}_1 = x_2 & x_1(0) = x_{10} \\ \dot{x}_2 = u + f(x_1, x_2, t) & x_2(0) = x_{20} \\ \dot{u} = v & u(0) = 0. \end{cases} \qquad (1.24)$$

We know that if the sliding variable (1.5) is constrained to zero in finite time $t = t_r$, then the state variables converge to zero asymptotically in accordance with Eq. (1.4) for all $t \geq t_r$. Here we assume $|f(x_1, x_2, t)| \leq L$ and in addition that it is smooth with bounded derivative $\left| \dot{f}(x_1, x_2, t) \right| \leq \bar{L}$.

In order to achieve chattering attenuation the following auxiliary sliding variable

$$s = \dot{\sigma} + \bar{c}\sigma \qquad (1.25)$$

is introduced. If we design a control law v that provides finite-time convergence of $s \to 0$, then the ideal sliding mode occurs in the sliding surface

$$s = \dot{\sigma} + \bar{c}\sigma = 0 \qquad (1.26)$$

and $\sigma, \dot{\sigma} \to 0$ together with $x_1, x_2 \to 0$, as time increases, even in the presence of the bounded disturbance $f(x_1, x_2, t)$. However, we will not have an ideal sliding mode, but instead an *asymptotic sliding mode* will occur in system (1.24) since the original sliding variable σ converges to zero only asymptotically. This is the price we are going to pay for the chattering attenuation. Using Eq. (1.21) for designing the SMC in terms of v, we obtain

$$s\dot{s} = s(v + c\bar{c}x_2 + (c + \bar{c})u + (c + \bar{c})f(x_1, x_2, t) + \dot{f}(x_1, x_2, t)) \qquad (1.27)$$

Choosing $v = -c\bar{c}x_2 - (c + \bar{c})u + v_1$ and substituting it into Eq. (1.27), we obtain

$$s\dot{s} = s(v_1 + (c+\bar{c})f(x_1, x_2, t) + \dot{f}(x_1, x_2, t)) \leq sv_1 + |s|(\bar{L} + (c+\bar{c})L) \qquad (1.28)$$

Selecting $v_1 = -\rho\,\text{sign}(s)$ with $\rho > 0$, and substituting it into Eq. (1.28) it follows that

$$s\dot{s} \leq |s|(-\rho + \bar{L} + (c + \bar{c})L) = -\frac{\alpha}{\sqrt{2}}|s|. \qquad (1.29)$$

Finally, if the control gain ρ is computed as

$$\rho = \bar{L} + (c + \bar{c})L + \frac{\alpha}{\sqrt{2}} \qquad (1.30)$$

then the control law v that drives s to zero in finite time $t_r \leq \frac{\sqrt{2}|s(0)|}{\alpha}$ is

$$v = -c\bar{c}x_2 - (c + \bar{c})u - \rho\,\text{sign}(s) \qquad (1.31)$$

Example 1.4. The results of the simulation of system (1.24) with the sliding mode control (1.25), (1.31), the initial conditions $x_1(0) = 1$, $x_2(0) = -2$, the control gain

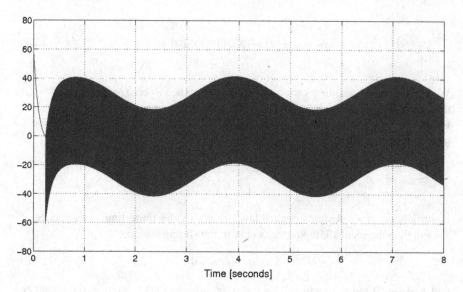

Fig. 1.15 Control v

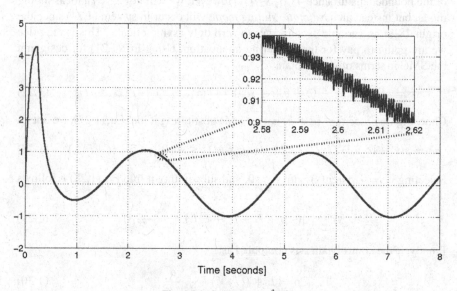

Fig. 1.16 Control $u = \int v \, dt$

$\rho = 30$, the parameters $c = 1.5$, $\bar{c} = 10$, and the disturbance $f(x_1, x_2, t) = \sin(2t)$ (which, again, is only used for simulation purposes) are presented in Figs. 1.15–1.20.

The control law (1.31) contains the high-frequency switching term $\rho \, \mathrm{sign}(s)$ that yields chattering (Fig. 1.15); however, chattering is attenuated in the physical

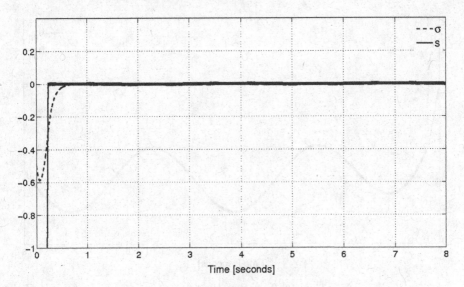

Fig. 1.17 Sliding variables

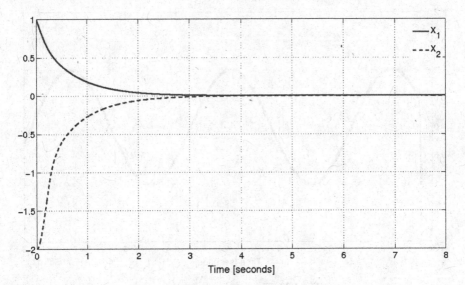

Fig. 1.18 State variables

control $u = \int v dt$ (Fig. 1.16). It can be observed from Fig. 1.17 that the auxiliary sliding variable s converges to zero in finite time and the original sliding variable σ converges to zero asymptotically. Therefore, the achieved sliding mode is called an *asymptotic sliding mode* with respect to the original sliding variable σ.

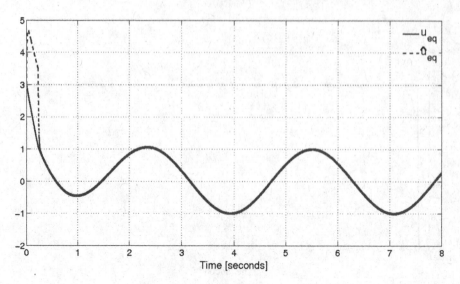

Fig. 1.19 Equivalent control estimation

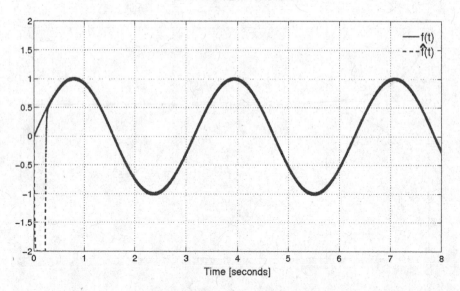

Fig. 1.20 The disturbance estimation

The state variables exhibit convergence to zero as time increases (see Fig. 1.18) in a similar way to the results from Fig. 1.5 that were achieved via the high-frequency switching sliding mode control u given by Eq. (1.19). Also, in order to implement the continuous sliding mode control $u = \int v dt$, with v given by Eqs. (1.25) and (1.31), it is necessary to differentiate σ. In the example, $\dot{\sigma}$ was computed numerically; however, it can be done using the sliding mode observers/differentiators that will be discussed later on.

1.3 Concept of Equivalent Control

Suppose that at time t_r the sliding surface $\sigma = x_2 + cx_1 = 0$ is reached and the trajectory $x_1(t), x_2(t)$ remains on the sliding surface thereafter by means of the SMC given by Eqs. (1.5), (1.19). This means that $\sigma = \dot{\sigma} = 0$ for all $t \geq t_r$. The condition $\dot{\sigma} = 0$ yields

$$\dot{\sigma} = cx_2 + f(x_1, x_2, t) + u = 0, \quad \sigma(t_r) = 0 \tag{1.32}$$

A control function that satisfies Eq. (1.32) can be easily computed as

$$u_{eq} = -cx_2 - f(x_1, x_2, t) \tag{1.33}$$

Definition 1.4. The control function (1.33), which needs to be applied to system (1.1) after reaching the sliding surface $\sigma = 0$, to ensure that the system trajectory stays on the surface thereafter, is called the *equivalent control*.

The following properties of the equivalent control can be established:

• The control function (1.33) is not the actual control that is applied to system (1.1) as soon as the sliding surface is reached. Furthermore, control (1.33) usually cannot be implemented, since the bounded disturbance $f(x_1, x_2, t)$ is not known and appears explicitly in Eq. (1.33). The equivalent control action describes the "average" effect of the high-frequency switching control (1.19) on system (1.1). The average can be achieved via low-pass filtering (LPF) of the high-frequency switching term $\rho \, \text{sign}(\sigma)$ in the control law (1.19). Therefore, the equivalent control can be estimated (online) as follows:

$$\hat{u}_{eq} = -cx_2 - \rho LPF \left(\text{sign}(\sigma) \right), t \geq t_r \tag{1.34}$$

For instance, the LPF can be implemented as a first-order differential equation

$$\begin{aligned} \tau \dot{z} &= -z + \text{sign}(\sigma) \\ \hat{u}_{eq} &= -cx_2 - \rho z \end{aligned} \tag{1.35}$$

where τ is a small positive scalar representing the time constant of the filter.

The signal u_{eq} can be estimated very accurately by $\hat{u}_{eq}$ by making τ as small as possible, but larger than the sampling time of the computer-implemented LPF.

• Comparing Eqs. (1.33) and (1.34) the disturbance term can be easily estimated:

$$\hat{f}(x_1, x_2, t) = \rho LPF \left(\text{sign}(\sigma) \right), t \geq t_r \tag{1.36}$$

Example 1.5. The system (1.1) with the sliding mode control (1.5), (1.19), the initial conditions $x_1(0) = 1$, $x_2(0) = -2$, the control gain $\rho = 2$, the parameter $c = 1.5$, and the disturbance $f(x_1, x_2, t) = \sin(2t)$, which is used for simulations purposes only, has been simulated. The equivalent control and the disturbance can be estimated using Eqs. (1.35) and (1.36) with $\tau = 0.01$. For comparison, the ideal equivalent control is plotted in accordance with Eq. (1.34). The results of the simulation are presented in Figs. 1.19 and 1.20.

Based on Figs. 1.19 and 1.20 we observe a high level of accuracy in terms of the estimation of the equivalent control and the disturbance.

1.4   Sliding Mode Equations

It was discussed earlier that system (1.1) with the SMC given by Eqs. (1.5), (1.19) exhibits a two-phase motion (Fig. 1.6), namely, the reaching phase (when the system trajectory moves towards the sliding surface) and the sliding phase (when the system trajectory moves along the sliding surface). The sliding variable (1.5) is supposed to be designed in order to provide a desired motion in the sliding mode. This design problem can be reduced to two tasks.

Task 1 is to find the system's equation in the sliding mode for all $t \geq t_r$; and task 2 is to parameterize the sliding variable (1.5) in order to ensure the desired/given compensated dynamics.

Substituting the SMC in Eqs. (1.5), (1.19) into Eq. (1.1) yields

$$\begin{cases} \dot{x}_1 = x_2 & x_1(t_r) = x_{1r} \\ \dot{x}_2 = -cx_2 - \rho\,\text{sign}(x_2 + cx_1) + f(x_1, x_2, t) & x_2(t_r) = -cx_{1r} \end{cases} \tag{1.37}$$

Equation (1.37) is not suitable for the sliding mode analysis, since the right-hand side of the system is a discontinuous high-frequency switching function, which loses its continuity in each point on the sliding surface $\sigma = x_2 + cx_1 = 0$. However, from Sect. 1.3, we know that the system's (1.1) dynamics in the sliding mode (when $\sigma = x_2 + cx_1 = 0$) are driven by the equivalent control (1.33). Therefore, in order to obtain the equations of system's compensated dynamics in the sliding mode, we substitute the equivalent control into Eq. (1.1). Bearing in mind that in the sliding mode $x_2 = -cx_1$ we obtain

$$\begin{cases} \dot{x}_1 = x_2 \\ \dot{x}_2 = \underbrace{(-cx_2 - f(x_1, x_2, t))}_{u_{eq}} + f(x_1, x_2, t) \end{cases} \Rightarrow \begin{cases} \dot{x}_1 = x_2 \\ \dot{x}_2 = -cx_2 \end{cases} \Rightarrow \begin{cases} \dot{x}_1 = -cx_1 \\ \dot{x}_2 = -cx_2 \end{cases}$$

$$\tag{1.38}$$

Finally the system's (1.1) compensated dynamics in the sliding mode are reduced to the form

$$\begin{cases} \dot{x}_1 = -cx_1 \\ x_2 = -cx_1 \end{cases}, x_1(t_r) = x_{1r} \tag{1.39}$$

for all $t \geq t_r$. A solution of Eq. (1.39) for all $t \geq t_r$ can be written as

$$x_1(t) = x_{1r} \exp[-c(t - t_r)]$$
$$x_2(t) = -cx_{1r} \exp[-c(t - t_r)]$$

(1.40)

It is clear that the parameter $c > 0$ can be selected to give a desired rate of convergence of x_1, x_2 to zero.

The following properties are exhibited by the system's dynamics in the sliding mode:

- The SMC controller design is reduced to two tasks. The first task consists of the design of the first-order sliding surface in Eq. (1.20). The second task is to design the control u to drive the sliding variable (1.5) to zero. Again, the first-order sliding variable dynamics given by Eq. (1.6) are employed.
- The original system's dynamics are of second-order while its compensated dynamics in the sliding mode (1.39) are of order equal to one. The reduction in order is due to the fact that Eq. (1.39) describes the "slow" motion only. The "fast" motion that is due to the high-frequency switching control (see Figs. 1.7 and 1.9) is of very small amplitude and is disregarded in Eq. (1.39).
- The system's dynamics in the sliding mode (1.39) do not depend on the bounded disturbance $f(x_1, x_2, t)$; however, its upper limit is taken into account in the SMC design [see Eqs. (1.16) and (1.17)].

1.5 The Matching Condition and Insensitivity Properties

It was discussed in Sect. 1.4 that the system's dynamics in the sliding mode do not depend on the bounded disturbance $f(x_1, x_2, t)$. We need to bear in mind that the disturbance $f(x_1, x_2, t)$ enters only the second equation of the system (1.1). The question is whether this *insensitivity property* of the system's dynamics in the sliding mode to the bounded disturbances/uncertainties can be extended to bounded disturbances/uncertainties entering the first equation of the system (1.1).

In order to address this issue consider the system

$$\begin{cases} \dot{x}_1 = x_2 + \varphi(x_1, x_2, t) & x_1(0) = x_{10} \\ \dot{x}_2 = u + f(x_1, x_2, t) & x_2(0) = x_{20} \end{cases}$$

(1.41)

where $|f(x_1, x_2, t)| \leq L$, $|\varphi(x_1, x_2, t)| \leq P$. Assume that an SMC u is designed to drive the trajectories of system (1.41) to the sliding surface $\sigma = x_2 + cx_1 = 0$ in finite time $t \leq t_r$ and to maintain motion on the surface thereafter. The dynamics of system in the sliding mode can be easily derived using the equivalent control approach presented in Sect. 1.5. In this example the reduced-order motion is described by

$$\begin{cases} \dot{x}_1 = x_2 + \varphi(x_1, x_2, t) \\ x_2 = -cx_1 \end{cases} \quad x_1(t_r) = x_{1r}$$

(1.42)

It can be observed from Eqs. (1.41)–(1.42) that the disturbance $f(x_1, x_2, t)$ does not affect the system's dynamics in the sliding mode while the disturbance $\varphi(x_1, x_2, t)$ that enters the first equation (where the control is absent) can prevent the state variable from converging to zero in the sliding mode (1.41). The disturbance $f(x_1, x_2, t)$ is called a disturbance *matched* by the control, and the disturbance $\varphi(x_1, x_2, t)$ is called an *unmatched* one.

Note that such a criterion for detecting *matched* and *unmatched* disturbances is valid only for SISO systems with the control u entering in only one equation. The matching condition will be generalized later on for nonlinear systems of an arbitrary order.

1.6 Sliding Mode Observer/Differentiator

So far, we have assumed that both state variables $x_1(t)$ and $x_2(t)$ are measured (available). In many cases only x_1 (a position) is measured, but x_2 (a velocity) must be estimated.

In order to estimate x_2 (assuming a bound on $|x_2|$ is known) the following observation algorithm is proposed:

$$\dot{\hat{x}}_1 = v \tag{1.43}$$

where v is an observer injection term that is to be designed so that the estimates $\hat{x}_1$, $\hat{x}_2 \to x_1$, x_2.

Let us introduce an estimation error (an auxiliary sliding variable)

$$z_1 = \hat{x}_1 - x_1 \tag{1.44}$$

Subtracting the first equation in Eq. (1.1) from Eq. (1.43) we obtain

$$\dot{z}_1 = -x_2 + v. \tag{1.45}$$

Let us design the injection term v that drives $z_1 = \hat{x}_1 - x_1 \to 0$ in finite time. In this case $\hat{x}_1$ will converge to x_1 in finite time. The following choice of injection term

$$v = -\rho \, \text{sign}(z_1), \quad \rho > |x_2| + \beta, \quad \beta > 0 \tag{1.46}$$

yields

$$z_1 \dot{z}_1 = z_1(-x_2 - \rho \, \text{sign}(z_1)) \leq |z_1|(|x_2| - \rho) \leq -\beta |z_1| \tag{1.47}$$

Inequality (1.47) mimics Eq. (1.21), which means that in a finite time $t_r \leq \frac{|z_1(0)|}{\beta}$, $z_1 \to 0$ or $\hat{x}_1 \to x_1$. Therefore, a sliding mode exists in the observer (1.43) for $t \geq t_r$. The sliding mode dynamics in Eq. (1.45) are computed using the concept of equivalent control studied in Sect. 1.4:

$$\dot{z}_1 = -x_2 + v_{eq} = 0 \tag{1.48}$$

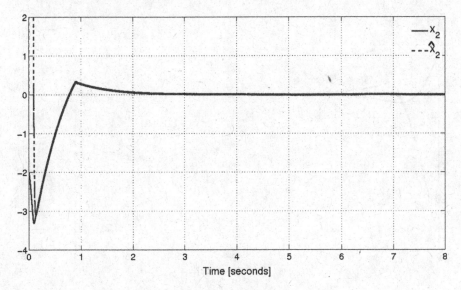

Fig. 1.21 Estimating x_2

What about estimating x_2? It is clear from (1.48) that the state variable x_2 can be exactly estimated as

$$x_2 = v_{eq}, t \geq t_r. \tag{1.49}$$

The equivalent injection v_{eq} can be estimated by LPF of the high-frequency switching control (1.46) as

$$\tau \dot{\hat{v}}_{eq} = -\hat{v}_{eq} - \rho \text{sign}(z_1) \tag{1.50}$$

where τ is a small positive constant, and finally

$$x_2 \approx \hat{x}_2 = \hat{v}_{eq}, t \geq t_r \tag{1.51}$$

Remark 1.3. The sliding mode observer given by Eqs. (1.43), (1.46), (1.50), and (1.51) also could be treated as a differentiator, since the variable it estimates is a derivative of the measured variable.

Example 1.6. The system (1.1) with the sliding mode control (1.5), (1.19), the initial conditions $x_1(0) = 1$, $x_2(0) = -2$, the control gain $\rho = 2$, the parameter $c = 1.5$, and the disturbance $f(x_1, x_2, t) = \sin(2t)$, which is used for simulation purposes only, is simulated. The variable x_1 is measured, and the variable x_2 is estimated using the *sliding mode observer* (1.43), (1.46), (1.50), and (1.51) with $\rho = 10$ and $\tau = 0.01$. The results of the simulations are shown in Figs. 1.21–1.24.

The sliding mode observer given by Eqs. (1.43), (1.46), (1.50), and (1.51) estimates x_2 very quickly and very accurately (Figs. 1.21 and 1.23). The sliding

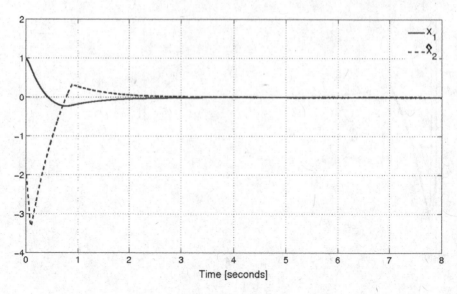

Fig. 1.22 State variables

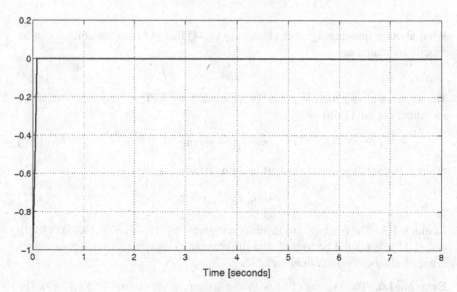

Fig. 1.23 Sliding variable z_1

variable (1.5) with x_2 replaced by its estimate $\hat{x}_2$, $\sigma = \hat{x}_2 + 1.5x_1$, converges to zero in finite time $t_r \approx 1\,\mathrm{s}$. Furthermore the state variables x_1, $x_2 \to 0$ as time increases (see Fig. 1.22) despite the presence of the bounded disturbance $f(x_1, x_2, t) = \sin(2t)$.

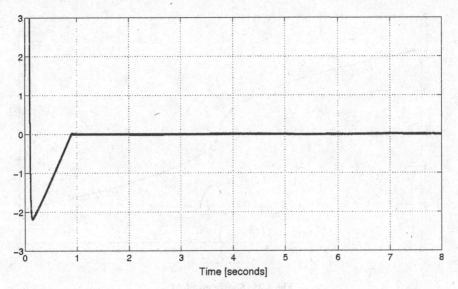

Fig. 1.24 Sliding variable σ

1.7 Second-Order Sliding Mode

As mentioned in Sect. 1.5, the compensated dynamics of system (1.1) in the sliding mode (1.39) are of the order one, while the system's uncompensated dynamics are of second order. This reduction of order is called a *partial dynamical collapse.*

Let us consider if a complete dynamical collapse is possible, which means that the second-order uncompensated dynamics in Eq. (1.1) are reduced to algebraic equations ($x_1 = x_2 = 0$) in finite time. Addressing this question is very important especially in cascade control systems, where dynamical collapse means elimination of inner loop dynamics and/or any parasitic dynamics if properly compensated by SMC.

The first problem is the sliding variable design. Let us try the following *nonlinear* sliding variable:

$$\sigma = \sigma(x_1, x_2) = x_2 + c\,|x_1|^{1/2}\,\mathrm{sign}(x_1), \quad c > 0 \tag{1.52}$$

Remark 1.4. The sliding *manifold* (it is not a straight line anymore due to its nonlinearity) that corresponds to the sliding variable (1.52)

$$x_2 + c\,|x_1|^{1/2}\,\mathrm{sign}(x_1) = 0, \quad c > 0 \tag{1.53}$$

is continuous (see Fig. 1.25).

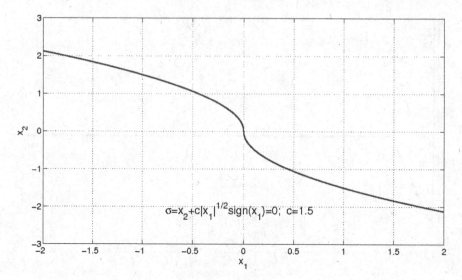

Fig. 1.25 Sliding manifold

The second problem is to design the control u that drives the sliding variable (1.52) to zero in finite time $t \leq t_r$. We will address the second problem rigorously later on in Chap. 4. Assume that this control is already available.

The sliding mode equations of the systems (1.1), (1.52) are defined for all $t \geq t_r$ as

$$\begin{cases} \dot{x}_1 = x_2 \\ \dot{x}_2 = -c\,|x_1|^{1/2} \operatorname{sign}(x_1) \end{cases} \quad x_1(t_r) = x_{1r} \tag{1.54}$$

Equation (1.54) can be rewritten as one nonlinear differential equation:

$$\dot{x}_1 = -c\,|x_1|^{1/2} \operatorname{sign}(x_1), x_1(t_r) = x_{1r} \tag{1.55}$$

Integrating Eq. (1.55) we obtain

$$|x_1(t)|^{1/2} - |x_{1r}|^{1/2} = -\frac{c}{2}(t - t_r) \tag{1.56}$$

We wish to identify a time instant $t = \bar{t}_r$ so that $x_1(\bar{t}_r) = x_2(\bar{t}_r) = 0$. This is

$$\bar{t}_r = \frac{2}{c}\,|x_{1r}|^{1/2}\,t_r \tag{1.57}$$

This result means that the state variables x_1, $x_2 \to 0$ in finite time equal to $\bar{t}_r - t_r$ while the system (1.1) is in the sliding mode, with dynamics described by Eq. (1.54). Obviously, the overall reaching time from the initial condition $x_1(0) = x_{10}$, $x_2(0) = x_{20}$ to zero will be $t \leq \bar{t}_r$, since $t \leq t_r$ is required to reach the sliding manifold (1.53) and it will take time $t = \bar{t}_r$ for the state variables to reach zero while

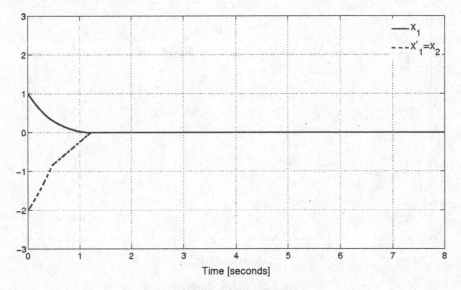

Fig. 1.26 Controlled output x_1 and its derivative $\dot{x}_1$

constrained to the *nonlinear sliding manifold* (1.53). This is a new phenomenon, since the variables in Eq. (1.1) reach zero asymptotically in the sliding mode (1.39) associated with a *linear sliding surface* (1.20).

The controller u design that drives the sliding variable (1.52) to zero in finite time $t \le t_r$ will be considered rigorously in Chap. 4. In this subsection we will design the controller in a similar fashion to the one given by Eq. (1.19):

$$u = -\rho\,\mathrm{sign}(\sigma) \Rightarrow u = -\rho\,\mathrm{sign}\left(x_2 + c|x_1|^{1/2}\mathrm{sign}(x_1)\right) \tag{1.58}$$

where the positive gain ρ is sufficiently large. The control law in Eq. (1.58) is called the control *with prescribed convergence law*.

Definition 1.5. The control $u = u(x_1, x_2)$ in Eq. (1.58) with a nonlinear sliding manifold (1.53) that drives the controlled output x_1 and its derivative $\dot{x}_1 = x_2$ to zero in finite time $t \le \bar{t}_r$ and keeps them there thereafter in the presence of a bounded disturbance $f(x_1, x_2, t)$ is called *second-order sliding mode (2-SM) control* and *an ideal 2-SM* is said to be taking place in system (1.1) for all $t > \bar{t}_r$.

The entire theory of *second-order sliding mode control* will be rigorously presented together with a variety of 2-SM controllers in Chap. 4 and more generalized *higher-order sliding mode control* will be discussed in Chap. 6.

Example 1.7. The results of the simulation of system (1.1) with the 2-SM control (1.58), the initial conditions $x_1(0) = 1$, $x_2(0) = -2$, the control gain $\rho = 2$, the parameter $c = 1.5$, and the disturbance $f(x_1, x_2, t) = \sin(2t)$, which is used for simulation purposes only, which illustrate the *second-order sliding mode control* concepts, are presented in Figs. 1.26–1.28.

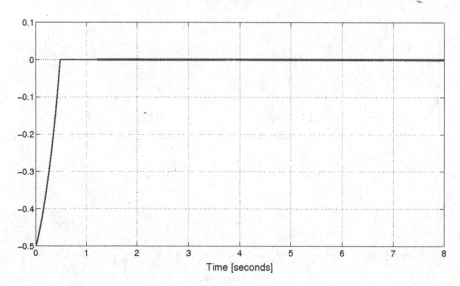

Fig. 1.27 Sliding variable σ

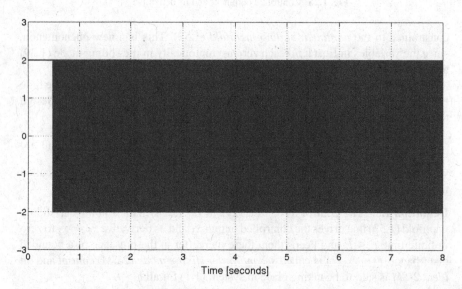

Fig. 1.28 Second-order sliding mode control

Discussion. The sliding variable σ reaches zero in finite time $t_r \approx 0.5$ s (see Fig. 1.27). It confirms the existence of the sliding mode in system (1.1) for all $t > 0.5$ s. The controlled output x_1 and its derivative $\dot{x}_1 = x_2$ reach zero in finite time $\bar{t}_r \approx 1.2$ s (Fig. 1.26). This confirms the existence of a second-order sliding mode in system (1.1) for all $t > 1.2$ s. Dynamical collapse of system (1.1) is achieved in the second-order sliding mode, since the system's dynamics are reduced to the algebraic equations $x_1(t) = x_2(t) = 0$ for all $t > 1.2$ s.

1.8 Output Tracking: Relative Degree Approach

The output tracking (servomechanism) control problem is a very common practical task. For instance, an aircraft flight control system makes the aircraft attitude (Euler) angles follow reference profiles, often generated in real time, by means of controlled deflection of the aerodynamic surfaces, while the state vector associated with the aircraft dynamics contains a number of other variables, which are the subject of control. Another example concerns controlling DC-to-DC electric power converters. The electric energy conversion is performed by means of high-frequency switching control of power transistor while maintaining the output voltage at a given level.

Let us revisit system (1.1):

$$\begin{cases} \dot{x}_1 = x_2 \\ \dot{x}_2 = u + f(x_1, x_2, t) & \quad x_1(0) = x_{10} \\ y = x_1 & \quad x_2(0) = x_{20} \end{cases} \tag{1.59}$$

where x_1, x_2 are position and velocity of the unit mass, y is a controlled output, u is a control force, and the disturbance term $f(x_1, x_2, t)$, which may comprise dry and viscous friction as well as any other unknown resistance forces, is assumed to be bounded, i.e., $|f(x_1, x_2, t)| \leq L > 0$.

The problem to be addressed now is to design an SMC control law $u = u(x_1, x_2, t)$ that makes the output y (the position of the unit mass) follow asymptotically a reference profile $y_c(t)$ given in current time. In other words, the control $u = u(x_1, x_2)$ is supposed to drive the output tracking error to zero: $\lim_{t \to \infty} (y_c(t) - y(t)) = 0$ in the presence of the bounded disturbance $f(x_1, x_2, t)$.

The proposed control design technique employs the concepts of input–output dynamics and relative degree.

Definition 1.6. Consider a SISO dynamic system with output $y \in \mathbb{R}$, state vector $x \in \Theta \subset \mathbb{R}^n$, and the control input $u \in \mathbb{R}$. If $y^{(i)}$ is independent of u for all $i = 1, 2, \ldots, k-1$, but $y^{(k)}$ is proportional to u with the coefficient of proportionality not equal to zero in a reasonable domain $\Omega \subset \Theta \subset \mathbb{R}^n$, then k is called the well-defined *relative degree*.

For the system (1.59) the *input–output dynamics* have relative degree $k = 2$ since

$$y^{(2)} = u + f(y, \dot{y}, t) \tag{1.60}$$

Remark 1.5. The relative degree of the system in Eq. (1.59) is equal to the system's order, which means that the system (1.59) does not have any *internal* dynamics.

1.8.1 Conventional Sliding Mode Controller Design

The desired input–output tracking error compensated dynamics for the system (1.59) or (1.60) can be defined as a linear differential equation of order equal to $k - 1$ with respect to the output tracking error $e = y_c(t) - y(t)$. Specifically, we define

$$\sigma = \dot{e} + ce, \quad c > 0 \tag{1.61}$$

Now we have to design a conventional SMC u that drives $\sigma \to 0$ in finite time and keeps it at zero thereafter, bearing in mind that as soon as the sliding variable σ reaches zero the sliding mode starts and the output tracking error e in the sliding mode will obey the desired reduced (first)-order differential equation

$$\sigma = \dot{e} + ce \doteq 0 \quad c > 0 \tag{1.62}$$

that yields convergence to zero as time increases.

The sliding variable dynamics are derived as

$$\dot{\sigma} = \underbrace{\ddot{y}_c + c\dot{y}_c - f(y, \dot{y}, t) - c\dot{y}}_{\varphi(y,\dot{y},t)} - u \quad \Rightarrow \quad \dot{\sigma} = \varphi(y, \dot{y}, t) - u \tag{1.63}$$

where y_c, $\dot{y}_c$, and $\ddot{y}_c$ are known in current time. The cumulative disturbance term $\varphi(y, \dot{y}, t)$ is assumed bounded, i.e., $|\varphi(y, \dot{y}, t)| \leq M$.

Conventional SMC u can be designed by using the sliding mode existence condition (1.21) rewritten in a form

$$\sigma\dot{\sigma} \leq -\bar{\alpha} |\sigma|, \quad \bar{\alpha} = \frac{\alpha}{\sqrt{2}} \tag{1.64}$$

Consequently

$$\sigma\dot{\sigma} = \sigma \left(\varphi(y, \dot{y}, t) - u \right) \leq |\sigma| M - \sigma u \tag{1.65}$$

and selecting

$$u = \rho\,\text{sign}(\sigma) \tag{1.66}$$

and substituting it into Eq. (1.66) we obtain

$$\sigma\dot{\sigma} \leq |\sigma| (M - \rho) = -\bar{\alpha} |\sigma| \tag{1.67}$$

The control gain ρ is computed as

$$\rho = M + \bar{\alpha} \tag{1.68}$$

Example 1.8. The results of the simulation of system (1.59) with a conventional SMC control (1.61), (1.66), (1.68), the initial conditions $x_1(0) = 1$, $x_2(0) = -2$, the control gain $\rho = 6$, the parameter $c = 1.5$, the output reference profile $y_c = 2\cos(t)$, and the disturbance $f(x_1, x_2, t) = \sin(2t)$, (which is used for simulation

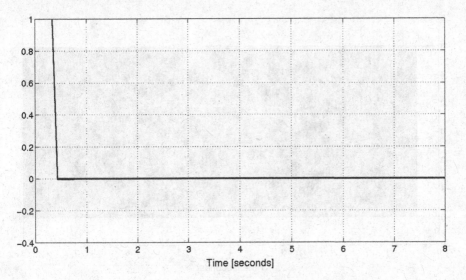

Fig. 1.29 Sliding variable σ

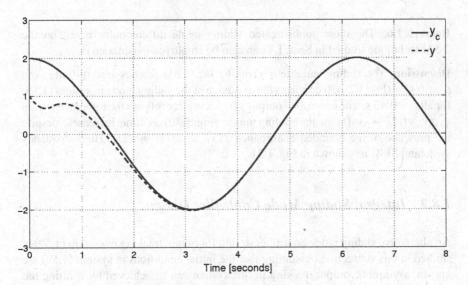

Fig. 1.30 The reference profile tracking

purposes only), illustrating the *output tracking SMC* concepts, are presented in Figs. 1.29–1.31. The component $\dot{e}$ of the sliding variable in Eq. (1.61) is computed using simple numerical differentiation during the simulation. In particular, the error e is fed to the input of the transfer function $\frac{s}{\tau s + 1}$ with $\tau = 0.01$, whose output gives an approximate $\dot{e}$.

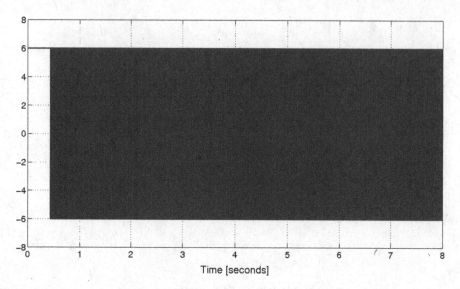

Fig. 1.31 Traditional sliding mode control

Remark 1.6. The more sophisticated sliding mode differentiator based on the 2-SM technique studied in Sect. 1.7 can also be employed to generate $\dot{e}$.

Discussion. The sliding variable σ given by Eq. (1.61) reaches zero in finite time $t_r \approx 0.45$ s (Fig. 1.29). It confirms the existence of the sliding mode in system (1.59) for all $t > 0.45$ s. The controlled output $y(t)$ accurately follows the reference profile $y_c(t)$: $y(t) \rightarrow y_c(t)$ in the sliding mode (Fig. 1.30) as time increases, despite the presence of the bounded-disturbance $f(x_1, x_2, t) = \sin(2t)$. High-frequency switching SMC u is shown in Fig. 1.31.

1.8.2 Integral Sliding Mode Controller Design

An alternative sliding mode controller design for output tracking in system (1.59) is studied in this subsection. Assuming that the initial conditions in system (1.59) are known, asymptotic output tracking in this system can be achieved by splitting the control function u into two parts:

$$u = u_1 + u_2 \tag{1.69}$$

and then:

- Designing the auxiliary sliding mode control law u_1 to compensate for the bounded disturbance $\varphi(y, \dot{y}, t)$ in Eq. (1.63) in the auxiliary sliding mode such that sliding starts right away (i.e., without a reaching phase)

- Designing the control u_2 to drive the sliding variable (1.61) to zero as time increases, bearing in mind that the sliding variable dynamics (1.63) are not perturbed anymore

Control u_1 Design. The auxiliary sliding variable is designed as

$$\begin{cases} s = \sigma - z \\ \dot{z} = -u_2 \end{cases} \tag{1.70}$$

The auxiliary sliding variable dynamics are given by

$$\begin{aligned} \dot{s} &= \dot{\sigma} - \dot{z} = \varphi(y, \dot{y}, t) - (u_1 + u_2) - (-u_2) \\ &= \varphi(y, \dot{y}, t) - u_1 \end{aligned} \tag{1.71}$$

Sliding mode control u_1 that drives the auxiliary sliding variable s to zero in finite time is designed as in Eq (1.66)

$$u_1 = \rho_1 \operatorname{sign}(s) \tag{1.72}$$

and thus the s-dynamics collapse in the auxiliary sliding mode. The original sliding variable dynamics (1.63) compensated by control u_1 given by Eq. (1.72) are

$$\begin{cases} \dot{\sigma} = \varphi(y, \dot{y}, t) - u_1 - u_2, \\ \dot{s} = \varphi(y, \dot{y}, t) - u_1, \quad u_1 = \rho_1 \operatorname{sign}(s) \end{cases} \tag{1.73}$$

In order to describe the σ-dynamics in the auxiliary sliding mode ($s = 0$) we have to find the equivalent control u_{1eq} that satisfies the condition $\dot{s} = 0$ and substitute it to Eq. (1.73). It is easy to see that

$$\begin{aligned} \dot{s} &= \varphi(y, \dot{y}, t) - u_{1eq} = 0 &\Rightarrow& \quad u_{1eq} = \varphi(y, \dot{y}, t) &\Rightarrow \\ \dot{\sigma} &= \varphi(y, \dot{y}, t) - \underbrace{\varphi(y, \dot{y}, t)}_{u_{1eq}} - u_2 &\Rightarrow& \quad \dot{\sigma} = -u_2 \end{aligned} \tag{1.74}$$

Therefore, the original sliding variable dynamics (1.74) do not depend on the disturbance $\varphi(y, \dot{y}, t)$ in the auxiliary sliding mode.

Now we will address the issue of starting the auxiliary sliding mode from the very beginning without any reaching phase. In order to achieve it we have to enforce the initial condition $s(0) = 0$ in Eq. (1.73). From Eq. (1.70) we obtain

$$\begin{aligned} s(0) &= \sigma(0) - z(0) = 0 &\Rightarrow& \quad z(0) = \sigma(0) &\Rightarrow \\ z(0) &= \dot{y}_c(0) + cy_c(0) - x_2(0) - cy(0) \end{aligned} \tag{1.75}$$

Therefore, the initial conditions for the variable z are identified from Eq. (1.70) that makes $s(0) = 0$.

The following properties of the auxiliary sliding mode control u_1 can be observed:

- The sliding mode control u_1 provides compensation for the disturbance $\varphi(y, \dot{y}, t)$ from the very beginning without any reaching phase.
- It was argued earlier that the use of sliding mode control reduces the system's order in the sliding mode. On the other hand the original sliding variable dynamics in Eq. (1.63) retain their order after being compensated by the auxiliary sliding mode control u_1 as in Eq. (1.74). So, a very specific type of sliding mode control has been designed.

Definition 1.7. Sliding mode control that retains the order of the compensated system's dynamics in the sliding mode is called *integral sliding mode control*.

In order to complete the controller design that drives the sliding variable (1.61), select the control function u_2 in Eq. (1.74) as

$$u_2 = k\sigma, \quad k > 0 \tag{1.76}$$

The σ-dynamics compensated by the control (1.69), (1.72), (1.76) become

$$\dot{\sigma} = -k\sigma, \quad \sigma(0) = \sigma_0 \tag{1.77}$$

and the desired convergence rate can be easily achieved by choosing the gain $k > 0$.

Example 1.9. The results of the simulation of system (1.59) with the integral SMC control (1.69), (1.70) (1.72), (1.75), (1.76), the initial conditions $x_1(0) = 1$, $x_2(0) = -2$, $z(0) = \dot{y}_c(0) + cy_c(0) - x_2(0) - cy(0)$, the control gains $\rho_1 = 8$, $k = 6$, the parameter $c = 1.5$, the output reference profile $y_c = 2\cos(t)$, and the disturbance $f(x_1, x_2, t) = \sin(2t)$ (which is used for simulation purposes only), which illustrate the *output tracking Integral SMC* concepts, are presented in Figs. 1.32–1.34. During the simulation, the component $\dot{e}$ of the sliding variable (1.61) is computed using a simple numerical differentiation. In particular, the error e is fed to the input of the transfer function $\frac{s}{\tau s + 1}$ with $\tau = 0.01$, which gives an approximation of $\dot{e}$.

Discussion. The auxiliary sliding variable s given by Eq. (1.62) is equal to zero from the very beginning [due to the selection of $z(0) = 3.5$ using Eq. (1.75)] and is kept at zero thereafter (Fig. 1.32) by means of *integral sliding mode control* (1.72). The disturbance $f(x_1, x_2, t) = \sin(2t)$ is compensated completely for all $t \geq 0$. The sliding variable σ converges to zero (Fig. 1.32) in accordance with Eq. (1.77) and the output $y(t)$ accurately follows the reference profile $y_c(t)$: $y(t) \to y_c(t)$ as time increases (Fig. 1.33). High-frequency switching SMC Eq. (1.69) is shown in Fig. 1.34.

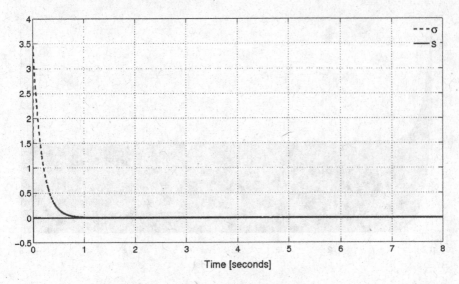

Fig. 1.32 Sliding variables σ and s

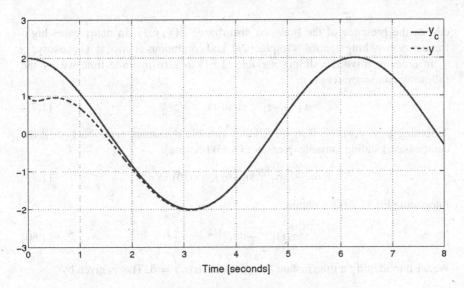

Fig. 1.33 The reference profile tracking

1.8.3 Super-Twisting Controller Design

In Sects. 1.8.1 and 1.8.2 the discontinuous high-frequency switching sliding mode controllers are designed to drive the sliding variable (1.61) to zero, which yields the solution to the output tracking problem, i.e., $y(t) \rightarrow y_c(t)$ as time increases,

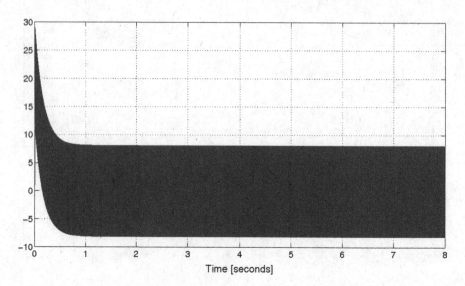

Fig. 1.34 Integral sliding mode control

despite the presence of the bounded disturbance $\varphi(y, \dot{y}, t)$. In many cases high-frequency switching control is impractical, and continuous control is a necessity.

In order to drive the sliding variable (1.61) to zero in finite time we try the following continuous control:

$$u = c\,|\sigma|^{1/2}\,\text{sign}(\sigma), \quad c > 0 \tag{1.78}$$

Assuming $\varphi(y, \dot{y}, t) = 0$ in the sliding variable dynamics equation (1.63), the compensated sliding variable dynamics (1.63) become

$$\dot{\sigma} = -c\,|\sigma|^{1/2}\,\text{sign}(\sigma), \quad \sigma(0) = \sigma_0 \tag{1.79}$$

Integrating Eq. (1.79) we obtain

$$|\sigma(t)|^{1/2} - |\sigma_0|^{1/2} = -\frac{c}{2}t \tag{1.80}$$

We wish to identify a time instant $t = t_r$ so that $\sigma(t_r) = 0$. This is given by

$$t_r = \frac{2}{c}\,|\sigma_0|^{1/2} \tag{1.81}$$

So the control (1.78) drives the sliding variable to zero in finite time (1.81). However, in the case of $\varphi(y, \dot{y}, t) \neq 0$, the compensated σ-dynamics become

$$\dot{\sigma} = \varphi(y, \dot{y}, t) - c\,|\sigma|^{1/2}\,\text{sign}(\sigma), \quad \sigma(0) = \sigma_0 \tag{1.82}$$

and convergence to zero does not occur.

If we could add a term to the control function (1.78) so that it will start following the disturbance $\varphi(y, \dot{y}, t) \neq 0$ in finite time, then the disturbance will be compensated for completely. As soon as the disturbance is canceled the sliding variable dynamics will coincide with Eq. (1.79) and $\sigma \to 0$ also in finite time.

Assuming $|\dot{\varphi}(y, \dot{y}, t)| \leq C$ the following control

$$\begin{cases} u = c\,|\sigma|^{1/2}\,\text{sign}(\sigma) + w \\ \dot{w} = b\,\text{sign}(\sigma) \end{cases} \quad c = 1.5\sqrt{C}; b = 1.1C \qquad (1.83)$$

makes the compensated σ-dynamics become

$$\begin{cases} \dot{\sigma} + c\,|\sigma|^{1/2}\,\text{sign}(\sigma) + w = \varphi(y, \dot{y}, t), \\ \dot{w} = b\,\text{sign}(\sigma) \end{cases} \qquad (1.84)$$

The control (1.83) meets our expectation, and the term w becomes equal to $\varphi(y, \dot{y}, t)$ in finite time, and therefore Eq. (1.84) becomes Eq. (1.79). Consequently $\sigma \to 0$ in finite time as well.

The control (1.83) is called *super-twisting control* and will be studied rigorously in Chap. 4.

The following properties are exhibited by the super-twisting control formulation:

- The super-twisting control (1.83) is a *second-order sliding mode control*, since it drives both σ, $\dot{\sigma} \to 0$ in finite time.
- The super-twisting control (1.83) is continuous since both $c\,|\sigma|^{1/2}\,\text{sign}(\sigma)$ and the term $w = b\int \text{sign}(\sigma)\,dt$ are continuous. Now, the high-frequency switching term $\text{sign}(\sigma)$ is "hidden" under the integral.

Example 1.10. The results of the simulation of the system (1.59) with the super-twisting control (1.61), (1.83), initial conditions $x_1(0) = 1$, $x_2(0) = -2$, the control gains $c = 13.5$, $b = 88$, the parameter $C = 80$, the output reference profile $y_c = 2\cos(t)$; and the disturbance $f(x_1, x_2, t) = \sin(2t)$, which is used for simulation purposes only, which illustrate the *super-twisting control* concept for *output tracking*, are presented in Figs. 1.35–1.38. During the simulation, the component $\dot{e}$ of the sliding variable (1.61) is computed using simple numerical differentiation, although the sliding mode differentiator studied in Sect. 1.6 could also be employed.

Discussion. The sliding variable σ is driven to zero in finite time (Fig. 1.35) by the *continuous* super-twisting control (Fig. 1.37). The high accuracy asymptotic output tracking (Fig. 1.36), which is achieved, is similar to that obtained with conventional SMC (Fig. 1.30) and integral SMC (Fig. 1.33), *but is obtained by means of continuous control* (Fig. 1.37) rather than high-frequency switching (Figs. 1.31 and 1.34). Including the attenuated (by integration) high frequency switching term $\text{sign}(\sigma)$ (Fig. 1.38) in the super-twisting control (1.83) is mandatory—it compensates for the disturbance while retaining a continuity of the control function (Fig. 1.37).

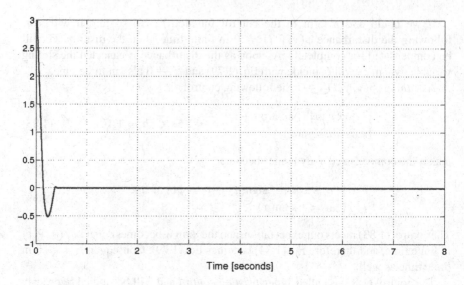

Fig. 1.35 Sliding variable σ

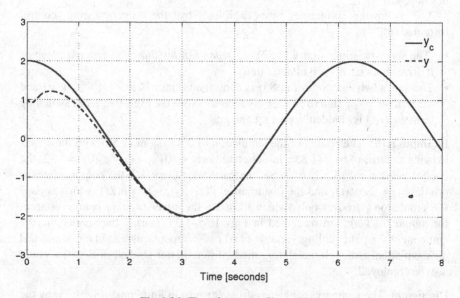

Fig. 1.36 The reference profile tracking

1.8.4 Prescribed Convergence Law Controller Design

Let us summarize the results on output tracking sliding mode control studied in the previous subsections:

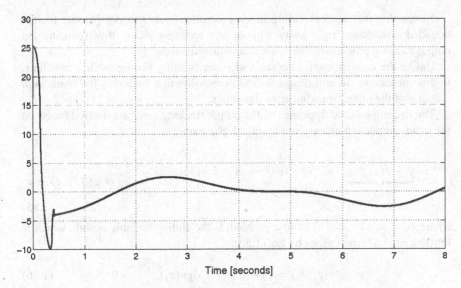

Fig. 1.37 Super-twisting control

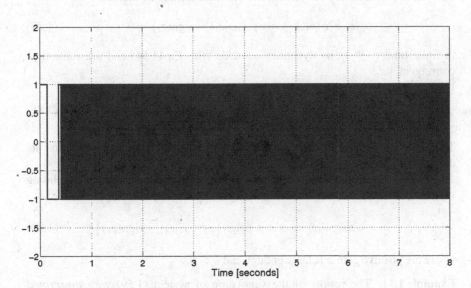

Fig. 1.38 Time history of sign (σ)

Sliding mode controller	Sliding variable convergence	Output tracking convergence	Type
Traditional	Finite time	Asymptotic	Discontinuous
Integral	Asymptotic	Asymptotic	Discontinuous
Super-twisting	Finite time	Asymptotic	Continuous

Notice that the integral sliding mode controller is able to compensate for the bounded disturbance right away without any reaching phase, thus reducing the output tracking problem to one without any disturbances.

Unlike the conventional, integral, and super-twisting sliding mode controllers, in this subsection we will design a 2-SM controller that ensures a *full dynamical collapse* of the output tracking error dynamics.

The uncompensated dynamics of the output tracking error are derived based on the input–output dynamics given by Eq. (1.60), namely

$$e^{(2)} = \underbrace{\ddot{y}_c - f(y, \dot{y}, t)}_{\varphi(y,\dot{y},t)} - u \quad \Rightarrow \quad e^{(2)} = \varphi(y, \dot{y}, t) - u \quad \Rightarrow \quad \begin{aligned} \dot{e}_1 &= e_2 \\ \dot{e}_2 &= \varphi(y, \dot{y}, t) - u \end{aligned}$$

$$(1.85)$$

where $e_1 = e$, $e_2 = \dot{e}$. The first problem is the sliding variable design, which is introduced in a format given by Eq. (1.52):

$$\sigma = \sigma(e_1, e_2) = e_2 + c\,|e_1|^{1/2}\,\mathrm{sign}(e_1), \quad c > 0 \qquad (1.86)$$

As soon as the trajectory of system (1.85) reaches the sliding manifold

$$\sigma = \sigma(e_1, e_2) = e_2 + c\,|e_1|^{1/2}\,\mathrm{sign}(e_1) = 0 \qquad (1.87)$$

in finite time $t \le t_r$ its dynamics coincide with Eq. (1.54) and are finite-time convergent, i.e.,

$$\begin{cases} \dot{e}_1 = e_2 \\ \dot{e}_2 = -c\,|e_1|^{1/2}\,\mathrm{sign}(e_1) \end{cases} \quad e_1(t_r) = e_{1r} \qquad (1.88)$$

Therefore (as studied in Sect. 1.8) e_1, e_2 (or the output tracking error and its derivative) converge to zero in finite time.

The 2-SM controller with the *prescribed convergence law* that drives the sliding variable (1.86) to zero in finite time is designed by analogy with Eq. (1.58):

$$u = -\rho\,\mathrm{sign}(\sigma) \Rightarrow u = -\rho\,\mathrm{sign}\left(e_2 + c\,|e_1|^{1/2}\,\mathrm{sign}(e_1)\right) \qquad (1.89)$$

where the positive gain ρ is large enough and $c > 0$.

Example 1.11. The results of the simulation of system (1.59) with *prescribed convergence law* control (1.89), which illustrate the *prescribed convergence law* control concepts for *output tracking*, are presented in Figs. 1.39–1.42. During the simulation, the component $e_2 = \dot{e}$ of the sliding variable in Eq. (1.86) is computed using simple numerical differentiation.

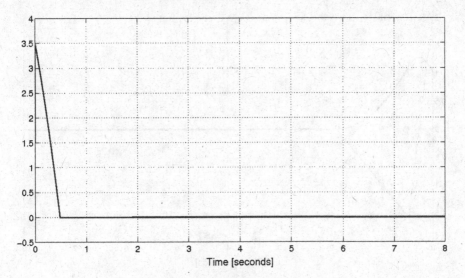

Fig. 1.39 Sliding variable σ

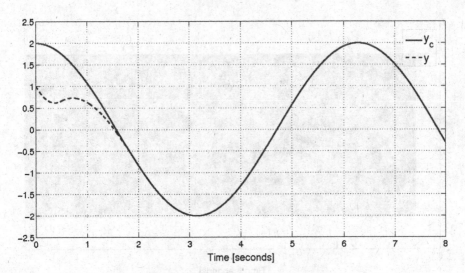

Fig. 1.40 The reference profile tracking

Discussion. The sliding variable σ is driven to zero in finite time (Fig. 1.39) by the *prescribed convergence law* (Fig. 1.42). The finite-time convergence of the output tracking error and its derivative to zero (or *dynamical collapse*) is achieved in the presence of the bounded disturbance $\varphi(y, \dot{y}, t)$ (Figs. 1.40 and 1.41).

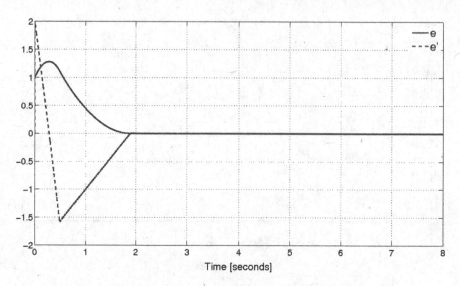

Fig. 1.41 The output tracking error and it derivative

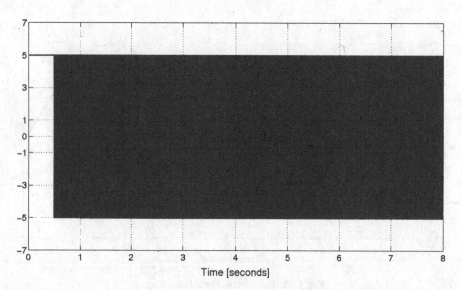

Fig. 1.42 The control function

1.9   Notes and References

The objective of this chapter is to provide an intuitive overview of conventional (first-order) and second-order sliding modes. The definitions and concepts are consequently taken from a wide range of sources and will be rigorously referred to and discussed in the subsequent chapters. The main terminology and techniques are taken from the established monographs and papers [57,61,64,67,74,91,92,113, 123, 132, 146, 147, 182, 183, 186, 195]. For an account of the early developments in VSS and SMC and the people involved, see the review in [184].

1.10   Exercises

Exercise 1.1. A simplified m longitudinal motion of an underwater vehicle can be described by

$$m\ddot{x} + k\dot{x}\,|\dot{x}| = u \qquad (1.90)$$

where x is position of the vehicle, u is the control input (a force that is provided by a propeller), m is the mass of the vehicle, and $k > 0$ is the drag coefficient. Assuming the value of m is known exactly, the drag coefficient is bounded $k_1 \le k \le k_2$ and the position and its derivative (velocity) x, $\dot{x}$ are measured:

(a) Obtain a state system model of the vehicle using $x_1 = x$, $x_2 = \dot{x}$ as the state variables.
(b) Design a conventional sliding mode control law u that drives x_1, $x_2 \to 0$ as time increases.
(c) Simulate the control system for $x_1(0) = 2\,\text{m}$, $x_2(0) = 0.5\,\text{m/s}$, $m = 4\,\text{kg}$, and $k = 1.5 + 0.4\sin(2t)\left[\frac{\text{kg}}{\text{ms}}\right]$. Plot the time histories of the sliding variable, the control function u, the position x_1, and the velocity x_2.
(d) Identify the quantities that reach zero in finite time and the ones that approach zero asymptotically.

Hint: The function $k\dot{x}\,|\dot{x}|$ can be bounded as $|k\dot{x}\,|\dot{x}|| \le k_2\dot{x}^2 = 1.9\dot{x}^2$.

Exercise 1.2. Repeat Exercise 1.1 approximating the *sign* function in the control law by the sigmoid function $\text{sign}(\sigma) \approx \frac{\sigma}{|\sigma|+\varepsilon}$ and separately by the saturation function

$$\text{sign}(\sigma) \approx \begin{cases} 1 & if\ \sigma > \varepsilon \\ \frac{\sigma}{\varepsilon} & if\ |\sigma| \le \varepsilon \\ -1 & if\ \sigma < -\varepsilon \end{cases}$$

for $\varepsilon = 0.01$. Compare the results of the simulations.

Exercise 1.3. Repeat Exercise 1.1 designing a conventional sliding mode control law in terms of $v = \dot{u}$ that drives to zero an auxiliary sliding variable $s = \dot{\sigma} + C_1\sigma$. The derivative of the original sliding variable σ may be obtained by using a sliding mode differentiator. Do you expect the original sliding variable σ to reach zero in finite time? Please explain why the original control function is continuous.

Exercise 1.4. Repeat Exercise 1.1 designing $u = u(x)$ using the super-twisting control law.

Exercise 1.5. Repeat Exercise 1.1 designing a second-order sliding mode control law with $v = \dot{u}$ using the prescribed convergence law technique.

Exercise 1.6. For the DC motor modeled by

$$J\frac{d\omega}{dt} = k_m i - T_L \tag{1.91}$$

$$L\frac{di}{dt} = -iR - k_b\omega + u \tag{1.92}$$

where J is the moment of inertia, i is the armature current, L and R are the armature inductance and resistance respectfully, ω is the motor angular speed, k_b is a constant of back electromotive force, k_m is a motor torque constant, T_L is an unknown load torque which is bounded and has bounded derivative, and u is a control function defined by the armature voltage, design a sliding mode control u, steering the angular speed ω to zero assuming both i and ω are measurable and all parameters are known. Simulate the control system with $R = 1$ Ohm, $L = 0.5$ H, $k_m = 5 \cdot 10^{-2}$ N m/A , $k_b = k_m$, $J = 10^{-3}$ N m s^2/rad, $T_L = 0.1 \sin(t)$ N m, $\omega(0) = 1$ rad/s, and $i(0) = 0$. Plot the time histories of the sliding variable, the control function u, the current i, and the angular speed ω.

Exercise 1.7. Repeat Exercise 1.6 assuming that only $\omega(t)$ is measured. Design a sliding mode observer for estimating $\hat{\imath}(t) \rightarrow i(t)$. Simulate the control system. Plot the time histories of the sliding variable, the control function u, the angular speed ω, the current i and its estimate $\hat{\imath}$.

Exercise 1.8. Repeat Exercise 1.6 and design a second-order sliding mode control law u in the form of super-twisting control. Simulate the control system. Plot the time histories of the sliding variable, the control function u, the angular speed ω, and the current i.

Exercise 1.9. For the DC motor given in Exercise 1.6 design a conventional sliding mode control u that provides asymptotic output tracking $\omega(t) \rightarrow \omega_c(t)$, where $\omega_c(t)$ is an angular speed command given in current time. Simulate the control system with $\omega_c(t) = 0.2 \sin(2t)$ rad/s. Plot the time histories of the sliding variable, the control function u, the angular speed ω, the current i, and the output tracking error $e_\omega = \omega_c(t) - \omega(t)$. The torque load profile $T_L = 0.1 \sin(t)$ N m is given for simulation purposes only and must be considered unknown when designing the controller.

Exercise 1.10. Repeat Exercise 1.9 using integral sliding mode control to compensate for the unknown external disturbance. Simulate the control system with $\omega_c(t) = 0.2 \sin(2t)$ rad/s. Plot the time histories of the sliding variable, the control function u, the angular speed ω, the current i, and the output tracking error $e_\omega = \omega_c(t) - \omega(t)$.

Exercise 1.11. Repeat Exercise 1.9 and design the second-order sliding mode control law u as a super-twisting control.

Chapter 2
Conventional Sliding Modes

This chapter considers the development of conventional sliding mode methods. The chapter describes the early work to define the notion of the solution of differential equations with discontinuous right-hand sides and the concept of "equivalent control" as a means to describe the reduced-order dynamics while a sliding motion is taking place. The main focus of the chapter is on the development of sliding mode design techniques for uncertain linear systems—specifically systems which can be thought of as predominantly linear in a characteristic, or nonlinear systems which can be modeled well (at least locally) by a linear system. For such systems, sliding surfaces formed from linear combinations of the states are considered (i.e., hyperplanes in the state space). In this chapter we consider different explicit design methods which can be used to synthesize hyperplanes which give appropriate closed-loop dynamics when a sliding motion is induced. Different classes of control law are then developed to guarantee the existence of a sliding motion in finite time and to ensure the sliding motion can be maintained in the face of uncertainty. The majority of the chapter is based on the assumption that state information is available for use in the control law. This is convenient and indeed mirrors the development of the ideas since their inception. However, the assumption that all the state, are available is somewhat impractical from an engineering perspective, and in the later sections we consider the case when only output information is available. The impact of this is studied both in terms of the constraints this imposes on the choice of sliding surfaces and the associated control laws.

2.1 Introduction

This chapter will discuss "conventional" sliding modes—or to be more precise first-order sliding modes when viewed in the context of higher-order sliding. Consider a general state-space system

$$\dot{x} = f(x, u, d) \tag{2.1}$$

Y. Shtessel et al., *Sliding Mode Control and Observation*, Control Engineering,
DOI 10.1007/978-0-8176-4893-0_2, © Springer Science+Business Media New York 2014

where $x \in \mathbb{R}^n$ is a vector which represents the state and $u \in \mathbb{R}^m$ is the control input. It is assumed that $f(\cdot)$ is differentiable with respect to x and absolutely continuous with respect to time. The quantity $d \in \mathbb{R}^q$ represents external bounded disturbances/uncertainties within the system. Consider a surface in the state space given by

$$S = \{x : \sigma(x) = 0\} \tag{2.2}$$

A formal definition of an ideal sliding mode will now be given generalizing Definition 1.3 from Chap. 1.

Definition 2.1. An ideal sliding mode is said to take place on Eq. (2.2) if the states $x(t)$ evolve with time such that $\sigma(x(t_r)) = 0$ for some finite $t_r \in \mathbb{R}^+$ and $\sigma(x(t)) = 0$ for all $t > t_r$.

During a sliding mode, $\sigma(t) = 0$ for all $t > t_r$. Intuitively this dynamical collapse implies the motion of the system when confined to S will be of reduced dynamical order. From a control systems perspective the capacity to analyze the dynamics of the reduced-order motion is important.

If the control action in $u = u(x)$ Eq. (2.1) is discontinuous, the differential equation describing the resulting closed-loop system written as

$$\dot{x}(t) = f^c(x) \tag{2.3}$$

is such that the function $f^c : \mathbb{R} \times \mathbb{R}^n \mapsto \mathbb{R}^n$ is discontinuous with respect to the state vector. The classical theory of differential equations is now not applicable since Lipschitz assumptions are usually employed to guarantee the existence of a unique solution. The solution concept proposed by Filippov for differential equations with discontinuous right-hand sides constructs a solution as the "average" of the solutions obtained from approaching the point of discontinuity from different directions.

2.1.1 Filippov Solution

Consider initially the case when the system has a single input and $\sigma : \mathbb{R}^n \mapsto \mathbb{R}$. Suppose x_0 is a point of discontinuity on S and define $f^c_-(x_0)$ and $f^c_+(x_0)$ as the limits of $f^c(x)$ as the point x_0 is approached from opposite sides of the tangent to S at x_0. The solution proposed by Filippov is given by

$$\dot{x}(t) = (1 - \alpha)f^c_-(x) + \alpha f^c_+(x) \tag{2.4}$$

where the scalar $0 < \alpha < 1$.

The scalar α is chosen so that

$$f^c_a := (1 - \alpha)f^c_- + \alpha f^c_+$$

is tangential to S (see Fig. 2.1).

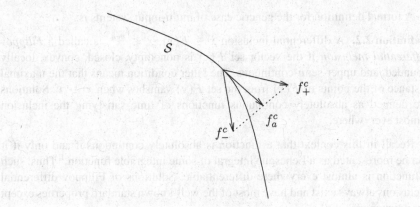

Fig. 2.1 A schematic of the Filippov construction

Remark 2.1. From the discussion above, it is clear the Filippov solution is an average solution of the two "velocity" vectors at the point x_0.

Equation (2.4) can be thought of as a differential equation whose right-hand side is defined as the convex set

$$F(x) = \{(1 - \alpha) f_-^c + \alpha f_+^c : \text{ for all } \alpha \in \begin{bmatrix} 0 & 1 \end{bmatrix}\}$$

and thus

$$\dot{x}(t) \in F(x)$$

The values of α which ensure $\dot{\sigma}(t) = 0$ can be computed explicitly from Eq. (2.4). For simplicity further suppose $\sigma = Sx$ where $S^T \in \mathbb{R}^n$. Then explicitly

$$\dot{\sigma} = S\dot{x} = (1 - \alpha)Sf_-^c + \alpha Sf_+^c$$

In order to maintain $\sigma = 0$, the scalar α must satisfy

$$(1 - \alpha)Sf_-^c + \alpha Sf_+^c = 0$$

and consequently (and uniquely)

$$\alpha = \frac{Sf_-^c}{Sf_-^c - Sf_+^c}$$

so that

$$\dot{x}(t) = \frac{Sf_-^c f_+^c - Sf_+^c f_-^c}{Sf_-^c - Sf_+^c}$$

A formal definition for the generic case of multi-input systems is:

Definition 2.2. A differential inclusion $\dot{x} \in F(x)$, $x \in \mathbb{R}^n$, is called a *Filippov differential inclusion* if the vector set $F(x)$ is nonempty, closed, convex, locally bounded, and upper-semi-continuous. The latter condition means that the maximal distance of the points of $F(x)$ from the set $F(y)$ vanishes when $x \to y$. Solutions are defined as absolutely continuous functions of time satisfying the inclusion almost everywhere.

Recall in this context that a function is absolutely continuous if and only if it can be represented as a Lebesgue integral of some integrable function.[1] Thus, such a function is almost everywhere differentiable. Solutions of Filippov differential inclusions always exist and have most of the well-known standard properties except the uniqueness.

Definition 2.3. It is said that a differential equation $\dot{x} = f(x)$ with a locally bounded Lebesgue-measurable right-hand side is understood in the Filippov sense, if it is replaced by a special Filippov differential inclusion $\dot{x} \in F(x)$, where

$$F(x) = \bigcap_{\delta > 0} \bigcap_{\mu N = 0} \overline{co}\, f(O_\delta(x) \backslash N) \tag{2.5}$$

Here μ is the Lebesgue measure, $O_\delta(x)$ is the δ-vicinity of x, and $\overline{co}\, M$ denotes the convex closure of M.

Note that any surface or curve has zero Lebesgue measure. Thus, values on any such set do not affect the Filippov solutions. In the most usual case, when f is continuous almost everywhere, the procedure is to take $F(x)$ being the convex closure of the set of all possible limit values of f at a given point x, obtained when its continuity point y tends to x. In the general case approximate-continuity points y are taken (one of the equivalent definitions by Filippov).[2] A solution of $\dot{x} = f(x)$ is defined as a solution of $\dot{x} \in F(x)$. Obviously, values of f on any set of measure 0 do not influence the Filippov solutions. Note that with continuous f the standard definition is obtained. The nonautonomous case is reduced to the considered one introducing the fictitious equation $\dot{t} = 1$.

In order to better understand the definition, consider the case when the number of limit values $f_1, \ldots, f_n$ at the point x is finite. Then any possible Filippov velocity has the form $\dot{x} = \lambda_1 f_1 + \cdots + \lambda_n f_n$, $\lambda_1 + \cdots + \lambda_n = 1$, $\lambda_i \geq 0$, and can be considered as a mean value of the velocity taking on the values f_i during the fraction of time $\lambda_i \Delta t$ of the current infinitesimal time interval Δt.

---

[1] For details see [158].
[2] For details see [158].

2.1.2 Concept of Equivalent Control

One way to undertake this analysis is by the so-called *equivalent control* method attributed to Utkin. This defines the equivalent control as the control action necessary to maintain an ideal sliding motion on $\mathcal{S}$. The idea is to exploit the fact that in conventional sliding modes both $\sigma = \dot{\sigma} = 0$. The constraint on the derivative of σ can be written as

$$\dot{\sigma} = \frac{\partial \sigma}{\partial x} \frac{dx}{dt} = \frac{\partial \sigma}{\partial x} f(x, u, d) = 0$$

This represents an algebraic equation in x, u, and d, and by definition, the equivalent control signal $u_{eq}(t)$, which is the continuous control function required to maintain sliding, is the solution to

$$\frac{\partial \sigma}{\partial x} f(x, u_{eq}, d) = 0 \tag{2.6}$$

For example, consider the affine system

$$\dot{x} = f(x) + g(x)u + d \tag{2.7}$$

The specific structure which has been imposed here ensures that for a given x the control input appears linearly. Consequently Eq. (2.6) simplifies to

$$\frac{\partial \sigma}{\partial x} f(x) + \frac{\partial \sigma}{\partial x} g(x)u_{eq} + \frac{\partial \sigma}{\partial x} d = 0 \tag{2.8}$$

and so, provided $\frac{\partial \sigma}{\partial x} g(x)$ is nonsingular, from Eq. (2.6)

$$u_{eq} = -\left(\frac{\partial \sigma}{\partial x} g(x) \right)^{-1} \frac{\partial \sigma}{\partial x} f(x) - \left(\frac{\partial \sigma}{\partial x} g(x) \right)^{-1} \frac{\partial \sigma}{\partial x} d \tag{2.9}$$

The closed-loop response is given by substituting the expression in Eq. (2.9) into Eq. (2.7) to yield

$$\dot{x} = \left(I - g(x) \left(\frac{\partial \sigma}{\partial x} g(x) \right)^{-1} \frac{\partial \sigma}{\partial x} \right) f(x) + \left(I - \left(\frac{\partial \sigma}{\partial x} g(x) \right)^{-1} \frac{\partial \sigma}{\partial x} \right) d \tag{2.10}$$

Example 2.1. Consider a system where the state vector is given by $x = [x_1, x_2]^T$ with the structure of (2.7) so that

$$f(x) = \begin{bmatrix} x_2 \\ 0 \end{bmatrix}, \quad g(x) = \begin{bmatrix} 0 \\ 1 \end{bmatrix}, \quad d = \begin{bmatrix} 0 \\ \phi(x_1, x_2, t) \end{bmatrix} \tag{2.11}$$

The scalar disturbance term $\phi(x_1, x_2, t)$ comprises dry and viscous friction as well as any other unknown resistance forces, and it is assumed to be bounded, i.e., $|\phi(x_1, x_2, t)| \leq L$. Now considering the surface $\sigma = x_2 + cx_1$, it implies

$$\frac{\partial \sigma}{\partial x} = \left[\frac{\partial \sigma}{\partial x_1}, \frac{\partial \sigma}{\partial x_2} \right] = [c, 1]$$

and consequently the equivalent control is given by Eq. (2.9), i.e.,

$$u_{eq} = -cx_2 - \phi(x_1, x_2, t)$$

Example 2.2. Consider the following multi-input multi-output (MIMO) linear system:

$$\begin{aligned} \dot{x}_1 &= x_1 + x_2 + x_3 + u_2 \\ \dot{x}_2 &= x_2 + 3x_3 + u_1 - u_2 \\ \dot{x}_3 &= x_1 + x_3 - u_1 \end{aligned}$$

and the corresponding outputs surfaces

$$\begin{aligned} \sigma_1 &= -x_1 + 10x_3 \\ \sigma_2 &= x_3 + x_2 \end{aligned}$$

Applying the concept of equivalent control we need to find the dynamic of sliding modes in the intersection of the output surfaces σ_1 and σ_2, i.e.,

$$\begin{aligned} \dot{\sigma}_1 &= 9x_1 - x_2 + 9x_3 - 10u_1 + u_2 \\ \dot{\sigma}_2 &= x_1 + x_2 + 4x_3 - u_2 \end{aligned}$$

From the invariance conditions $\dot{\sigma}_1 = 0$, $\sigma_1 = 0$, and $\dot{\sigma}_2 = 0$, $\sigma_2 = 0$ we obtain:

$$\begin{aligned} u_{1eq} &= -x_1 - 2x_2 - 4x_3 \\ u_{2eq} &= x_1 + 1.3x_3 \end{aligned}$$

and the reduced dynamics of the original system is given by

$$\begin{aligned} x_1 &= 10x_3 \\ x_2 &= -x_3 \\ \dot{x}_3 &= -0.3x_3 \end{aligned}$$

Remark 2.2. There are several points to note from the analysis given above:

- The expression for the equivalent control u_{eq} in Eq. (2.9) comes from formally solving Eq. (2.6), considered as an algebraic equation. It is therefore quite independent of the control signal which is actually applied. The control signal which is physically applied to the plant can be discontinuous in nature. However, the solution to Eq. (2.9) will always be smooth.

- The equivalent control u_{eq} in Eq. (2.9) depends on the disturbance $d(t)$ which will generally be unknown. Consequently Eq. (2.9) will not be physically implementable.
- The control signal in Eq. (2.9) is best thought of as an abstract concept to facilitate the creation of an expression for the reduced-order system in Eq. (2.10), hence establishing a differential equation from which the stability of the closed-loop system can be studied.

Another crucial property of sliding mode control systems can now be demonstrated, namely its robustness—or more precisely, its invariance to a certain class of uncertainty. Suppose the disturbance d acts in the channels of the inputs so that

$$d(t) = g(x)\xi(t) \tag{2.12}$$

for some (unknown) signal $\xi(t)$. Then it is easy to see that in Eq. (2.10)

$$\left(I - g(x)\left(\frac{\partial \sigma}{\partial x}g(x)\right)^{-1}\frac{\partial \sigma}{\partial x}\right)d = \left(I - g(x)\left(\frac{\partial \sigma}{\partial x}g(x)\right)^{-1}\frac{\partial \sigma}{\partial x}\right)g(x)\xi$$

$$= g(x)\xi - g(x)\xi$$

$$= 0$$

and so Eq. (2.10) collapses to

$$\dot{x} = \left(I - g(x)\left(\frac{\partial \sigma}{\partial x}g(x)\right)^{-1}\frac{\partial \sigma}{\partial x}\right)f(x) \tag{2.13}$$

The closed-loop (reduced-order) system given in Eq. (2.13) is completely independent of ξ. This *invariance property* has motivated research in sliding mode control.

Clearly from Eq. (2.10) the choice of the surface affects the dynamics of the reduced-order motion. In terms of control system design, the selection of the surface is one of the key design choices. Later in this chapter an alternative viewpoint and design framework will be given which is more amenable from the perspective of synthesizing choices for S.

Example 2.3. Consider a second-order system representing a DC motor:

$$\dot{\theta}(t) = w(t) \tag{2.14}$$

$$\dot{w}(t) = \frac{F(t)}{J} + \frac{K_t}{J}u(t) \tag{2.15}$$

where θ represents the shaft position and w is the angular rotation speed. The scalar J represents the inertia of the shaft, $F(t)$ represents the effects of dynamic friction and K_t represents the motor constant. Assume all the coefficients are unknown but bounded so that $0 < \underline{J} \leq J \leq \bar{J}$, $|F(t)| \leq \bar{F}$ and the motor constant $\underline{K}_t \leq K_t \leq \bar{K}_t$.

Suppose a switching function σ is defined as

$$\sigma = w + m\theta \tag{2.16}$$

where m is a positive design scalar. During the sliding motion if $\sigma \equiv 0$, then combining Eqs. (2.14) and (2.16) gives

$$\dot{\theta}(t) = w(t) = -m\theta(t)$$

and a first-order system is obtained which is *independent of the uncertainty* associated with $F(t)$, J, and K_t. The closed-loop solution is given by

$$\theta(t) = \theta_0 e^{-m(t-t_s)} \tag{2.17}$$

where θ_0 represents the value of $\theta(\cdot)$ at the time instant t_s at which sliding is achieved. Clearly in Eq. (2.17) the effect of the uncertainty has been totally rejected and robust closed-loop performance has been achieved.

Figures 2.2–2.4 are associated with simulations where $\frac{K_t}{J} = 0.45$ and $m = 1$. The controller regulates the shaft position back to zero from an initial displacement of 1 rad.

The next section focuses on linear (in fact uncertain linear) system representations which have been more well studied in the literature and yield systematic tractable methods for the design of $\mathcal{S}$.

2.2 State-Feedback Sliding Surface Design

Consider the nth-order linear time-invariant system with m inputs given by

$$\dot{x}(t) = Ax(t) + Bu(t) \tag{2.18}$$

where $A \in \mathbb{R}^{n \times n}$ and $B \in \mathbb{R}^{n \times m}$ with $1 \leq m \leq n$. Without loss of generality it can be assumed that the input distribution matrix B has full rank. Define a switching function $\sigma : \mathbb{R} \to \mathbb{R}^m$ to be

$$\sigma(x) = Sx(t) \tag{2.19}$$

where $S \in \mathbb{R}^{m \times n}$ is of full rank and let $\mathcal{S}$ be the hyperplane defined by

$$\mathcal{S} = \{x \in \mathbb{R}^n \ : \ Sx = 0\} \tag{2.20}$$

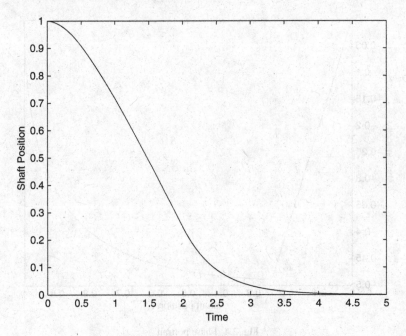

Fig. 2.2 Evolution of shaft position

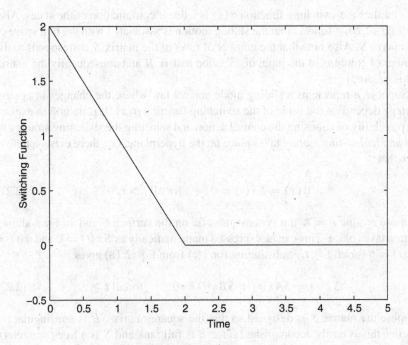

Fig. 2.3 Evolution of switching function

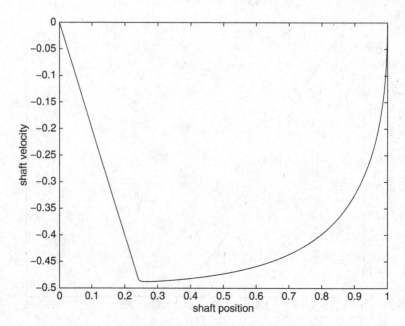

Fig. 2.4 Phase portrait

This implies the switching function $\sigma(x)$ is a linear combination of the states. Also from Eq. (2.20) it follows that the sliding motion is associated with the null space of the matrix S. Also note that the number of rows of the matrix S corresponds to the number of columns of the input distribution matrix B and consequently the matrix SB is square.

Suppose u represents a sliding mode control law where the changes in control strategy depend on the value of the switching function $\sigma(x)$. It is natural to explore the possibility of choosing the control action and selecting the switching strategy so that an *ideal sliding motion* takes place on the hyperplane, i.e., there exists a time t_r such that

$$\sigma(x) = Sx(t) = 0 \qquad \text{for all } t > t_r \qquad (2.21)$$

Suppose at time $t = t_r$ the system states lie on the surface S and an ideal sliding motion takes place. This can be expressed mathematically as $Sx(t) = 0$ and $\dot{\sigma}(t) = S\dot{x}(t) = 0$ for all $t \geq t_r$. Substituting for $\dot{x}(t)$ from Eq. (2.18) gives

$$S\dot{x}(t) = SAx(t) + SBu(t) = 0 \qquad \text{for all } t \geq t_r \qquad (2.22)$$

Suppose the matrix S is designed so that the square matrix SB is nonsingular (in practice this is easily accomplished since B is full rank and S is a free parameter).

The equivalent control, written as u_{eq}, as argued above, is the unique solution to the algebraic equation (2.22), namely

$$u_{eq}(t) = -(SB)^{-1} SAx(t) \tag{2.23}$$

This represents the control action which is required to maintain the states on the switching surface. Substituting the expression for the equivalent control into Eq. (2.18) results in a free motion

$$\dot{x}(t) = \left(I_n - B(SB)^{-1}S\right) Ax(t) \qquad \text{for all } t \geq t_r \text{ and } Sx(t_r) = 0 \tag{2.24}$$

It can be seen from Eq. (2.24) that the sliding motion is a control independent free motion which depends on the choice of sliding surface.

Now suppose the system (2.18) is uncertain:

$$\dot{x}(t) = Ax(t) + Bu(t) + B\xi(t, x, u) \tag{2.25}$$

where $\xi : \mathbb{R} \times \mathbb{R}^n \times \mathbb{R}^m \mapsto \mathbb{R}^m$ is unknown but bounded and encapsulates any nonlinearities or uncertainties in the system. Uncertainty which acts in the channel of the inputs is often referred to as *matched uncertainty*. Suppose a control law can be found for the system in Eq. (2.25) which induces a sliding motion on Eq. (2.20). Arguing as before, the equivalent control is in this case given by

$$u_{eq}(t) = -(SB)^{-1} SAx(t) - \xi(t, x, u) \tag{2.26}$$

The closed-loop sliding motion is given by substituting Eq. (2.26) in Eq. (2.25) and yields

$$\dot{x}(t) = \left(I_n - B(SB)^{-1}S\right) Ax(t) \tag{2.27}$$

This motion is completely independent of the uncertainty. Although the sliding motion is clearly dependent on the matrix S, how to select S to achieve a specific design goal is not transparent. One way to see the effect is to first transform the system into a suitable *regular form*.

2.2.1 Regular Form

In this section a coordinate transformation is introduced to create a special structure in the input distribution matrix. Since by assumption $\text{rank}(B) = m$ there exists an orthogonal matrix $T_r \in \mathbb{R}^{n \times n}$ such that

$$T_r B = \begin{bmatrix} 0 \\ B_2 \end{bmatrix} \tag{2.28}$$

where $B_2 \in \mathbb{R}^{m \times m}$ and is nonsingular. The matrix T_r can be obtained via so-called QR factorization.[3] This means a design algorithm can be created to deliver a change of coordinates in which a specific structure is imposed on the input distribution matrix.

Remark 2.3. This is not the only way to achieve the partition in Eq. (2.28). In principle, any nonsingular matrix which partitions the input distribution matrix can be employed—indeed later in the chapter a different approach based on orthogonal complements will be used. One advantage of the method of QR factorization is that the method generates an orthogonal matrix T_r. Consequently the associated coordinate transformation is orthogonal which means it has good numerical conditioning and also Euclidean distance is preserved.

Let $z = T_r x$ and partition the new coordinates so that

$$z = \begin{bmatrix} z_1 \\ z_2 \end{bmatrix} \tag{2.29}$$

where $z_1 \in \mathbb{R}^{n-m}$ and $z_2 \in \mathbb{R}^m$. The system matrices (A, B) in the original coordinates become $A \leftrightarrow T_r A T_r^T$ and $B \leftrightarrow T_r B$ in the "z" coordinates. Now the linear system (2.18) can be written as

$$\dot{z}_1(t) = A_{11} z_1(t) + A_{12} z_2(t) \tag{2.30}$$

$$\dot{z}_2(t) = A_{21} z_1(t) + A_{22} z_2(t) + B_2 u(t) \tag{2.31}$$

in which

$$T_r A T_r^T = \begin{bmatrix} A_{11} & A_{12} \\ A_{21} & A_{22} \end{bmatrix}$$

The representation in Eqs. (2.30) and (2.31) is referred to as *regular form*. Suppose the matrix defining the switching function in the new coordinate system is compatibly partitioned as

$$S T_r^T = \begin{bmatrix} S_1 & S_2 \end{bmatrix} \tag{2.32}$$

where $S_1 \in \mathbb{R}^{m \times (n-m)}$ and $S_2 \in \mathbb{R}^{m \times m}$. Since $SB = S_2 B_2$ it follows that a necessary and sufficient condition for the matrix SB to be nonsingular is that $\det(S_2) \neq 0$ since $\det(SB) = \det(S_2 B_2) = \det(S_2) \det(B_2)$ and therefore

$$\det(SB) \neq 0 \Leftrightarrow \det(S_2) \neq 0$$

since $\det(B_2) \neq 0$ by construction. By design assume this to be the case. During an ideal sliding motion

$$S_1 z_1(t) + S_2 z_2(t) = 0 \qquad \text{for all } t > t_s \tag{2.33}$$

---

[3]For details of QR factorization methods see [177].

and therefore exploiting the non-singularity of S_2, the relationship in Eq. (2.33), can be rewritten as

$$z_2(t) = -M z_1(t) \tag{2.34}$$

where $M = S_2^{-1} S_1$. Substituting in Eq. (2.30) gives

$$\dot{z}_1(t) = (A_{11} - A_{12} M) z_1(t) \tag{2.35}$$

This equation is a straightforward expression describing the reduced-order dynamics in terms of the design freedom associated with the sliding surface.

If Eq. (2.30) is considered in isolation with z_1 thought of as the state vector and z_2 as a "virtual" control input, then Eq. (2.34) can be thought of as a state-feedback control law for Eq. (2.30). Consequently the dynamics describing the sliding motion in Eq. (2.35) can be thought of as the closed-loop system applying the feedback control law (2.34)–(2.30). It can be seen from Eq. (2.35) that S_2 has no direct effect on the dynamics of the sliding motion and acts only as a scaling factor for the switching function. A common choice for S_2, however, which stems from the so-called *hierarchical* design procedure, is to let $S_2 = \Lambda B_2^{-1}$ for some diagonal design matrix $\Lambda \in \mathbb{R}^{m \times m}$ which implies $SB = \Lambda$. By selecting M and S_2 the switching function in Eq. (2.32) is completely determined.

Remark 2.4.   The matrix S of the switching function $\sigma(x) = Sx(x)$ has the form:

$$S = S_2 \begin{bmatrix} M & I_{m \times m} \end{bmatrix} T_r. \tag{2.36}$$

There exist two major techniques for the design of the matrix M; these are

- Eigenvalue placement method
- Linear-quadratic minimization

2.2.2   Eigenvalue Placement

Single-input systems represented by the pair (A, B) where $B \in \mathbb{R}^n$ can be written in the so-called controllability canonical form[4]

$$A = \begin{bmatrix} 0 & 1 & 0 & \cdots & 0 \\ \vdots & 0 & 1 & & \vdots \\ \vdots & & \ddots & \ddots & 0 \\ 0 & & & 0 & 1 \\ -a_1 & -a_2 & \cdots & \cdots & -a_n \end{bmatrix} \qquad B = \begin{bmatrix} 0 \\ \vdots \\ \vdots \\ 0 \\ 1 \end{bmatrix} \tag{2.37}$$

---

[4]See for example [47].

where the scalars a_i are the coefficients of the characteristic equation of the A matrix:

$$\lambda^n + a_n\lambda^{n-1} + \cdots + a_2\lambda + a_1 = 0$$

For this general system an appropriate switching function is

$$\sigma(x) = s_1 x_1 + s_2 x_2 + \ldots + s_{n-1} x_{n-1} + x_n \qquad (2.38)$$

where the scalars s_i are to be chosen. Partition the state space associated with Eq. (2.37) into the first $n - 1$, and the last equation, so that

$$
\begin{bmatrix} \dot{x}_1 \\ \dot{x}_2 \\ \vdots \\ \vdots \\ \dot{x}_{n-1} \end{bmatrix} =
\begin{bmatrix} 0 & 1 & 0 & \cdots & 0 \\ \vdots & 0 & 1 & & \vdots \\ \vdots & & \ddots & \ddots & 0 \\ \vdots & & & 0 & 1 \\ 0 & \cdots & \cdots & \cdots & 0 \end{bmatrix}
\begin{bmatrix} x_1 \\ x_2 \\ \vdots \\ \vdots \\ x_{n-1} \end{bmatrix} +
\begin{bmatrix} 0 \\ \vdots \\ \vdots \\ 0 \\ 1 \end{bmatrix} x_n
$$

During the sliding motion $\sigma(x) = 0$ and so from Eq. (2.38) the last coordinate can be expressed as

$$x_n = -s_1 x_1 - s_2 x_2 - \ldots - s_{n-1} x_{n-1}$$

Now substituting for x_n in Eq. (2.2.2) yields a description of the sliding motion as

$$
\begin{bmatrix} \dot{x}_1 \\ \dot{x}_2 \\ \vdots \\ \vdots \\ \dot{x}_{n-1} \end{bmatrix} =
\begin{bmatrix} 0 & 1 & 0 & \cdots & 0 \\ \vdots & 0 & 1 & & \vdots \\ \vdots & & \ddots & \ddots & 0 \\ \vdots & & & 0 & 1 \\ -s_1 & \cdots & \cdots & \cdots & -s_{n-1} \end{bmatrix}
\begin{bmatrix} x_1 \\ x_2 \\ \vdots \\ \vdots \\ x_{n-1} \end{bmatrix}
$$

Therefore the characteristic equation of the sliding motion is

$$\lambda^{n-1} + s_{n-1}\lambda^{n-2} + \ldots + s_2\lambda + s_1 = 0$$

The scalars $s_1, \ldots s_n$ should therefore be chosen to make the polynomial above Hurwitz. More generally for multi-input systems, the regular form from Eqs. (2.30) and (2.31) must be relied upon. For a hyperplane parameterized as in Eq. (2.36) the sliding motion is governed by the system matrix $(A_{11} - A_{12}M)$ in Eq. (2.35) where the matrices A_{11} and A_{12} are associated with the regular form in Eqs. (2.30) and (2.31). In the context of designing a regulatory system, the matrix $(A_{11} - A_{12}M)$ must have stable eigenvalues. The switching surface design problem can therefore be considered to be one of choosing a state-feedback matrix M to stabilize the reduced-order system associated with the pair (A_{11}, A_{12}). Because of the special structure of the regular form, it can be shown that the pair (A_{11}, A_{12}) is controllable if and only if (A, B) is controllable.[5] Consequently if the original pair (A, B) is controllable then the problem of synthesizing the matrix S associated with the switching function can always be solved to ensure the associated hyperplane $\mathcal{S}$ yields a stable sliding motion. This is important; otherwise, the method would lack credibility. Any eigenvalue pole placement methods can be employed.

The eigenvalue placement algorithm is demonstrated on the following example.

Example 2.4. Consider the linear system:

$$\dot{x} = Ax + Bu = \begin{bmatrix} 1 & 1 & 1 \\ 0 & 1 & 3 \\ 1 & 0 & 1 \end{bmatrix} x + \begin{bmatrix} 0 & 1 \\ 1 & -1 \\ -1 & 0 \end{bmatrix} \begin{bmatrix} u_1 \\ u_2 \end{bmatrix}$$

Apply the transformation $z = T_r x$, where T_r is obtained by QR factorization, such that

$$\bar{B} = T_r B = \begin{bmatrix} \frac{1}{3}\sqrt{3} & \frac{1}{3}\sqrt{3} & \frac{1}{3}\sqrt{3} \\ \frac{1}{3}\sqrt{6} & -\frac{1}{6}\sqrt{6} & -\frac{1}{6}\sqrt{6} \\ 0 & \frac{1}{2}\sqrt{2} & -\frac{1}{2}\sqrt{2} \end{bmatrix} B = \begin{bmatrix} 0 & 0 \\ 0 & \frac{1}{2}\sqrt{6} \\ \sqrt{2} & -\frac{1}{2}\sqrt{2} \end{bmatrix}$$

to create the equivalent system:

$$\begin{bmatrix} \dot{z}_1 \\ \dot{z}_2 \end{bmatrix} = \bar{A} \begin{bmatrix} z_1 \\ z_2 \end{bmatrix} + \bar{B}u = \begin{bmatrix} \bar{A}_{11} & \bar{A}_{12} \\ \bar{A}_{21} & \bar{A}_{22} \end{bmatrix} \begin{bmatrix} z_1 \\ z_2 \end{bmatrix} + \begin{bmatrix} \bar{0} \\ B_2 \end{bmatrix} \begin{bmatrix} u_1 \\ u_2 \end{bmatrix}$$

where $\bar{A} = T_r A T_r^T$ and $T_r T_r^T = I_{n \times n}$. After some computations,

$$\bar{A}_{11} = [3], \quad \bar{A}_{12} = \begin{bmatrix} -\frac{1}{6}\sqrt{18} & -\frac{1}{2}\sqrt{6} \end{bmatrix}, \quad \bar{A}_{21} = \begin{bmatrix} 0 \\ \frac{1}{3}\sqrt{6} \end{bmatrix}, \quad \bar{A}_{22} = \begin{bmatrix} \frac{1}{2} & \frac{1}{4}\sqrt{12} \\ -\frac{5}{12}\sqrt{12} & -\frac{1}{2} \end{bmatrix}$$

"Now setting $z_2 = -Mz_1$ where $M = [M_{11} \ M_{12}]^T$, the dynamics for z_1 are given by:

$$\dot{z}_1 = (\bar{A}_{11} - A_{12}M)z_1 = (3 + \frac{1}{6}\sqrt{18}M_{11} + \frac{1}{2}\sqrt{6}M_{12})z_1$$

Provided the eigenvalue $\alpha = 3 + \frac{1}{6}\sqrt{18}M_{11} + \frac{1}{2}\sqrt{6}M_{12} < 0$, stability of z_1 is ensured. For example setting $\alpha = -1$, implies $M = [-\frac{18}{\sqrt{18}} \ -\frac{2}{\sqrt{6}}]^T$. Finally with $S_2 = I_{m \times m} = 1$, the matrix S is given by:

$$S = S_2[M \ I_{m \times m}]T_r = \begin{bmatrix} -3.1162 & -1.409 & -2.8233 \end{bmatrix}"$$

---

[5]See, for example, Sect. 3.4 in [67].

2.2.3 Quadratic Minimization

Consider the problem of minimizing the quadratic performance index

$$J = \frac{1}{2} \int_{t_s}^{\infty} x(t)^T Q x(t) \, dt \tag{2.39}$$

where Q is both symmetric and positive definite, and t_s is the time at which the sliding motion commences. The objective is to minimize Eq. (2.39) subject to the system equation (2.18) under the assumption that sliding takes place. Notice this is quite different from the "classical" LQR problem formulation which includes a penalty weighting on the control effort. Here no penalty cost on the control is imposed, and this represents a so-called cost-free control problem.

It is assumed that the state of the system at time t_s, given by $x(t_s)$, is a known initial condition, and the closed-loop system is stable such that $x(t) \rightarrow 0$ as $t \rightarrow \infty$. To solve this problem, the matrix Q from Eq. (2.39) is transformed and partitioned compatibly with the z coordinates from Eq. (2.29) so that

$$T_r Q T_r^T = \begin{bmatrix} Q_{11} & Q_{12} \\ Q_{12}^T & Q_{22} \end{bmatrix} \tag{2.40}$$

In the "z" coordinates, the cost J in Eq. (2.39) can be written as

$$J = \frac{1}{2} \int_{t_s}^{\infty} z_1(t)^T Q_{11} z_1(t) + 2 z_1(t)^T Q_{12} z_2(t) + z_2(t)^T Q_{22} z_2(t) \, dt \tag{2.41}$$

If the component z_1 is considered to be the state vector and z_2 the "virtual control" input then this expression represents a "traditional" mixed cost LQR problem associated with the state-space representation in Eq. (2.30) since the term $z_1^T Q_{12} z_2$ involves a mix of the state vector and the virtual control. To avoid this complication, a trick can be employed to "eliminate" the cross term. Define a new "virtual control" input as

$$v := z_2 + Q_{22}^{-1} Q_{12}^T z_1 \tag{2.42}$$

After algebraic manipulation, Eq. (2.41) may then be written as

$$J = \frac{1}{2} \int_{t_s}^{\infty} z_1^T \hat{Q} z_1 + v^T Q_{22} v \, dt \tag{2.43}$$

where

$$\hat{Q} := Q_{11} - Q_{12} Q_{22}^{-1} Q_{12}^T \tag{2.44}$$

Note that $\hat{Q}$ represents part of the Schur complement of Eq. (2.40). Recall the constraint equation (the null-space dynamics associated with the regular form) may be written as

$$\dot{z}_1(t) = A_{11} z_1(t) + A_{12} z_2(t) \tag{2.45}$$

Eliminating the z_2 contribution from Eq. (2.45) and using Eq. (2.42), after writing the differential equations in terms of the virtual control, the modified constraint equation becomes

$$\dot{z}_1(t) = \hat{A} z_1(t) + A_{12} v(t) \tag{2.46}$$

where

$$\hat{A} = A_{11} - A_{12} Q_{22}^{-1} Q_{12}^{T} \tag{2.47}$$

The positive definiteness of Q ensures from Schur complement arguments that $Q_{22} > 0$, so that Q_{22}^{-1} exists, and also that $\hat{Q} > 0$. Furthermore, the controllability of the original (A, B) pair ensures the pair $(\hat{A}, A_{12})$ is controllable. Consequently, the problem becomes one of minimizing the functional (2.43) subject to the system (2.46). This can be interpreted as a standard LQR optimal state-regulator problem. A necessary condition to ensure a solution to the LQR problem is that the pair $(\hat{A}, \hat{Q}^{1/2})$ is detectable[6] and then after solving the Riccati equation

$$\hat{P} \hat{A}^{T} + \hat{A} \hat{P} + \hat{Q} - \hat{P} A_{12} Q_{22}^{-1} A_{12}^{T} \hat{P} = 0 \tag{2.48}$$

the matrix parameterizing the hyperplane is

$$M = Q_{22}^{-1} Q_{12}^{T} + Q_{22}^{-1} A_{12}^{T} \hat{P} \tag{2.49}$$

Robustness of the LQR Sliding Surface Design

An advantage of this approach compared to pole placement is that the LQR optimization method inherits robustness. Suppose in fact *unmatched uncertainty* is present in the system so that Eq. (2.30) becomes

$$\dot{z}_1(t) = A_{11}(I + \Delta_1) z_1(t) + A_{12}(I + \Delta_1) z_2(t) \tag{2.50}$$

where Δ_1 and Δ_2 represent (unknown) multiplicative perturbations. Suppose Q is chosen so that $Q_{12} = 0$ and $Q_{22} = q I_m$ where q is a positive scalar. If a sliding mode is enforced, the reduced-order sliding motion will be given by

$$\dot{z}_1(t) = \left(A_{11}(I + \Delta_1) - \frac{1}{q} A_{12}(I + \Delta_2) A_{12}^{T} P_1 \right) z_1(t) \tag{2.51}$$

where P_1 is the symmetric positive definite solution to

$$P_1 A_{11}^{T} + A_{11} P_1 + Q_{11} - \frac{1}{q} P_1 A_{12} A_{12}^{T} P_1 = 0 \tag{2.52}$$

---

[6]A more detailed description of this approach is given in Sect. 4.2.2 in [67] and details about LQR methods appear in Appendix C.

Under some conditions on the perturbations it will be shown that Eq. (2.50) is stable. Assume Δ_1 is sufficiently small that

$$P_1\Delta_1 + \Delta_1^T P_1 < Q_{11} \tag{2.53}$$

and consider $V_1 = z_1^T P_1 z_1$ as a candidate Lyapunov function. Then

$$\dot{V} \le z_1^T \frac{1}{q}\left(P_1 A_{12} A_{12}^T P_1 - P_1 A_{12}(I+\Delta_2)A_{12}^T P_1 - P_1 A_{12}(I+\Delta_2^T)A_{12}^T P_1\right) z_1$$

$$= -z_1^T \frac{1}{q} P_1 A_{12}\left(\Delta_2 + \Delta_2^T + I\right) A_{12}^T P_1 z_1 \tag{2.54}$$

after substituting from the Riccati equation from Eq. (2.52) and the bound on the Δ_1 inequality from Eq. (2.53). Consequently if the uncertainty Δ_2 satisfies

$$\Delta_2 + \Delta_2^T + I > 0 \tag{2.55}$$

then $\dot{V} < 0$ for $z_1 \ne 0$ and the sliding motion remains stable. Some special cases can be considered:

• Unstructured Perturbations: Consider the expression

$$\Theta := 2(\Delta_2 + \frac{1}{2}I)^T(\Delta_2 + \frac{1}{2}I)$$

Clearly $\Theta \ge 0$ for all Δ_2. Expanding the right hand side it follows

$$2\Delta_2^T \Delta_2 + \Delta_2 + \Delta_2^T + \frac{1}{2}I \ge 0$$

which implies

$$\Delta_2 + \Delta_2^T + I \ge -2\Delta_2^T\Delta_2 + \frac{1}{2}I = 2(\frac{1}{4} - \Delta_2^T\Delta_2) \tag{2.56}$$

for all Δ_2. However if $\|\Delta_2\| < \frac{1}{2}$, the right hand side of (2.56) is positive and $\Delta_2 + \Delta_2^T + I > 0$.
• Structured Perturbations: Suppose the uncertainty is structured and has the special form $\Delta = \text{diag}(\delta_1,\ldots\delta_m)$ where the δ_i are scalars. Then Eq. (2.55) is equivalent to

$$2\delta_i + 1 > 0 \quad \text{for } i = 1\ldots m$$

and stability is ensured if $\delta_i > -\frac{1}{2}$ for $i = 1\ldots m$.

The analysis confirms the robustness of the LQR sliding surface design.

Example 2.5. Consider Example 2.4, now we will design the sliding surface with LQ minimization. Consider the matrix

$$Q = \begin{bmatrix} 1 & 0 & 0 \\ 0 & 2 & 0 \\ 0 & 0 & 3 \end{bmatrix}$$

After transformation we have

$$T_r Q T_r^T = \begin{bmatrix} 2 & -\frac{1}{6}\sqrt{18} & -\frac{1}{6}\sqrt{6} \\ -\frac{1}{6}\sqrt{18} & \frac{3}{2} & \frac{1}{12}\sqrt{12} \\ -\frac{1}{6}\sqrt{6} & \frac{1}{12}\sqrt{12} & \frac{5}{2} \end{bmatrix}$$

the elements of the new matrix Q are given by

$$Q_{11} = [2], \; Q_{12} = \begin{bmatrix} -\frac{1}{6}\sqrt{18} & -\frac{1}{6}\sqrt{6} \end{bmatrix},$$
$$Q_{21} = \begin{bmatrix} -\frac{1}{6}\sqrt{18} \\ -\frac{1}{6}\sqrt{6} \end{bmatrix}, \; Q_{22} = \begin{bmatrix} \frac{3}{2} & \frac{1}{12}\sqrt{12} \\ \frac{1}{12}\sqrt{12} & \frac{5}{2} \end{bmatrix}$$

and after direct computations we have $\hat{A} = 2$ and $\hat{Q} = 1.6364$. Now we can use the *lqr* command of MATLAB in order to obtain the matrix M:

$$M = [-2.032 - 2.3464]$$

Finally we obtain the matrix S:

$$S = \begin{bmatrix} -2.032 & -2.3464 & 1 \end{bmatrix} T_r = \begin{bmatrix} -3.089 & 0.492 & -0.922 \end{bmatrix}$$

2.3   State-Feedback Relay Control Law Design

In this section, although previously multi-input systems were considered at the outset, here the development of control laws for single-input systems will be considered first before multi-input generalizations are considered.

2.3.1   Single-Input Nominal Systems

Using the nomenclature developed in the previous section, suppose $m = 1$, i.e., the system is single input in nature. Assume that the switching function $\sigma(x) = Sx$ has already been defined. In this situation the matrix S is a row vector of the same order as the states. The objective is to force $\sigma \rightarrow 0$ in *finite time* and to ensure $\sigma \equiv 0$ for all subsequent time. From the nominal representation in Eq. (2.18) it follows

$$\dot{\sigma}(t) = S\dot{x}(t) = SAx(t) + SBu(t) \tag{2.57}$$

The objective is, through feedback control, to turn the equation above into the differential equation

$$\dot{\sigma}(t) = -\eta \text{sign}(\sigma(t)) \tag{2.58}$$

or equivalently

$$\sigma\dot{\sigma}(t) = -\eta|\sigma(t)| \tag{2.59}$$

For $\sigma(0) \neq 0$ the solution to the equation above becomes zero in finite time. This can easily be seen by the change of variable $V = \frac{1}{2}\sigma^2$. Clearly $\dot{V} = \sigma\dot{\sigma}$ and $|\sigma| = \sqrt{2V}$. Consequently the equation above becomes

$$\dot{V} = -\sqrt{2}\eta V^{1/2}$$

This implies

$$V^{1/2}(t) = V^{1/2}(0) - \sqrt{2}\eta t$$

and therefore at time $t_r = V^{1/2}(0)/(\sqrt{2}\eta)$, it follows $V^{1/2}(t_r) = 0$. Notice that $V = \frac{1}{2}\sigma^2$ can be viewed as a Lyapunov function for the system (2.59) because

$$\dot{V} = \sigma\dot{\sigma} = -\eta\sigma\mathrm{sign}(\sigma) = -\eta|\sigma| < 0$$

when $\sigma \neq 0$. Comparing Eqs. (2.57) and (2.59) it is clear that choosing

$$u(t) = \underbrace{-(SB)^{-1}SAx(t)}_{u_{eq}(t)} - \eta(SB)^{-1}\mathrm{sign}(\sigma) \tag{2.60}$$

as the control law in Eq. (2.57) creates in closed loop the system in (2.59). The simple control law in Eq. (2.60) thus ensures σ is driven to zero in finite time—in fact in $|\sigma(0)|/\eta$ units of time.

2.3.2 Single-Input Perturbed Systems

This can be easily extended to the case of systems with bounded matched uncertainty. Now consider the uncertain linear system

$$\dot{x}(t) = Ax(t) + Bu(t) + B\xi(t, x) \tag{2.61}$$

where the (unknown) function $\xi : \mathbb{R}^+ \times \mathbb{R}^n \mapsto \mathbb{R}^m$ represents matched uncertainty. With the same choice of switching function

$$\dot{\sigma}(t) = S\dot{x}(t) = SAx(t) + SB\xi(t, x) + SBu(t) \tag{2.62}$$

In this situation consider the control law

$$u(t) = -(SB)^{-1}SAx(t) - (\eta + \rho(t, x)|SB|)(SB)^{-1}\mathrm{sign}(\sigma(t)) \tag{2.63}$$

where the scalar function $\rho(t, x)$ represents *a known* upper bound on the (unknown) uncertainty $\xi(t, x)$. Substituting for Eq. (2.60) in Eq. (2.62) gives

$$
\begin{aligned}
\dot{\sigma}(t) &= SAx(t) + SB\xi(t, x) + SBu(t) \\
&= SAx(t) + SB\xi(t, x) - SAx(t) - (\eta + \rho(t, x)|SB|)\text{sign}(\sigma) \\
&= SB\xi(t, x) - \eta\text{sign}(\sigma) - \rho(t, x)|SB|\text{sign}(\sigma) \qquad (2.64)
\end{aligned}
$$

However since

$$
SB\xi(t, x) \le |SB\xi(t, x)| = |SB||\xi(t, x)| \le |SB|\rho(t, x)
$$

it follows from Eq. (2.64) that

$$
\begin{aligned}
\sigma\dot{\sigma}(t) &= \sigma SB\xi(t, x) - \eta\sigma\text{sign}(\sigma) - \rho(t, x)|SB|\sigma\text{sign}(\sigma) \\
&\le |\sigma||SB|\rho(t, x) - \eta|\sigma| - \rho(t, x)|SB||\sigma| \\
&= -\eta|\sigma| \qquad (2.65)
\end{aligned}
$$

and once again σ is driven to zero in less than $|\sigma(0)|/\eta$ units of time.

Consider next a system with what might be described as multiplicative uncertainty:

$$
\dot{x}(t) = Ax(t) + B(1 + \delta)u(t) + B\xi(t, x) \qquad (2.66)
$$

where $\delta \in (-\delta_0, \delta_1)$ with known scalars $0 < \delta_0 < 1$ and $\delta_1 > 0$. Compared to Eq. (2.61) it is clear there is now uncertainty associated with the control signal. Note the limitation that $\delta_0 < 1$ prevents a "change of polarity" with respect to the control. Proceeding as before

$$
\dot{\sigma}(t) = S\dot{x}(t) = SAx(t) + SB\xi(t, x) + SB\delta u(t) + SBu(t) \qquad (2.67)
$$

Consider the control law

$$
u(t) = u_l(t) + u_n(t), \qquad (2.68)
$$

where as before the linear term $u_l(t) = -(SB)^{-1}SAx(t)$ and the nonlinear term (with a new modulation function) is $u_n(t) = -(SB)^{-1}\bar{\rho}(t, x)\text{sign}(\sigma(t))$ where

$$
\bar{\rho}(t) = (\eta + |SB|(\rho(t, x) + \delta_1|u_l(t)|))(1 - \delta_0)^{-1} \qquad (2.69)
$$

As before the scalar function $\rho(t, x)$ represents a known upper bound on the (unknown) uncertainty $\xi(t, x)$. Note that the modulation function in the nonlinear term also depends on δ_0. Substituting for Eq. (2.68) in Eq. (2.62) gives

$$
\begin{aligned}
\dot{\sigma}(t) &= SAx(t) + SB\xi(t, x) + SBu(t) \\
&= SB\xi(t, x) + SB(1 + \delta)u_n(t) + SB\delta u_l(t) \qquad (2.70)
\end{aligned}
$$

Consequently

$$\sigma\dot\sigma(t) = \sigma SB\xi(t,x) + \sigma SB(1+\delta)u_n(t) + \sigma SB\delta u_l(t)$$
$$\leq |\sigma|\,||SB||\,|\xi(t,x)| - (1+\delta)\bar\rho(t,x)\mathrm{sign}(\sigma) + |\sigma|\,||SB||\,|\delta|\,|u_l(t)|$$
$$\leq |\sigma|\,||SB||\,|\xi(t,x)| - (1+\delta)\bar\rho(t,x)|\sigma| + |\sigma|\,||SB|\delta_1|u_l(t)|$$
$$\leq |\sigma|\,||SB||\,|\xi(t,x)| - (1-\delta_0)\bar\rho(t,x)|\sigma| + |\sigma|\,||SB|\delta_1|u_l(t)| \tag{2.71}$$

Substituting for $\bar\rho(t,x)$ from Eq. (2.69) yields

$$\sigma\dot\sigma(t) \leq -\eta|\sigma| \tag{2.72}$$

and a sliding mode is guaranteed to be achieved in finite time, and subsequently maintained.

Example 2.6. Consider the state-space model

$$\dot x(t) = Ax(t) + bu(t) \tag{2.73}$$

where

$$A = \begin{bmatrix} 0 & 1 & 0 \\ 0 & -2 & 1 \\ 0 & 0 & -1 \end{bmatrix} \quad b = \begin{bmatrix} 0 \\ 0 \\ 10 \end{bmatrix}$$

These represent the equations of motion of a hot-air balloon where the control input is the fuel flow into the burner and the first component represents the altitude of the balloon.[7] The open-loop poles are $\{0, -1, -2\}$.

The aim is to select a switching function defined by

$$S = \begin{bmatrix} s_1 & s_2 & 1 \end{bmatrix}$$

or equivalently

$$\sigma(x) = s_1 x_1 + s_2 x_2 + x_3$$

to ensure the reduced-order sliding motion confined to S is stable, and meets any design specifications. While sliding, when $\sigma = 0$,

$$x_3 = -\begin{bmatrix} s_1 & s_2 \end{bmatrix}\begin{bmatrix} x_1 \\ x_2 \end{bmatrix} \tag{2.74}$$

Because of the special form of the state space

$$\begin{bmatrix} \dot x_1 \\ \dot x_2 \end{bmatrix} = \begin{bmatrix} 0 & 1 \\ 0 & -2 \end{bmatrix}\begin{bmatrix} x_1 \\ x_2 \end{bmatrix} + \begin{bmatrix} 0 \\ 1 \end{bmatrix}x_3 \tag{2.75}$$

$$x_3 = -\begin{bmatrix} s_1 & s_2 \end{bmatrix}\begin{bmatrix} x_1 \\ x_2 \end{bmatrix} \tag{2.76}$$

[7]Taken from [86].

Simplifying Eqs. (2.74) and (2.75)

$$\begin{bmatrix} \dot{x}_1 \\ \dot{x}_2 \end{bmatrix} = \begin{bmatrix} 0 & 1 \\ -s_1 & -2-s_2 \end{bmatrix} \begin{bmatrix} x_1 \\ x_2 \end{bmatrix} \qquad (2.77)$$

Equations (2.75) and (2.74) represent a second-order system in which x_3 has the role of the control variable and $\begin{bmatrix} s_1 & s_2 \end{bmatrix}$ is a full state-feedback matrix. The characteristic equation of Eq. (2.77) is

$$\det \begin{bmatrix} \lambda & -1 \\ s_1 & \lambda + 2 + s_2 \end{bmatrix} = 0$$

in other words

$$\lambda^2 + (2 + s_2)\lambda + s_1 = 0 \qquad (2.78)$$

Choosing the required sliding mode poles to be $\{-1 \pm j\}$ gives a desired characteristic equation

$$\lambda^2 + 2\lambda + 2 = 0$$

Comparing coefficients with Eq. (2.78) gives $s_1 = 2$ and $s_2 = 0$ and the resulting switching function

$$\sigma(x) = 2x_1 + x_3 \qquad (2.79)$$

A control law must be developed such that the reachability condition (2.59) is satisfied. It follows (in this case) that

$$\dot{\sigma} = 2\dot{x}_1 + \dot{x}_3$$

Now substituting from the original equations

$$\dot{\sigma} = 2 \underbrace{x_2}_{\dot{x}_1} + \underbrace{-x_3 + 10u}_{\dot{x}_3}$$

Now choose

$$u = -\frac{1}{5}x_2 + \frac{1}{10}x_3 - \frac{\eta}{10}\,\text{sign}(\sigma) \qquad (2.80)$$

where η is a positive scalar. It follows that

$$\dot{\sigma} = -\eta\,\text{sign}(\sigma) \quad \Rightarrow \quad \sigma\dot{\sigma} = -\eta|\sigma|$$

Hence Eq. (2.80) is an appropriate variable structure controller which induces a sliding motion. The plot of the switching function is shown in Fig. 2.5.

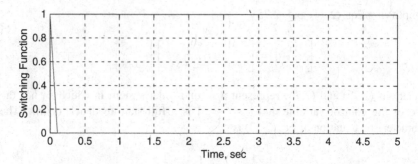

Fig. 2.5 Switching function

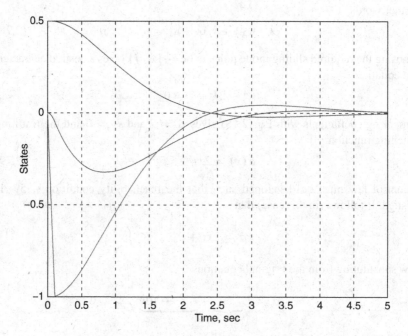

Fig. 2.6 States time history

Note that in finite time—approximately 0.1 s—the switching function has become zero. Furthermore once zero (at which point the hyperplane has been reached) the states are forced to remain on the surface. The state's evolution is presented in Fig. 2.6. It is clear from Fig. 2.6 that x_1 and x_2 approach zero as time increases in the sliding mode.

2.3.3   *Relay Control for Multi-input Systems*

In this subsection these ideas are extended to multi-input systems. Many different multivariable control structures exist and fundamentally the key thing that must be achieved is a multivariable version of the sign function.

In the multi-input case, once again

$$\dot{\sigma}(t) = S\dot{x}(t) = SAx(t) + SBu(t) \tag{2.81}$$

The "complication" now is the fact that SB is a square *matrix* and not a scalar. A simple way to circumvent this is to enforce a structure on SB through choice of S while at the same time ensuring appropriate properties for the reduced-order sliding motion. It will be argued that in fact S can always be chosen so that $SB = I_m$ while simultaneously ensuring appropriate dynamics for the sliding mode.

Recall from Eq. (2.32) that the sliding surface hyperplane matrix can be parameterized as

$$S = \begin{bmatrix} S_1 & S_2 \end{bmatrix} T_r \tag{2.82}$$

where the matrices S_2 and S_1 were linked via the parameter $M = S_2^{-1} S_1$ and T_r was the orthogonal matrix used as the basis of the coordinate transformation to achieve regular form. Equivalently Eq. (2.82) above can be written as

$$S = S_2 \begin{bmatrix} M & I_m \end{bmatrix} T_r \tag{2.83}$$

In the above, for a given pair (A, B), the matrix T_r is established from QR reduction based in B. Furthermore the dynamics of the reduced-order motion depend solely on the choice of matrix M which may be viewed as a state-feedback gain for the null-space system in (2.30). Clearly the choice of S_2 does not affect the dynamics of the sliding motion and it is this design freedom which is exploited to ensure that $SB = I_m$. For a given (A, B) and design parameter M, it follows

$$
\begin{aligned}
SB &= \begin{bmatrix} S_1 & S_2 \end{bmatrix} T_r B \\
&= \begin{bmatrix} S_1 & S_2 \end{bmatrix} \begin{bmatrix} 0 \\ B_2 \end{bmatrix} \quad \text{from Eq. (2.28)} \\
&= S_2 B_2 \tag{2.84}
\end{aligned}
$$

Consequently choosing $S_2 = B_2^{-1}$ ensures the sliding motion dynamics are specified as $(A_{11} - A_{12}M)$ and simultaneously $SB = I_m$.

Now the multi-input structure in Eq. (2.81) can be decomposed into m independent equations. Specifically exploiting the fact that $SB = I_m$, Eq. (2.81) can be written componentwise as

$$\dot{\sigma}_i(t) = (SA)_i x(t) + u_i(t) \tag{2.85}$$

where $(SA)_i$ is the ith row of the $m \times n$ matrix SA and u_i and σ_i are the ith components of the vectors u and σ, respectively. Now as discussed in the previous section, m independent single-input controllers can be designed for each of the components u_i.

2.4   State-Feedback Unit-Vector Control

Of the many different multivariable sliding mode control structures which exist, the one that will be considered here is the *unit-vector* approach. This has the advantage of being an inherently multi-input approach. Consider an uncertain system of the form

$$\dot{x}(t) = Ax(t) + Bu(t) + f(t, x, u) \tag{2.86}$$

where the function $f : \mathbb{R} \times \mathbb{R}^n \times \mathbb{R}^m \to \mathbb{R}^m$, which represents the uncertainties or nonlinearities, satisfies the so-called *matching condition*, i.e.,

$$f(t, x, u) = B\xi(t, x, u) \tag{2.87}$$

where $\xi : \mathbb{R} \times \mathbb{R}^n \times \mathbb{R}^m \to \mathbb{R}^m$ and is unknown but satisfies

$$\|\xi(t, x, u)\| \le k_1 \|u\| + \alpha(t, x), \tag{2.88}$$

where $1 > k_1 \ge 0$ is a known constant and $\alpha(\cdot)$ is a known function.

2.4.1   Design in the Presence of Matched Uncertainty

The proposed control law comprises two components: a linear component to stabilize the nominal linear system and a discontinuous component. Specifically

$$u(t) = u_l(t) + u_n(t) \tag{2.89}$$

where the linear component is given by

$$u_l(t) = -\Lambda^{-1} (SA - \Phi S) x(t) \tag{2.90}$$

where $\Phi \in \mathbb{R}^{m \times m}$ is any stable design matrix and $\Lambda = SB$. The nonlinear component is defined to be

$$u_n(t) = -\rho(t, x)\Lambda^{-1} \frac{P_2\sigma(t)}{\|P_2\sigma(t)\|}, \tag{2.91}$$

where $P_2 \in \mathbb{R}^{m \times m}$ is a symmetric positive definite matrix satisfying the Lyapunov equation

$$P_2 \Phi + \Phi^T P_2 = -I \qquad (2.92)$$

and the scalar function $\rho(t, x)$, which depends only on the magnitude of the uncertainty, is any function satisfying

$$\rho(t, x) \geq (\|\Lambda\|(k_1\|u_l\| + \alpha(t, x)) + \gamma) / (1 - k_1\|\Lambda\|\|\Lambda^{-1}\|) \qquad (2.93)$$

where $\gamma > 0$ is a design parameter. In this equation it is assumed that the scaling parameter S_2 has been chosen so that $\Lambda = SB$ has the property that

$$k_1\|\Lambda\|\|\Lambda^{-1}\| < 1 \qquad (2.94)$$

A necessary condition for Eq. (2.94) to hold is that $k_1 < 1$ because $\|\Lambda\|\|\Lambda^{-1}\| > 1$ for all choices of S.

Before demonstrating that the above controller induces a sliding motion, it will first be established that any scalar modulation function satisfying Eq. (2.93) bounds the uncertain term $\xi(t, x, u)$.

Rearranging Eq. (2.93) gives

$$\begin{aligned}
\rho(t, x) &\geq \|\Lambda\|(k_1\|u_l\| + \alpha(t, x)) + \gamma + k_1\|\Lambda\|\|\Lambda^{-1}\|\rho(t, x) \\
&\geq \|\Lambda\|(k_1\|\Lambda^{-1}\|\rho(t, x) + k_1\|u_l\| + \alpha(t, x)) + \gamma \\
&\geq \|\Lambda\|(k_1\|u\| + \alpha(t, x)) + \gamma \\
&\geq \|\Lambda\|\|\xi(t, x, u)\| + \gamma \qquad (2.95)
\end{aligned}$$

In obtaining the third inequality the fact that

$$u = u_l - \rho(t, x)\Lambda^{-1}\frac{P_2\sigma}{\|P_2\sigma\|} \quad \Rightarrow \quad \|u\| < \|u_l\| + \rho(t, x)\|\Lambda^{-1}\|$$

is used. Inequality (2.95) demonstrates $\rho(t, x)$ is greater in magnitude than the matched uncertainty occurring in Eq. (2.87). Substituting for the control law in Eq. (2.86) yields

$$\begin{aligned}
\dot{\sigma} &= SAx(t) + \Lambda u + \Lambda\xi(t, x, u) \\
&= \Phi\sigma - \rho(t, x)\frac{P_2\sigma}{\|P_2\sigma\|} + \Lambda\xi(t, x, u) \qquad (2.96)
\end{aligned}$$

It will now be shown that $V(\sigma) = \sigma^T P_2 \sigma$ guarantees quadratic stability for the switching states σ, and in particular

$$\begin{aligned}
\dot{V} &= \sigma^T(P_2\Phi + \Phi^T P_2)\sigma - 2\rho\|P_2\sigma\| + 2\sigma^T P_2\Lambda\xi \\
&\leq \sigma^T(P_2\Phi + \Phi^T P_2)\sigma - 2\rho\|P_2\sigma\| + 2\|P_2\sigma\|\|\Lambda\|\|\xi\| \\
&= -\sigma^T\sigma - 2\|P_2\sigma\|(\rho(t, x) - \|\Lambda\|\|\xi\|) \\
&\leq -\sigma^T\sigma - 2\gamma\|P_2\sigma\| \qquad (2.97)
\end{aligned}$$

Assuming that the closed-loop system has no finite-escape time during the reaching phase, then this control law guarantees that the switching surface is reached in finite time despite the disturbance or uncertainty. Once the sliding motion is attained, it is completely independent of the uncertainty.

Example 2.7. Consider the satellite dynamics given by

$$
\begin{aligned}
I_1\dot{\omega}_1 &= (I_2 - I_3)\omega_2\omega_3 + u_1 \\
I_2\dot{\omega}_2 &= (I_3 - I_1)\omega_3\omega_1 + u_2 \\
I_3\dot{\omega}_3 &= (I_1 - I_2)\omega_1\omega_2 + u_3
\end{aligned}
\tag{2.98}
$$

where I_1, I_2, and I_3 represent the moments of inertia around the principal axes of the body. The variables ω_1, ω_2, and ω_3 are the angular velocities, which are measurable. The variables u_1, u_2, and u_3 are the control input torques. Defining $x_1 = \omega_1$, $x_2 = \omega_2$, $x_3 = \omega_3$, and $x = [x_1, x_2, x_3]^T$, the system in (2.98) has the representation:

$$
\dot{x} = Bu + B\xi(x) =
\begin{bmatrix}
\frac{1}{I_1} & 0 & 0 \\
0 & \frac{1}{I_2} & 0 \\
0 & 0 & \frac{1}{I_3}
\end{bmatrix}
\begin{bmatrix}
u_1 \\ u_2 \\ u_3
\end{bmatrix}
+
\begin{bmatrix}
\frac{1}{I_1} & 0 & 0 \\
0 & \frac{1}{I_2} & 0 \\
0 & 0 & \frac{1}{I_3}
\end{bmatrix}
\begin{bmatrix}
(I_2 - I_3)x_2x_3 \\
(I_3 - I_1)x_3x_1 \\
(I_1 - I_2)x_1x_2
\end{bmatrix}
$$

$$
\sigma = Sx
$$

where σ is the sliding output and $\xi(x)$ represents the nonlinearities which satisfy the *matching condition*. Furthermore $\xi(x)$ satisfies $||\xi(x)|| \leq \alpha(x) \leq c$ in a domain $x \in \Omega \subset \mathbb{R}^3$ that includes the origin, where c is a known constant. Then the problem is to stabilize the equilibrium point $x = 0$ of the satellite in finite time. Based on Eq. (2.89), the proposed control law is:

$$
u = (SB)^{-1}(\Phi S)x - \rho(x)(SB)^{-1}\frac{P_2\sigma}{||P_2\sigma||}
$$

Choosing $S = \mathrm{diag}(l_1, l_2, l_3)$ it follows $SB = I_{3\times3}$. If $\Phi = -\frac{1}{2}I_{3\times3}$, $P_2 = I_{3\times3}$ is the solution of the Lyapunov equation:

$$
P_2\Phi + \Phi^T P_2 = -I
\tag{2.99}
$$

Now choosing the Lyapunov function $V = \sigma^T P_2\sigma$ we have

$$
\dot{V} \leq -||\sigma||^2 - 2\gamma||P_2\sigma|| < -2\gamma||\sigma|| < 0
$$

and finally

$$
\dot{V} < -2\gamma V^{\frac{1}{2}}
$$

where $\gamma = \rho - c > 0$. For simulation purposes we consider $\rho = 3$. Therefore the proposed control law guarantees that the sliding surface is reached in finite time,

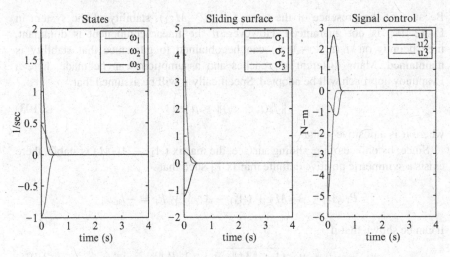

Fig. 2.7 Stabilization of the equilibrium point of the satellite

which in this case means that x equals zero in finite time. The system in Eq. (2.98) has been simulated with the control law (2.89) using the initial conditions $x(0) = [0.5, -1, 2]^T$ and the parameters $I_1 = 1\,\text{kgm}^2$, $I_2 = 0.8\,\text{kgm}^2$, and $I_3 = 0.4\,\text{kgm}^2$. The results obtained with the proposed control law are presented in Fig. 2.7. In order to attenuate chattering the discontinuous portion of the control signal is approximated by $\frac{P_2\sigma}{\|P_2\sigma\|} \approx \frac{P_2\sigma}{\|P_2\sigma\|+\epsilon}$, with $\epsilon \ll 1$.

2.4.2   Design in the Presence of Unmatched Uncertainty

If the uncertainty does not meet the matching requirements, after transformation into regular form, in the "z" coordinates, a system of the form

$$\dot{z}_1(t) = A_{11}z_1(t) + A_{12}z_2(t) + f_u(t, z_1, z_2) \tag{2.100}$$

$$\dot{z}_2(t) = A_{21}z_1(t) + A_{22}z_2(t) + B_2u(t) + f_m(t, z_1, z_2) \tag{2.101}$$

is obtained where $f_m(t, z_1, z_2)$ and $f_u(t, z_1, z_2)$ represent the matched and unmatched components of the uncertainty, respectively. As argued in the earlier sections, the effects of the matched uncertainty $f_m(t, z_1, z_2)$ can be canceled. This section considers the null-space dynamics in Eq. (2.100). If a sliding motion can be induced on $\mathcal{S}$, then $z_2 = -Mz_1$ and the reduced-order motion is governed by

$$\dot{z}_1(t) = (A_{11} - A_{12}M)z_1(t) + f_u(t, z_1, -Mz_1) \tag{2.102}$$

Because of the presence of the term $f_u(t, z_1, -Mz_1)$, stability of the system in Eq. (2.102) is not guaranteed. However, if the linear component is dominant, then bounds on $f_u(t, z_1, -Mz_1)$ can be obtained to guarantee that stability is maintained. Many different approaches and assumptions can be made: here a Lyapunov approach will be adopted. Specifically it will be assumed that

$$\| f_u(t, z_1, z_2) \| \le \mu \|z\| \tag{2.103}$$

where μ is a positive scalar.

Since, by choice of the sliding surface, the matrix $(A_{11} - A_{12}M)$ is stable, there exists a symmetric positive definite matrix P_1 such that

$$P_1(A_{11} - A_{12}M) + (A_{11} - A_{12}M)^T P_1 = -I_{n-m}$$

It can be shown that if

$$\mu \sqrt{(1 + \|M\|^2} < 1/(2\|P_1\|) \tag{2.104}$$

then Eq. (2.102) is stable while sliding. To establish this, first the constraint in Eq. (2.103) will be written in terms of $\|z_1\|$. Since during the sliding motion $z_2 = -Mz_1$ it follows

$$\|z\|^2 = \left\| \begin{bmatrix} z_1 \\ z_2 \end{bmatrix} \right\|^2 \le \|z_1\|^2 + \|z_2\|^2 \le \|z_1\|^2 + \|Mz_1\|^2 \le (\|M\|^2 + 1)z_1^2$$

and consequently

$$\|z\| \le \sqrt{(\|M\|^2 + 1)} \|z_1\|$$

Create from the symmetric positive definite matrix P_1 a candidate Lyapunov function $V(z_1) = z_1^T P_1 z_1$. It follows

$$\begin{aligned}
\dot{V} &= z_1^T P_1 \dot{z}_1 + \dot{z}_1^T P_1 z_1 \\
&= z_1^T \left(P_1(A_{11} - A_{12}M) + (A_{11} - A_{12}M)^T P_1 \right) z_1 + 2z_1^T P_1 f_u(t, z_1, -Mz_1) \\
&\le -z_1^T z_1 + 2\|z_1\| \|P_1\| \| f_u(t, z_1, -Mz_1)\| \\
&\le -z_1^T z_1 + 2\|z_1\| \|P_1\| \mu \|z\| \\
&\le -z_1^T z_1 + 2\|z_1\| \|P_1\| \mu \sqrt{(\|M\|^2 + 1)} \|z_1\| \\
&\le \|z_1\|^2 (2\mu \|P_1\| \sqrt{(\|M\|^2 + 1)} - 1) \tag{2.105}
\end{aligned}$$

If the inequality in Eq. (2.104) holds then

$$\dot{V} \le \|z_1\|^2 (2\|P_1\| \mu \sqrt{(\|M\|^2 + 1)} - 1) < 0 \tag{2.106}$$

and so the reduced-order motion is stable.

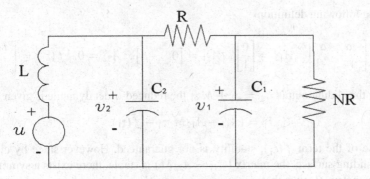

Fig. 2.8 Chua's circuit

Some modifications need to be made to the control law gain (2.93) to ensure a sliding motion can be achieved and maintained. Now

$$\dot{\sigma}(t) = SA(t) + SBu(t) + Sf(t, x) \tag{2.107}$$

It can be shown using similar arguments to those deployed earlier that

$$\rho(t, x) \geq \frac{\|S_2\| \left(\|M\| \mu \|x(t)\| \right) \|u_l(t)\| + \alpha(t, x)) + \gamma}{(1 - k_3 \|\Lambda\| \|\Lambda^{-1}\| \|B_2^{-1}\|)} \tag{2.108}$$

guarantees the existence of a sliding motion.

Example 2.8. Chua's circuit consists of one inductor, two capacitors, and one piecewise-linear nonlinear resistor; see Fig. 2.8. The normalized dynamic equations of the circuit are:

$$\begin{array}{rcl}
\dot{x}_1 & = & \alpha(x_2 - x_1 - f(x_1)) \\
\dot{x}_2 & = & x_1 - x_2 + x_3 \\
\dot{x}_3 & = & -\beta x_2 + u
\end{array} \tag{2.109}$$

where $x_1 = v_1$, $x_2 = v_2$, and x_3 is the current through the inductor; α and β are known parameters and $f(x_1)$ is a function that depends on the nonlinear resistor; this function represents the *unmatched uncertainty*. We consider the nonlinear function $f(x_1) = -\frac{1}{5}x_1(1 - x_1^2)$. Here it is assumed that $|x_1| < 1$, which implies $|f(x_1)| \leq \frac{1}{5}|x_1| + \frac{1}{5}|x_1^2| \leq \frac{2}{5}|x_1|$; then inequality (2.103) is satisfied with $\mu = \frac{2}{5}$. The goal of this example is to stabilize the equilibrium point $x = [0, 0, 0]^T$ of Chua's circuit. First define $z_1 = [x_1, x_2]^T$ and $z_2 = x_3$; then Chua's circuit has the following representation:

$$\begin{bmatrix} \dot{z}_1 \\ \dot{z}_2 \end{bmatrix} = \begin{bmatrix} A_{11} & A_{12} \\ A_{21} & A_{22} \end{bmatrix} \begin{bmatrix} z_1 \\ z_2 \end{bmatrix} + \begin{bmatrix} \bar{0} \\ 1 \end{bmatrix} u - \begin{bmatrix} \bar{f}(z_1) \\ 0 \end{bmatrix}$$

with the following definitions:

$$A_{11} = \begin{bmatrix} -\alpha & \alpha \\ 1 & -1 \end{bmatrix}, \ A_{12} = \begin{bmatrix} 0 \\ 1 \end{bmatrix}, \ A_{21} = \begin{bmatrix} 0 & -\beta \end{bmatrix}, \ A_{22} = 0, \ \bar{f}(z_1) = \begin{bmatrix} f(x_1) \\ 0 \end{bmatrix}$$

Setting the sliding output $\sigma_1 = z_2 + M z_1$ the reduced-order dynamic is given by:

$$\dot{z}_1 = (A_{11} - A_{12}M)z_1 - \bar{f}(z_1).$$

Because of the term $\bar{f}(z_1)$, stability is not guaranteed. However since by choice of the sliding surface, the matrix $(A_{11} - A_{12}M)$ is stable, there exists a symmetric positive matrix P_1 such that

$$P_1(A_{11} - A_{12}M) + (A_{11} - A_{12}M)^T P_1 = -I_{2 \times 2}$$

Now the candidate Lyapunov function $V = z_1^T P_1 z_1$ is proposed. It follows:

$$\dot{V} \leq ||z_1||^2 (2||P_1||\mu\sqrt{||M||^2 + 1} - 1)$$

Then if the inequality $(2||P_1||\mu\sqrt{||M||^2 + 1} - 1) < 0$ is satisfied the reduced-order dynamic is stable. Typical system parameters are chosen to be $\alpha = 9.1/7$ and $\beta = -8/7$. For $M = [2, 2]$ we obtain

$$P_1 = \begin{bmatrix} 0.3553 & -0.0293 \\ -0.0293 & 0.1764 \end{bmatrix}, \ (2||P_1||\mu\sqrt{||M||^2 + 1} - 1) = -0.1854$$

and inequality (2.104) is satisfied. Now with the control law

$$u = -(A_{21} - A_{22}M)z_1 - M(A_{11} - A_{12}M)z_1 - \rho(z)\frac{\sigma_1}{||\sigma_1||}$$

and the candidate Lyapunov function $V = \frac{1}{2}\sigma_1^2$, we have

$$\dot{V} \leq -(\rho(z) - \mu||M||||z||)||\sigma_1||$$

If the inequality $\rho(z) > \mu||M||||z||$ is satisfied, finite time convergence is ensured, provided during the reaching phase $|x_1(t)| < 1$. For simulation purposes $\rho = 4 > \mu||M||||z|| = 1.14$. Using the initial conditions $x_1(0) = 0.4$, $x_2(0) = 0.2$, and $x_3(0) = 0.5$, simulations were performed with the proposed control law and the results are presented in Fig. 2.8. In order to attenuate chattering the discontinuous portion of the control signal is approximated by $\rho(z)\frac{\sigma_1}{||\sigma_1||} \approx \rho(z)\frac{\sigma_1}{||\sigma_1||+\epsilon}$, with $\epsilon \ll 1$ (Fig. 2.9).

In engineering situations, tracking problems are often encountered whereby (usually) the output of the system is required to follow a predefined reference signal. In the following section an integral action based method is considered for output tracking.

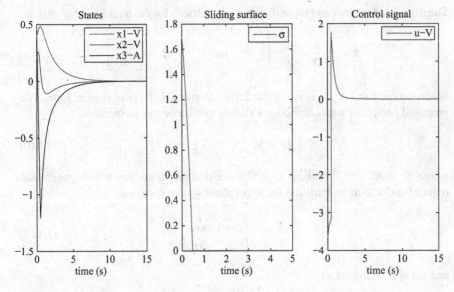

Fig. 2.9 Stabilization of the equilibrium point of Chua's circuit

2.5 Output Tracking with Integral Action

Consider the development of a tracking control law for the nominal linear system

$$\dot{x}(t) = Ax(t) + Bu(t) \tag{2.110}$$

$$y(t) = Cx(t) \tag{2.111}$$

which is assumed to be square, i.e., it has the same number of inputs and outputs. In addition, for convenience, assume the matrix pair (A, B) is in regular form. The control law described here utilizes an *integral action* methodology. Consider the introduction of additional states $x_r \in \mathbb{R}^m$ satisfying

$$\dot{x}_r(t) = r(t) - y(t) \tag{2.112}$$

where the differentiable signal $r(t)$ satisfies

$$\dot{r}(t) = \Gamma\left(r(t) - R\right) \tag{2.113}$$

with $\Gamma \in \mathbb{R}^{m \times m}$ a stable design matrix and R a constant demand vector. Augment the states with the integral action states and define

$$\tilde{x} = \begin{bmatrix} x_r \\ x \end{bmatrix} \tag{2.114}$$

The associated system and input distribution matrices for the *augmented system* are

$$\tilde{A} = \begin{bmatrix} 0 & -C \\ 0 & A \end{bmatrix} \quad \text{and} \quad \tilde{B} = \begin{bmatrix} 0 \\ B \end{bmatrix} \tag{2.115}$$

assuming the pair (A, B) is in regular form, the pair $(\tilde{A}, \tilde{B})$ is in regular form. The proposed controller seeks to induce a sliding motion on the surface

$$S = \{\tilde{x} \in \mathbb{R}^{n+m} \ : \ S\tilde{x} = S_r r\} \tag{2.116}$$

where $S \in \mathbb{R}^{m \times (n+p)}$ and $S_r \in \mathbb{R}^{m \times m}$ are design parameters which govern the reduced-order motion. Partition the hyperplane system matrix as

$$S = \begin{matrix} \overset{n}{\leftrightarrow} & \overset{m}{\leftrightarrow} \\ [\ S_1 & S_2\ ] \end{matrix} \tag{2.117}$$

and the system matrix as

$$\tilde{A} = \begin{matrix} \overset{n}{\leftrightarrow} & \overset{m}{\leftrightarrow} \\ \begin{bmatrix} \tilde{A}_{11} & \tilde{A}_{12} \\ \tilde{A}_{21} & \tilde{A}_{22} \end{bmatrix} & \begin{matrix} \updownarrow n \\ \updownarrow m \end{matrix} \end{matrix} \tag{2.118}$$

and assume $\Lambda = S\tilde{B}$ is nonsingular. If a controller exists which induces an ideal sliding motion on S and the augmented states are partitioned as

$$\tilde{x} = \begin{bmatrix} x_1 \\ x_2 \end{bmatrix} \tag{2.119}$$

where $x_1 \in \mathbb{R}^n$ and $x_2 \in \mathbb{R}^m$, then the ideal sliding motion is given by

$$\dot{x}_1(t) = (\tilde{A}_{11} - \tilde{A}_{12}M)x_1(t) + (\tilde{A}_{12}S_2^{-1}S_r + B_r)r(t) \tag{2.120}$$

where $M = S_2^{-1}S_1$ and $B_r = \begin{bmatrix} I_m & 0_{n \times m} \end{bmatrix}^T$. In order for the design methods described earlier to be valid, it is necessary for the matrix pair $(\tilde{A}_{11}, \tilde{A}_{12})$ to be completely controllable.

Remark 2.5. Necessary conditions on the original system are that (A, B, C) is completely controllable and has no invariant zeros at the origin.[8]

The development that follows mirrors the approach in Sect. 2.3.3 where Φ is any stable design matrix. The overall control law is then given by

---

[8]For details see Sect. 4.4.2 in [67] and Appendix C.

$$u = u_l(\tilde{x}, r) + u_n(\tilde{x}, r) \qquad (2.121)$$

where the continuous control

$$u_l(\tilde{x}, r) = -\Lambda^{-1} (SA - \Phi S) \tilde{x}(t) \qquad (2.122)$$

and the discontinuous control vector

$$u_n(\sigma, r) = -\rho_c(u_l, y)\Lambda^{-1} \frac{P_2(S\tilde{x} - S_r r)}{\| P_2(S\tilde{x} - S_r r) \|} \qquad (2.123)$$

where P_2 is a symmetric positive definite matrix satisfying

$$P_2\Phi + \Phi^T P_2 = -I \qquad (2.124)$$

The positive scalar function which multiplies the unit-vector component can be obtained from arguments similar to those in Sect. 2.3. It follows that, in terms of the original coordinates,

$$u_n(\tilde{x}, r) = L\tilde{x} + L_r r + L_{\dot{r}} \dot{r} \qquad (2.125)$$

with the gains defined as

$$L = -\Lambda^{-1}(S\tilde{A} - \Phi S) \qquad (2.126)$$

$$L_r = -\Lambda^{-1} (\Phi S_r + S_1 B_r) \qquad (2.127)$$

$$L_{\dot{r}} = \Lambda^{-1} S_r \qquad (2.128)$$

The parameter S_r can take any value and does not affect the stability of the closed-loop system. One common choice is to let $S_r = 0$ for simplicity. Another option, which has been found to give good results with practical applications, is to choose S_r so that at steady state the integral action states are zero in the absence of any uncertainty.

Up to this point it has been assumed that all the states are available for use in the control law. This hypothesis will be dropped in the remaining sections of Chap 2.

2.6 Output-Based Hyperplane Design

Consider the linear system in Eq. (2.18) and suppose that only the measured outputs

$$\dot{x}(t) = Ax(t) + Bu(t) \qquad (2.129)$$

$$y(t) = Cx(t) \qquad (2.130)$$

where $C \in \mathbb{R}^{p \times n}$ are available. The methods described earlier in this chapter are now no longer directly applicable since the state vector is not directly available. One approach is to use an observer (this will be discussed in later chapters) to estimate the states and then to use the estimate in place of the real states. This is conceptually straightforward but has potential pitfalls—particularly if linear observers are used. It is well understood that in linear systems (with linear control laws and linear observers), the robustness associated with the feedback law can be destroyed by the introduction of an observer as part of the feedback loop. Also the introduction of an observer will add significant computational costs in terms of implementation. Instead let us consider the possibility of introducing controllers in the spirit of Sect. 2.5 subject to the constraint that only output information is available.

2.6.1 Static Output-Feedback Hyperplane Design

The state-feedback control strategies described earlier are not immediately employable here. Firstly, it is intuitively likely that since the hyperplane design problem resolved itself into a state-feedback control paradigm for the fictitious triplé (A_{11}, A_{12}), in the situation when only output information is available this will become some form of restricted state-feedback design problem: in fact a static output-feedback design problem for a certain triple (A_{11}, A_{12}, C_1). Secondly the control laws described in the previous section involved a linear state-feedback control component. This is unlikely to be achievable in an output-feedback context.

Here, the situation when there are more outputs than inputs is considered, since in the square case, no design freedom exists in terms of selecting the dynamics of the sliding motion. Two assumptions will be made:

(A1) The parameter CB is full rank.
(A2) Any invariant zeros of (A, B, C) have negative real parts.

Remark 2.6. The dependence on invariant zeros is not perhaps surprising and the presence of zeros plays an important role in linear systems theory. It is also worth noting that for appropriate initial conditions associated with a particular zero direction $y(t)$ can be made zero for all time with an appropriate control linear control input although the state itself is not zero. This has clear links with the concept of a sliding motion (without the robustness properties).

These assumptions will be central to the output-feedback-based sliding mode control used here. The following lemma provides a canonical form for the system triple (A, B, C) which will be used in the subsequent analysis:

Lemma 2.1. *Let (A, B, C) be a linear system with $p > m$ and $rank(CB) = m$. Then a change of coordinates exists so that the system triple with respect to the new coordinates has the following structure:*

(a) *The system matrix can be written as*

$$A = \begin{bmatrix} A_{11} & A_{12} \\ A_{21} & A_{22} \end{bmatrix} \tag{2.131}$$

where $A_{11} \in \mathbb{R}^{(n-m)\times(n-m)}$ and the (fictitious) pair (A_{11}, C_1) is detectable where

$$C_1 = \begin{bmatrix} 0 & I_{p-m} \end{bmatrix} \tag{2.132}$$

(b) *The input distribution matrix has the form*

$$B = \begin{bmatrix} 0 \\ B_2 \end{bmatrix} \quad \text{where } B_2 \in \mathbb{R}^{m\times m} \text{ and is nonsingular} \tag{2.133}$$

(c) *The output distribution matrix has the form*

$$C = \begin{bmatrix} 0 & T \end{bmatrix} \quad \text{where } T \in \mathbb{R}^{p\times p} \text{ and is orthogonal} \tag{2.134}$$

Remark 2.7. It can be shown that the unobservable modes of (A_{11}, C_1) are in fact the invariant zeros of the triple (A, B, C).

The idea is to first achieve a regular form structure for the input distribution matrix and then to exploit the fact that CB is full rank to create another transformation which preserves the structure of B while enforcing the partition nature of the output distribution matrix C, which is a feature of the canonical form.

Remark 2.8. Note, this can be viewed as a special case of the traditional regular form discussed earlier which was used as the basis for switching function design in the state-feedback case.

Remark 2.9. Clearly the existence of unstable zeros renders difficulties. In fact unstable invariant zeros means the techniques described in this section are not applicable. While this is not ideal, it must be remembered in linear systems the presence of right half plane zeros limits the performance that can be imposed on the closed-loop system.[9]

Under the premise that only output information is available, the switching function must be of the form

$$\sigma(x) = FCx(t) \tag{2.135}$$

where $F \in \mathbb{R}^{m\times p}$. Suppose a controller exists which induces a stable sliding motion on

$$S = \{x \in \mathbb{R}^n \ : \ FCx = 0\} \tag{2.136}$$

---

[9]This situation is discussed further in "Notes and References" in Chap. 8.

For a first-order sliding motion to exist on $\mathcal{S}$, the equivalent control will be given by solving

$$\dot{\sigma} = FCAx(t) + FCBu_{eq}(t) = 0$$

For an unique equivalent control to exist, the matrix $FCB \in \mathbb{R}^{m \times m}$ must have full rank: this implies that $\text{rank}(CB) = m$ since $\text{rank}(FCB) \leq \text{rank}(CB)$. In all the analysis which follows it is assumed without loss of generality that the system is already in the canonical form of Lemma 2.1. Define matrices F_1 and F_2 such that

$$\begin{array}{cc} \overset{p-m}{\leftrightarrow} & \overset{m}{\leftrightarrow} \\ \left[\begin{array}{cc} F_1 & F_2 \end{array} \right] & = FT \end{array} \tag{2.137}$$

where $F_2 \in \mathbb{R}^{m \times m}$ is assumed to be nonsingular. Notice this structure has a relationship to the hyperplane matrix parametrization given in Eq. (2.83) since Eq. (2.137) can be re-written as

$$F = F_2 \left[\begin{array}{cc} K & I_m \end{array} \right] T^T \tag{2.138}$$

where $K = F_2^{-1} F_1$. The structure in Eq. (2.138) may not be particularly intuitive, but it nicely isolates the design freedom present in the problem. The matrix F_2 is essentially a scaling matrix which is square and invertible and it plays no role in determining the dynamics of the sliding motion. Furthermore it is analogous to the role the matrix S_2 plays in the switching function expansion in Eq. (2.83). The other design parameter is K. This is analogous to the matrix M from Eq. (2.83). It is clear from the dimension of the matrices F and M that the former has only $(p - m) \times (n - m)$ elements while the latter has $m \times (n - m)$. This reduction in parametrization results in less design flexibility—which is the price to pay for only having output rather than state information.

Based on Eq. (2.138) the matrix which defines the switching function can then be written as

$$FC = \left[\begin{array}{cc} F_1 C_1 & F_2 \end{array} \right]$$

where

$$C_1 = \left[\begin{array}{cc} 0_{(p-m) \times (n-p)} & I_{(p-m)} \end{array} \right] \tag{2.139}$$

In this way $FCB = F_2 B_2$ and $\det(F_2) \neq 0 \Leftrightarrow \det(FCB) \neq 0$. It follows during sliding $\sigma = FCx(t) = 0$ and

$$F_1 C_1 z_1 + F_2 z_2 = 0$$

which substituting in the null-space equations yields

$$\dot{z}_1(t) = (A_{11} - A_{12} F_2^{-1} F_1 C_1) z_1(t)$$

$$= (A_{11} - A_{12} K C_1) z_1(t) \tag{2.140}$$

since $K = F_2^{-1} F_1$. Consequently the problem of designing a suitable hyperplane is equivalent to an output-feedback problem for the system (A_{11}, A_{12}, C_1).

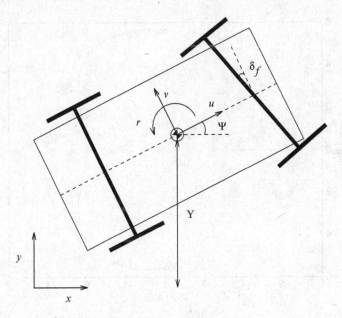

Fig. 2.10 Schematic of the vehicle

Remark 2.10. If the pair (A_{11}, C_1) is observable and the triple (A_{11}, A_{12}, C_1) satisfies the Kimura–Davison conditions[10]

$$m + p \geq n + 1 \tag{2.141}$$

output-feedback pole placement methods can be used to place the poles appropriately.

Example 2.9. Consider the fourth-order system

$$A = \begin{bmatrix} -3.9354 & 0 & 0 & -14.7110 \\ 0 & 0 & 0 & 1.0000 \\ 1.0000 & 14.9206 & 0 & 1.6695 \\ 0.7287 & 0 & 0 & -2.1963 \end{bmatrix} \quad B = \begin{bmatrix} 0 \\ 0 \\ 0 \\ 0.8116 \end{bmatrix} \tag{2.142}$$

$$C = \begin{bmatrix} 0 & 0 & 1 & 0 \\ 0 & 0 & 0 & 1 \end{bmatrix} \tag{2.143}$$

This represents a linearization of the rigid body dynamics of a passenger vehicle (Fig. 2.10). The first state is an average of the lateral velocity v and yaw rate r; the second state represents Ψ, the vehicle orientation; the third state, Y, is the lateral

---

[10]See "Notes and References" at the end of the chapter.

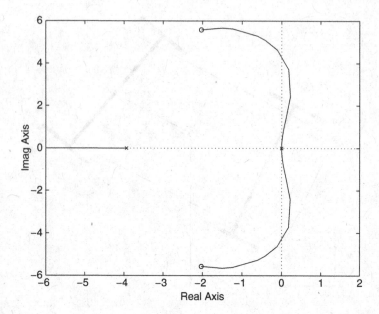

Fig. 2.11 Sliding mode poles as a root locus

deviation from the intended lane position, and the fourth state, r, is the yaw rate. The input to the system, δ_f, is the angular position of the front wheels relative to the chassis.

Notice this is already in the canonical form of Eqs. (2.131)–(2.134), and so

$$A_{11} = \begin{bmatrix} -3.9354 & 0 & 0 \\ 0 & 0 & 0 \\ 1.0000 & 14.9206 & 0 \end{bmatrix} \qquad A_{12} = \begin{bmatrix} -14.7110 \\ 1.0000 \\ 1.6695 \end{bmatrix}$$

$$C_1 = \begin{bmatrix} 0 & 0 & 1 \end{bmatrix} \tag{2.144}$$

The so-called Kimura–Davison conditions are not met for this example since $m + p = 3 < 4 = n$. In this case $p - m = 1$ and $m = 1$ and so the hyperplane matrix can be parameterized as

$$F = \begin{bmatrix} k & 1 \end{bmatrix}$$

where k is a scalar design parameter. The sliding motion is determined by a classical root locus of the "plant" $G_p = C_1(sI - A_{11})^{-1}A_{12}$ in series with a gain "k" in a unity feedback configuration. The root locus plot is given in Fig. 2.11.

For all values of $k > 2.58$, all the sliding mode poles lie in the LHP, and so, this constitutes an appropriate solution to the existence problem.

2.6.2 Static Output-Feedback Control Law Development

Having designed the surface, it is necessary to develop a controller to induce and
sustain a sliding motion. Perhaps a natural choice would be a control structure of
the form

$$u(t) = -Gy(t) - \rho(t, y) \frac{FCx(t)}{\|FCx(t)\|} \tag{2.145}$$

where the quantity $\rho(t, y)$ must upper bound the uncertainty. A common design
methodology is based on synthesizing a static output-feedback gain G numerically
to ensure the so-called reachability condition is satisfied.

   To facilitate the analysis, an additional, switching function dependent coordinate
transformation will be made. Let $z \mapsto T_K z = \hat{z}$ where

$$T_K := \begin{bmatrix} I_{(n-m)} & 0 \\ KC_1 & I_m \end{bmatrix} \tag{2.146}$$

with C_1 defined in (2.139). In this new coordinate system, the system triple
$(\hat{A}, \hat{B}, F\hat{C})$ has the property that

$$\hat{A} = \begin{bmatrix} \hat{A}_{11} & \hat{A}_{12} \\ \hat{A}_{21} & \hat{A}_{22} \end{bmatrix} \qquad \hat{B} = \begin{bmatrix} 0 \\ B_2 \end{bmatrix} \qquad F\hat{C} = \begin{bmatrix} 0 & I_m \end{bmatrix} \tag{2.147}$$

where B_2 is defined in (2.133). The matrix $\hat{A}_{11} = A_{11} - A_{12}KC_1$ which is assumed
to be stable by choice of K.
Furthermore

$$\hat{C} = \begin{bmatrix} 0_{p \times (n-p)} & \hat{T} \end{bmatrix} \tag{2.148}$$

where

$$\hat{T} := \begin{bmatrix} (T_1 - T_2 K) & T_2 \end{bmatrix} \tag{2.149}$$

and T_1 and T_2 represent the first $p - m$ and last m columns of the matrix T
from Eq. (2.134). Notice that $\hat{T}$ is nonsingular. Define a partition of $\hat{A}_{21}$ from
Eq. (2.147) as

$$\hat{A}_{21} = \begin{bmatrix} \overset{n-p}{\longleftrightarrow} & \overset{p-m}{\longleftrightarrow} \\ \hat{A}_{211} & \hat{A}_{212} \end{bmatrix} \tag{2.150}$$

Ideally the degrees of freedom in selecting the controller should be determined
numerically so that the reachability condition

$$\dot{\sigma}^T \sigma < 0 \tag{2.151}$$

is satisfied where $\sigma(x) = FCx(t)$ is the switching function. If Eq. (2.151) can be
satisfied, then the sliding surface $\mathcal{S}$ is *globally attractive*.

Assuming an appropriate switching surface has been designed to solve the existence problem, the linear part of the control law can be chosen as

$$G = -\gamma F, \qquad \gamma > 0 \tag{2.152}$$

For a large enough scalar γ it can be shown that a sliding motion is obtained in finite time from any initial condition. However the reachability condition $\dot{\sigma}^T \sigma < 0$, where $\sigma(t) = Fy(t)$, only holds in a compact domain around the origin. Outside this domain the controller behaves as a variable structure controller with the property that it forces the state trajectories into the invariant domain (sometimes referred to as the "sliding patch") in finite time. Inside this domain the reachability condition $\dot{\sigma}^T \sigma < 0$ holds and so sliding occurs in finite time. *Provided the existence problem can be solved, no additional structural or system conditions need to be imposed.*

In practical situations the shortcoming of this controller is that γ must be large and hence the controller takes on a "high gain" characteristic.

Without loss of generality, write the linear feedback gain as

$$G = [\; G_1 \quad G_2 \;]\hat{T}^{-1} \tag{2.153}$$

where $\hat{T}$ is from Eq. (2.149) and $G_1 \in \mathbb{R}^{m \times (p-m)}$ and $G_2 \in \mathbb{R}^{m \times m}$. Define a symmetric positive definite block diagonal matrix

$$P := \begin{bmatrix} P_1 & 0 \\ 0 & P_2 \end{bmatrix} > 0 \tag{2.154}$$

where $P_1 \in \mathbb{R}^{(n-m) \times (n-m)}$ and $P_2 \in \mathbb{R}^{m \times m}$. Then it is possible to find a matrix P as in Eq. (2.154), and a gain matrix G so that

$$PA_c + A_c^T P < 0 \tag{2.155}$$

where $A_c = \hat{A} - \hat{B}G\hat{C}$; then the control law will induce a sliding motion on the surface S inside the domain (the sliding patch)

$$\Omega = \{(\hat{z}_1, \hat{z}_2) \; : \; \|\hat{z}_1\| < \eta \gamma_0^{-1}\}$$

where $\gamma_0 = \|P_2(\hat{A}_{21} - G_1 C_1)\|$ and $\hat{z}_1 \in \mathbb{R}^{(n-m)}$, $\hat{z}_2 \in \mathbb{R}^m$ represent a partition of the state $\hat{z}$.

From the point of view of control law design, a requirement is to make

$$\|P_2(\hat{A}_{21} - G_1 C_1)\| \tag{2.156}$$

small to make the sliding patch Ω large.

The block diagonal structure in Eq. (2.154), together with the canonical form in Eq. (2.147), effectively guarantees a solution to the structural constraint

$P\hat{B} = (F\hat{C})^T$, which in turn ensures the transfer function matrix $F\hat{C}(sI - \hat{A})^{-1}\hat{B}$ is strictly positive real.[11]

Example 2.10. Consider the example from Sect. 2.9. From the root locus in Fig. 2.11 it can be seen that high gain is needed to improve the damping of the dominant complex conjugate pair. With a value of $\gamma = 100$ the sliding motion poles are governed by

$$\{-1.9241 \pm 5.6081i, \ -167.0320\}$$

and

$$\hat{A}_{21} = \begin{bmatrix} 100.7287 & 1492.0600 & -16474.8680 \end{bmatrix}$$

It can be shown for

$$G = \begin{bmatrix} 1095.7134 & 208.2982 \end{bmatrix}$$

that

$$\mathrm{eig}(A - BGC) = \{-162.0054, \ -12.4248, \ -0.3736 \pm 5.0794i\}$$

and an associated Lyapunov matrix can be found which also satisfies the structural constraint.

Remark 2.11. Although from a control theory point of view this example demonstrates the theory is valid and that a static output-feedback controller does exist, the resulting scheme may not be practical. Here the gain G is large and the sliding motion will be governed by two quite poorly damped dominant complex eigenvalues. This motivates the consideration of compensator-based output-feedback sliding mode controller design.

2.6.3 Dynamic Output-Feedback Hyperplane Design

So far, only the static output feedback case has been considered. In certain circumstances, the subsystem triple (A_{11}, A_{12}, C_1) is known to be infeasible with respect to static output-feedback stabilization. In such situations, consider the introduction of a dynamic compensator. Specifically, let

$$\dot{z}_c(t) = H z_c(t) + D y(t) \tag{2.157}$$

where the matrices $H \in \mathbb{R}^{q \times q}$ and $D \in \mathbb{R}^{q \times p}$ are to be determined. Define a new hyperplane in the augmented state space, formed from the plant and compensator state spaces, as

$$S_c = \{(z, z_c) \in \mathbb{R}^{n+q} \ : \ F_c z_c + FC z = 0\} \tag{2.158}$$

---

[11]For details and definitions see [175]. This is also discussed in Appendix B.

where $F_c \in \mathbb{R}^{m \times q}$ and $F \in \mathbb{R}^{m \times p}$. Define $D_1 \in \mathbb{R}^{q \times (p-m)}$ and $D_2 \in \mathbb{R}^{q \times m}$ as

$$\begin{bmatrix} D_1 & D_2 \end{bmatrix} = DT \tag{2.159}$$

then the compensator can be written as

$$\dot{z}_c(t) = Hz_c(t) + D_1 C_1 z_1(t) + D_2 z_2(t) \tag{2.160}$$

where C_1 is defined in Eq. (2.139). Assume that a control action exists which forces and maintains motion on the hyperplane $\mathcal{S}_c$ given in Eq. (2.158). As before, in order for a unique equivalent control to exist, the square matrix F_2 must be invertible. By writing $K = F_2^{-1} F_1$ and defining $K_c = F_2^{-1} F_c$ then the system matrix governing the reduced-order sliding motion, obtained by eliminating the coordinates z_2, can be written as

$$\dot{z}_1(t) = (A_{11} - A_{12} K C_1) z_1(t) - A_{12} K_c z_c(t) \tag{2.161}$$

$$\dot{z}_c(t) = (D_1 - D_2 K) C_1 z_1(t) + (H - D_2 K_c) z_c(t) \tag{2.162}$$

It follows that stability of the sliding motion depends only on the matrix

$$\begin{bmatrix} A_{11} - A_{12} K C_1 & -A_{12} K_c \\ (D_1 - D_2 K) C_1 & H - D_2 K_c \end{bmatrix} \tag{2.163}$$

As in the uncompensated case, it is necessary for the pair (A_{11}, C_1) to be detectable. To simplify the design problem, at the expense of removing some of the design flexibility, one can specifically choose $D_2 = 0$ in Eq. (2.163).

The resulting matrix in Eq. (2.163) can be viewed as the negative feedback interconnection of the "plant" $G_p(s) = C_1 (sI - A_{11})^{-1} A_{12}$ and the "compensator"

$$K(s) = K + K_c (sI - H)^{-1} D_1 \tag{2.164}$$

Note that this still has a very generalized structure and any linear design paradigm that creates an internally stabilizing closed loop can be employed to synthesize the sliding mode compensator matrices D_1 and H, and a hyperplane, represented by the matrices K and K_c,

In certain situations it is advantageous to consider the feedback configuration in Fig. 2.12 and to design a compensator $K(s)$ using any suitable paradigm to yield appropriate closed-loop performance. From the state-space realization of $K(s)$ in Eq. (2.164), the parameters K, K_c, D_1, and H can be identified. If the quantity $p - m$ is small, using "classical control" ideas, very simple compensators may be found which give good closed-loop performance to the fictitious feedback system in Fig. 2.12 which governs the sliding mode performance of the real system.

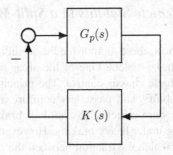

Fig. 2.12 A general linear feedback configuration

2.6.4   Dynamic Output-Feedback Control Law Development

Example 2.11. The third-order "plant" from Eq. (2.144) has a double pole at the origin. In terms of classical control, this suggests the use of a lead compensator. Choosing

$$K(s) = \frac{(s+0.5)}{(s+10)} \qquad (2.165)$$

in unity feedback with $G_p(s)$ gives closed-loop poles at

$$\{-12.1413, \ -0.9656, \ -1.2490 \pm 0.9718\}$$

This has improved the damping ratio of the dominant complex pair. A realization of the compensator in Eq. (2.165) is $H = -10$, $D_1 = 1$, $K_c = -9.5$, and $K = 1$. Using these values (and $D_2 = 0$) in the formulae in Eqs. (2.138) and (2.159) gives

$$F_a := \begin{bmatrix} F_c & F \end{bmatrix} = \begin{bmatrix} -11.7059 & | & 1.2322 & 1.2322 \end{bmatrix} \qquad (2.166)$$

and

$$D = \begin{bmatrix} 1 & 0 \end{bmatrix}$$

Using these matrices, an appropriate controller to induce a sliding motion may be obtained by using the method described earlier for the augmented system

$$A_a = \begin{bmatrix} H & DC \\ 0 & A \end{bmatrix} \qquad B_a = \begin{bmatrix} 0 \\ B \end{bmatrix} \qquad C_a = \begin{bmatrix} 1 & 0 \\ 0 & C \end{bmatrix} \qquad (2.167)$$

where the switching function matrix F_a is defined in Eq. (2.166). Choosing

$$G = \begin{bmatrix} -45.9050 & 4.7749 & 0.3392 \end{bmatrix}$$

gives

$$\text{eig}(A_a - B_a G C_a) = \{-8.6605, \ -3.0494 \pm 1.9424i, \ -0.8238 \pm 0.3124\}$$

2.6.5   Case Study: Vehicle Stability in a Split-Mu Maneuver

This section will show how these output-feedback sliding mode ideas can be applied to a realistic control problem. One of the areas of active research in the automotive industry is vehicle chassis control. The principal aims are to increase vehicle safety, maneuverability, and passenger comfort while reducing the work load on the driver. Most modern vehicles have antilock brake systems (ABS) which prevent the wheels locking under heavy braking. However, if a so-called split-mu surface is encountered, in which the traction between the road surface and the tire on one wheel becomes significantly reduced compared to the others, for instance, if a patch of ice is encountered, then any braking maneuver will introduce a sudden unexpected yawing moment. At high speed the driver will not have sufficient time to react, and a potentially extremely dangerous spin may occur.

This case study will consider a steer-by-wire system which will aim to maintain heading and vehicle stability in such a situation. An eighth-order nonlinear model of the vehicle, wheels, and road/tire interaction has been developed.[12] This model has been tuned to be representative of a typical family saloon. A linearization has been performed about an operating condition which represents a straight line trajectory at $15\,\mathrm{ms}^{-1}$ longitudinal velocity, corresponding wheel velocities, and zero lateral velocity, yaw rate, and yaw angle. The linearization assumes a uniform friction coefficient on each wheel. Considering yaw rate and lateral deviation as measured variables, a reduced-order linear model of the rigid body states is given in Eqs. (2.142) and (2.143).

The scenario that will be considered is an emergency stop on a split-mu surface. A simple ABS system has been incorporated into the nonlinear vehicle model. This is designed to bring the vehicle to a standstill in as short a time as possible. The steer-by-wire system will alter the front wheel position (the control input) in an effort to brake in a straight line. The key requirement is, therefore, to keep the third state, Y, in Eq. (2.143) as near to zero as possible.

From the canonical form in Eq. (2.144) the output of the fictitious system $G_p(s) = C_1(sI - A_{11})^{-1}A_{12}$ (as far as the switching function design is concerned) is, in the real system, the output of interest, Y. Because $G_p(s)$ has a double integrator characteristic no integral action is needed to achieve a steady-state error of zero.

The closed-loop simulation obtained from a fully nonlinear model is shown in Fig. 2.13. The controller manages to stop the vehicle from developing an excessive yaw angle. Also, the lateral deviation, Y, is halted with a peak of 17 cm and is regulated to zero. The controller develops and maintains sufficient yaw angle as is necessary to counteract the yawing moment induced by the asymmetric ABS braking. The input signal which the controller utilizes to perform this is shown in Fig. 2.14. It is also demonstrated that a sliding motion is maintained throughout the maneuver.

---

[12]For details of the model see [72, 108].

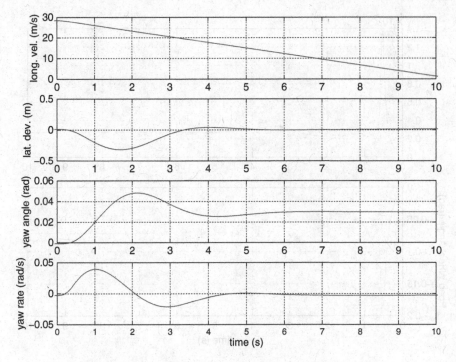

Fig. 2.13 Vehicle states and braking torques for closed-loop simulation

2.7 Integral Sliding Mode Control

As discussed earlier in this chapter, sliding mode control techniques are very useful for the controller design in systems with matched disturbances/parametric uncertainties. The system's compensated dynamics become insensitive to these matched disturbances and uncertainties under sliding mode control. This property of insensitivity is only achieved when the sliding surface is reached and the sliding motion is established. In this section we explore a method to compensate for the disturbance without the presence of a reaching phase.

For known linear systems, traditional controllers, including proportional-plus-derivative (PD), proportional-plus-integral-plus-derivative (PID), and optimal linear-quadratic regulator (LQR), can be successfully designed to achieve ideal closed-loop dynamics. Also known, nonlinear systems (with certain structures) can be controlled using, for instance, feedback linearization, backstepping, or any other Lyapunov-based nonlinear technique. Once a system is subjected to external bounded disturbances, it is natural to try to compensate such effects by means of an auxiliary (sliding mode) control while the original controller compensates for the unperturbed system.

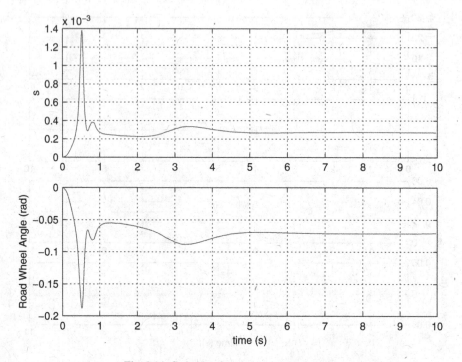

Fig. 2.14 Switching function and control signal

In this section we will design a so-called integral sliding mode (ISM) auxiliary controller compensating for the disturbance from $t \geq 0$, while retaining the order of the uncompensated system. This can be achieved assuming that the initial conditions are known.

2.7.1 Problem Formulation

Let us consider the following controlled uncertain system represented by the state-space equation

$$\dot{x} = f(x) + B(x)u + \phi(x,t), \qquad (2.168)$$

where $x \in \mathbb{R}^n$ is the state vector, $u \in \mathbb{R}^m$ is the control input vector, and $x(0) = x_0$. The function $\phi(x,t)$ represents the uncertainties affecting the systems, due to parameter variations, unmodeled dynamics, and/or exogenous disturbances.

Let $u = u_0$ be a nominal control designed for Eq. (2.168) assuming $\phi = 0$, where u is designed to achieve a desired task, whether stabilization, tracking, or an optimal control problem. Thus, the trajectories of the ideal system ($\phi = 0$) are given by the solutions of

$$\dot{x}_0 = f(x_0) + B(x_0)u_0. \qquad (2.169)$$

When $\phi \neq 0$ the trajectories of Eqs. (2.168) and (2.169) are different. The trajectories of Eq. (2.169) satisfy some specified requirements, whereas the trajectories of Eq. (2.168) might have a quite different and possibly even undesirable performance (depending on ϕ).

In order to design the controller assume that:

(A1) Rank $(B(x)) = m$ for all $x \in \mathbb{R}^n$.
(A2) The disturbance $\phi(x,t)$ is matched and there exists a vector $\xi(x,t) \in \mathbb{R}^m$ such that $\phi(x,t) = B(x)\xi(x,t)$.
(A3) A known upper bound for $\xi(x,t)$ can be found, i.e.,

$$\|\xi(x,t)\| \leq \xi^+(x,t). \tag{2.170}$$

Obviously, the second restriction is needed to exactly compensate ϕ. If ϕ were known exactly, it would be enough to chose $u = -\xi$. However, since ξ is uncertain, some other restrictions are needed in order to eliminate the influence of ϕ. Here the sliding mode approach is used to replace exact knowledge of ϕ.

2.7.2 Control Design Objective

Now the problem is *to design a control law* such that $x(0) = x_0(0)$, and *guarantees the identity* $x(t) = x_0(t)$ for all $t \geq 0$. Comparing Eqs. (2.168) and (2.169), it can be seen that the objective is achieved only if the equivalent control is equal to minus the uncertainty (i.e., $u_{1_{eq}} = -\xi$). Thus, the control objective can be reformulated in the following terms: *design the control* $u = u(t)$ as

$$u(t) = u_0(t) + u_1(t), \tag{2.171}$$

where $u_0(t)$ is the nominal control part designed for Eq. (2.169) and $u_1(t)$ is the ISM control part compensating for the unmeasured matched uncertainty $\phi(x,t)$, starting from $(t = 0)$.

2.7.3 Linear Case

Let us consider the linear time-invariant case:

$$\dot{x}(t) = Ax(t) + B(u_0(t) + u_1(t)) + \phi(t,x), \qquad \phi(t,x) = B\xi(t,x) \tag{2.172}$$

In this case the vector function σ can be defined as

$$\sigma(x) = G(x(t) - x(0)) - G \int_0^t (Ax(\tau) + Bu_0(\tau))\,d\tau, \tag{2.173}$$

where $G \in \mathbb{R}^{m \times n}$ is a projection matrix satisfying the condition

$$\det(GB) \neq 0$$

The time derivative of σ has the form

$$\dot{\sigma}(x) = GB(u_1 + \xi)$$

The control u_1 is taken as the unit vector

$$u_1 = -\rho(t, x) \frac{(GB)^T \sigma}{\|(GB)^T \sigma\|} \tag{2.174}$$

where $\rho(t, x) \geq \|\xi^+(t, x)\|$. Taking $V = \frac{1}{2} \sigma^T \sigma$ and in view of Eq. (2.170) the derivative of V on time is

$$\dot{V} = \sigma^T GB(u_1 + \gamma)$$
$$\leq -\|(GB)^T \sigma\|(M - \xi^+) < 0$$

Hence, the ISM is guaranteed.

Example 2.12. Let us consider the following system:

$$\dot{x} = Ax + B(u_0 + u_1) + \phi \tag{2.175}$$

representing a linearized model of an inverted pendulum on a cart, where x_1 and x_2 are the cart and pendulum positions, respectively, and x_3 and x_4 are the respective velocities. The matrices A and B are taken with the following values:

$$A = \begin{bmatrix} 0 & 0 & 1 & 0 \\ 0 & 0 & 0 & 1 \\ 0 & 1.25 & 0 & 0 \\ 0 & 7.55 & 0 & 0 \end{bmatrix}, B = \begin{bmatrix} 0 \\ 0 \\ 0.19 \\ 0.14 \end{bmatrix},$$

and the control $u_0 = u_0^*$ is designed for the nominal system, where u_0^* solves the following optimal problem subject to an LQ performance index:

$$J(u_0) = \int_0^\infty x_0^T(t) Q x_0(t) + u_0^T(t) R u_0(t) \, dt$$

where $u_0^* = \arg \min J(u_0)$. It is known (see Appendix C) that the solution of this problem is the following:

$$u_0^*(x) = -R^{-1} B^T P x = -K x,$$

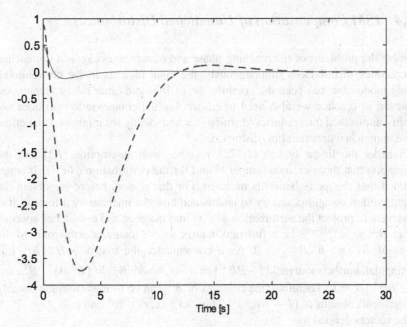

Fig. 2.15 States x_1 (*dashed*) and x_2 (*solid*) using ISM in the presence of matched uncertainties

where P is a symmetric positive definite matrix that is the solution of the algebraic Riccati equation

$$A^T P + PA - PBR^{-1} B^T P = -Q.$$

For the considered matrices A and B and taking $Q = I$ and $R = 1$, we have

$$K = \begin{bmatrix} -1 & 131.36 & -4.337 & 48.47 \end{bmatrix}.$$

In the simulation $\phi = B\xi$ with $\xi = 2\sin(0.5t) + 0.1\cos(10t)$ and the ISM control is

$$u_1 = -5\operatorname{sign}(\sigma),$$

where σ is designed according to Eq. (2.173). Now, the only restriction on the matrix G is $\det GB \neq 0$. One simple choice is $G = \begin{bmatrix} 0 & 0 & 1 & 0 \end{bmatrix}$; thus we obtain $GB = 0.19$, which obviously is different from zero.

Figure 2.15 shows the position of the cart and the pendulum. We can see that there is no influence of the disturbance ξ due to the compensation effect caused by the ISM control part u_1.

2.7.4 ISM Compensation of Unmatched Disturbances

To solve the problems of the reaching phase and of robustness against unmatched uncertainties/disturbances simultaneously, the main idea, as in the conventional sliding mode case, has been the combination of ISM and other robust techniques. However, in practice we also need to ensure that the compensation (a) does not amplify unmatched uncertainties/disturbances and ideally (b) minimizes the effect of the unmatched uncertainties/disturbances.

Consider the linear system (2.172), together with assumption (A1) and the assumption that there exists an upper bound for the perturbation $\phi(x,t)$. It is not assumed that this perturbation is matched. For that reason, before we obtain the sliding motion equations and try to understand how the uncertainty affects it, it is convenient to project the perturbation $\phi(x,t)$ into matched and unmatched spaces.

Let $B^\perp \in \mathbb{R}^{n \times (n-m)}$ be a full rank matrix whose image is orthogonal to the image of B, i.e., $B^T B^\perp = 0$. As a consequence the matrix $\begin{bmatrix} B & B^\perp \end{bmatrix}$ is nonsingular. Furthermore $\mathrm{rank}(I - BB^+) = n - m$, where $B^+ = (B^T B)^{-1} B^T$, and $(I - BB^+)B = 0$. Therefore, the columns of $B^\perp$ can be formed from the linearly independent columns of $(I - BB^+)^T$. Thus, let $\xi(x,t) \in \mathbb{R}^m$ and $\mu(x,t) \in \mathbb{R}^{n-m}$ be the vectors defined by

$$\begin{bmatrix} \xi(x,t) \\ \mu(x,t) \end{bmatrix} = \begin{bmatrix} B & B^\perp \end{bmatrix}^{-1} \phi(x,t) \tag{2.176}$$

Thus, Eq. (2.172) takes the following form:

$$\dot{x} = Ax + B(u_1 + u_0) + B\xi + B^\perp \mu \tag{2.177}$$

Then selecting σ as in Eq. (2.173), we have

$$\dot{\sigma} = GB(u_1 + \xi) + GB^\perp \mu$$

The control component u_1 should be designed as in Eq. (2.174) under the assumption GB is nonsingular. Let the modulation function $\rho \geq \xi^+ + (GB)^+ GB^\perp \mu$. The equivalent control obtained from solving $\dot{\sigma} = 0$ is given by the equation

$$u_{1_{eq}} = -\xi - (GB)^{-1} GB^\perp \mu$$

Substitution of $u_{1_{eq}}$ in Eq. (2.177) yields the sliding motion equation:

$$\dot{x} = Ax + Bu_0 + (I - B(GB)^{-1}G)B^\perp \mu$$

Define $\bar{d} := (I - B(GB)^{-1}G)B^{\perp}\mu$. Taking $G = B^T$ or $G = B^+$, we get $\bar{d} = B^{\perp}\mu$, i.e., the application of the sliding mode controller has not affected the unmatched disturbance part.

Now the question is, is it possible to select G to ensure the norm of $\bar{d}$ is less than the norm of $B^{\perp}\mu$? This is addressed in the following proposition.

Proposition 2.1 *The set of matrices* $\{G = QB^T : Q \in \mathbb{R}^{m \times m} \text{ and } \det(Q) \neq 0\}$ *is the solution of the optimization problem*

$$G^* = \arg\min_{G \in \bar{G}} \|(I - B(GB)^{-1}G)B^{\perp}\mu, \mu \neq 0\|$$

where $\bar{G} = \{G \in \mathbb{R}^{m \times n} : \det(GB) \neq 0\}$.

Proof. Since $B^{\perp}\mu$ and $B(GB)^{-1}GB^{\perp}\mu$ are orthogonal vectors, the norm of the vector $\|(I - B(GB)^{-1}G)B^{\perp}\mu\|$ is always greater than $\|B^{\perp}\mu\|$. Indeed,

$$\|(I - B(GB)^{-1}G)B^{\perp}\mu\|^2 = \|B^{\perp}\mu\|^2 + \|B(GB)^{-1}GB^{\perp}\mu\|^2$$

That is,

$$\|(I - B(GB)^{-1}G)B^{\perp}\mu\| \geq \|B^{\perp}\mu\| \tag{2.178}$$

If identity (2.178) is achieved, then the norm of $\|(I - B(GB)^{-1}G)B^{\perp}\mu\|$ is minimized with respect to G. The identity is obtained, if and only if $B(GB)^{-1}GB^{\perp}\mu=0$. Or equivalently, since rank$(B) = m$, $GB^{\perp}\mu = 0$, i.e., $G = QB^T$, where Q is nonsingular. This completes the proof. $\square$

Remark 2.12. Notice that the control law itself is not modified in order to optimize the effect of the unmatched uncertainties, and moreover, an optimal solution for G^* is simple: the simplest choice is $G^* = B^T$, but $B^+ = (B^T B)^{-1} B^T$ is another possibility.

Proposition 2.2 *For an optimal matrix* G^*, *the Euclidean norm of the disturbance is not amplified, that is,*

$$\|\phi(t)\| \geq \|(I - B(G^*B)^{-1}G^*)B^{\perp}\mu(t)\| \tag{2.179}$$

Proof. From Proposition 2.1 it follows that

$$\|(I - B(G^*B)^{-1}G^*)B^{\perp}\mu(t)\| = \|(I - BB^+)B^{\perp}\mu(t)\| = \|B^{\perp}\mu(t)\| \tag{2.180}$$

Now, since $\phi(t) = B\xi + B^{\perp}\mu$, and $B^T B^{\perp} = 0$, we obtain the equation

$$\|\phi(t)\|^2 = \|B\xi(t) + B^{\perp}\mu(t)\|^2 = \|B\xi(t)\|^2 + \|B\mu(t)\|^2 \geq \|B\mu(t)\|^2 \tag{2.181}$$

Hence, comparing Eqs. (2.180) and (2.181), we can obtain Eq. (2.2). $\square$

Example 2.13. Consider system Eq. (2.175) with the uncertainty $\phi(x,t)$ shown below:

$$\phi(x,t) = \begin{bmatrix} 0 & 0 & 2\sin(0.5t) + 0.1\cos(10t) & 0.1\sin(1.4t) \end{bmatrix}^T$$

The first step is to project the perturbation $\phi(x,t)$ into the matched and unmatched spaces using the expression in Eq. (2.176). Note the selection of the matrix $B^{\perp}$ is not unique: one simple choice is given by

$$B^{\perp} = \begin{bmatrix} 1 & 0 & 0 \\ 0 & 1 & 0 \\ 0 & 0 & -0.14 \\ 0 & 0 & 0.19 \end{bmatrix}$$

In this way ξ and μ in system (2.177) become

$$\xi = 3.41(2\sin(0.5t) + 0.1\cos(10t)) + 2.51(0.1\sin(1.4t)),$$

$$\mu = \begin{bmatrix} 0 \\ 0 \\ 3.41\,[0.1\sin(1.4t)] - 2.51\,[2\sin(0.5t) + 0.1\cos(10t)] \end{bmatrix}$$

The control law is the same as in Example 2.12, except for the choice of matrix G, which according to Proposition 2.1 is optimal if $G = B^T = \begin{bmatrix} 0 & 0 & 0.19 & 0.14 \end{bmatrix}$ or $G = B^+$. Here we consider three cases: the case when we use $G = B^T$, the case when the ISM control is not applied, and the case when G is not selected in an optimal manner. For this last case, we use $G = \begin{bmatrix} 0 & 0 & 1 & 0 \end{bmatrix}$ as in Example 2.12. For the first case, the states x_1 and x_2 (the positions) are depicted in Fig. 2.16; there we can see that the uncertainties do not significantly affect the trajectories of the system. To compare the effect of the ISM in presence of unmatched disturbances, Fig. 2.17 shows the trajectories of x_1 and x_2 when the ISM control part is omitted ($u = u_0$). It is clear that in this case, the disturbances affect the system considerably compared with the trajectories of Fig. 2.17 where a well-designed ISM control (with an optimal G) reduces significantly the effect of the disturbances. Figure 2.18 shows the effect of the matrix G in the design of the sliding surface. In Fig. 2.18 we compare the behavior of the variable x_2 when G is badly chosen.

2.8 Notes and References

A very readable concise treatment of some of the material in this chapter appears in *The Control Handbook* [58].

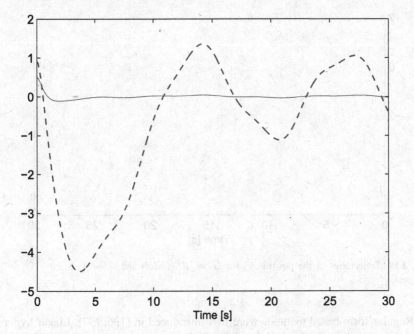

Fig. 2.16 States x_1 (*dashed*) and x_2(*solid*) without ISM compensator

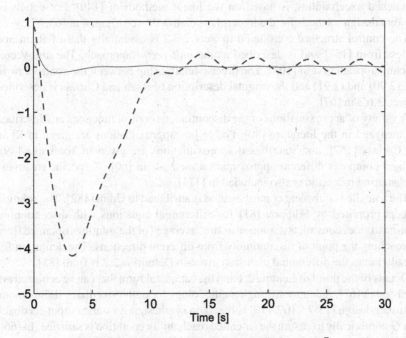

Fig. 2.17 States x_1 (*dashed*) and x_2 (*solid*) for $G = B^T$

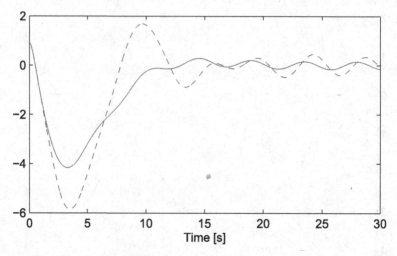

Fig. 2.18 Trajectories of the position x_2 for $G = B^T$ (*solid*) and $G = \begin{bmatrix} 0 & 0 & 1 & 0 \end{bmatrix}$ (*dashed*)

Regular form-based methods were first introduced in [136, 137]. Linear hyperplane design methods for a class of single-input systems are given in [2, 171]. Another approach to the hyperplane design, which can be used for treating the unmatched uncertainties, is based on the linear methods of [149]. For details of different design methods for the hyperplane matrix M see Chap. 4 in [63, 67].

The control structure considered in Sect. 2.3.2 is essentially that of Ryan and Corless from [157] and is described as the *unit-vector* approach. The unit-vector structure appears first in [103]. The precise relationship between the control law in Eqs. (2.90) and (2.91) and the original description of Ryan and Corless is described in Sect. 3.6.3 in [67].

A variety of approximations of the discontinuous control functions are described and analyzed in the literature [40]. Power law approximations are given in Ryan and Corless [157], and state based approximations are given in Tomizuka [49]. DeJager compares different approximation methods in [60]. A specific treatment on chattering reduction is also included in [171].

The so-called equivalent control method is attributed to Utkin [182]. The solution concept proposed by Filippov [81] for differential equations with discontinuous right-hand sides constructs a solution as the "average" of the solutions obtained from approaching the point of discontinuity from different directions. The definition for the solution to the differential inclusion given in Definition 2.2 is from [81].

Details of the proof of Lemma 2.1 and the canonical form that can be achieved are discussed in [67]. A common design methodology for output-feedback sliding mode controller design [13, 73, 109, 110] is based on synthesizing a static output-feedback gain G numerically to ensure the so-called reachability condition is satisfied. In [66], assuming an appropriate switching surface has been designed to solve the existence problem, the linear part of the control law was chosen as a scaling of the switching function, thought of as a feedback gain. The static output-feedback control law development in Sect. 2.6.2 is based on [71]. Example 2.9 is taken from the sliding mode output-feedback paper [72]. In the case study relating to vehicle control, the aims of increasing vehicle safety, maneuverability, and passenger comfort, while reducing the work load on the driver are discussed in detail in [1]. The limitations of some static output-feedback sliding mode controllers are discussed in [68].

The ISM concept was proposed independently by Matthews and DeCarlo [140] and Utkin and Shi [185]. In this chapter we followed the approach of Matthews and DeCarlo [140]. The ISM approach described in Sect. 2.7 can be easily extended to the class of affine nonlinear systems given by

$$\dot{x} = f(x) + B(x)(u_0 + u_1 + \xi)$$

For a detailed analysis of this case see [43]. Additional material for the advance study of the ISM approach can be found in many publications including [156, 192]. In particular, various combinations of ISM with $\mathcal{H}_\infty$ are studied in [43–45, 191]. The use of ISM schemes for "robustification" of solutions of LQR problems can be found in [93, 154] (a multi-model optimization problem); see also the design of robust output LQR control in [21, 26] and multiplant LQR control in [27]. The ISM controllers are widely used in robotic applications [50, 151] when it is necessary to track the reference trajectories. The corresponding references can be found in the books [186, 187] and the paper [59] and references therein.

An interesting application avenue exploiting the robustness properties of sliding modes with respect to matched uncertainties is the area of fault tolerant control. In such scenarios, actuator faults appear naturally within the control channels of the plant and can be accommodated "automatically" by sliding mode controllers [6, 7, 105, 162].

Although not discussed in this book, the main ideas and techniques of discrete-time sliding mode control can be found [3, 14, 97, 141, 187].

2.9 Exercises

Exercise 2.1. Consider the linear system

$$\dot{x} = Ax + Bu$$

$$\sigma = Gx$$

with σ as the sliding variable. Find the equivalent control u_{eq} and the sliding mode equations when

(a) $A = \begin{bmatrix} 2 & 19 \\ 3 & 29 \end{bmatrix}$, $B = \begin{bmatrix} 2 & 3 \end{bmatrix}^T$ and $G = \begin{bmatrix} 9 & 12 \end{bmatrix}$

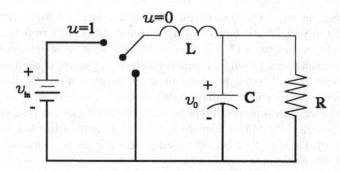

Fig. 2.19 DC–DC buck converter

(b) $A = \begin{bmatrix} 1 & 1 & 1 \\ 0 & 1 & 3 \\ 1 & 0 & 1 \end{bmatrix}$, $B = \begin{bmatrix} 3 & 9 \\ 1 & -2 \\ -1 & 0 \end{bmatrix}$ and $G = \begin{bmatrix} 1 & 29 & 0 \\ 1 & 12 & 0 \end{bmatrix}$

Exercise 2.2.
Consider the system given by

$$\dot{x}_1 = x_1 + x_2 + x_3 + u_1 + 10u_2$$
$$\dot{x}_2 = x_2 + 3x_3 + u_1 - 2u_2$$
$$\dot{x}_3 = x_1 + x_3 - u_1$$
$$\sigma_1 = x_1 + 10x_2, \; \sigma_2 = x_1 + 5x_2$$

Find the system dynamics in the sliding mode $\sigma_1 = \sigma_2 = 0$.

Exercise 2.3. Consider the DC–DC buck converter in Fig. 2.19 which belongs to the class of attenuation circuits; the corresponding dynamic equations are given by:

$$L\frac{di}{dt} = -v + uV_{in}$$
$$C\frac{dv}{dt} = i - \frac{v}{R}$$

where i is the current through the inductor L, v is the voltage across the capacitor C, V_{in} is the input voltage, and $u \in \{0, 1\}$ is the switching control signal. The goal is to stabilize the output voltage v at the desired level v_d. This goal is to be achieved via stabilization of the inductor current i at the desired level $i_d = \frac{v_d}{R}$ using sliding mode control, for this purpose:

(a) Setting $\sigma = i - i_d$ and using the control input $u = \frac{1}{2}(1 - \text{sign}(\sigma))$, find the equivalent control and the sliding mode dynamics.
(b) Considering $L = 20\,\text{mH}$, $C = 20\,\mu\text{F}$, $R = 30\,\Omega$, $V_{in} = 15\,\text{V}$, $v_d = 10\,\text{V}$, and the initial conditions $i(0) = 0.1\,\text{A}$ and $v(0) = 5\,\text{V}$, confirm the efficacy of the controller by simulations.

Exercise 2.4. Consider the linear system

$$\dot{x} = Ax + B(u + f)$$

with:

$$A = \begin{bmatrix} 3 & 5 \\ -1 & 2 \end{bmatrix}, B = \begin{bmatrix} 1 \\ 2 \end{bmatrix}$$

that is stabilized by the sliding mode controller $u = -(6||x|| + 3)\text{sign}(\sigma)$ in the presence of the bounded disturbance $|f| \le 2$. Design the sliding variable σ assuming $f = 0$, considering the two cases:

(a) LQR minimization, with $Q = I$ and $R = 1$
(b) Eigenvalue assignment, considering the eigenvalue $\lambda = -2$

Confirm your design by simulations, using $f = 2\sin(t)$ and the initial conditions $x_1(0) = 2$ and $x_2(0) = 1$.

Exercise 2.5. Using the sliding variable $\sigma = x_1$ and the control $u = -\text{sign}(\sigma)$, find the sliding mode equation for the systems given below, by using the equivalent control method and Filippov method for the cases:

(a)

$$\dot{x}_1 = u$$
$$\dot{x}_2 = (2u^2 - 1)x_2$$

(b)

$$\dot{x}_1 = u$$
$$\dot{x}_2 = (u - 2u^3)x_2$$

(c)

$$\dot{x}_1 = u$$
$$\dot{x}_2 = (u - 2u^2)x_2$$

Exercise 2.6. Consider the linear system given by

$$\dot{x} = Ax + Bu$$
$$y = Cx$$

with

$$A = \begin{bmatrix} 0 & 1 & 0 \\ 0 & 0 & 1 \\ -1 & \frac{1}{3} & -1 \end{bmatrix}, B = \begin{bmatrix} 0 \\ 1 \\ 1 \end{bmatrix}, C = \begin{bmatrix} 1 & 8/3 & 1 \\ 4 & 2 & -2 \\ 0 & 0 & 1 \end{bmatrix}$$

Design an output-feedback sliding mode control to stabilize the system at the origin using $u = -K\text{sign}(\sigma)$; use the eigenvalue assignment algorithm in order to design the sliding variable σ, and consider the eigenvalues, $\lambda_{1,2} = -2 \pm j5$. Considering

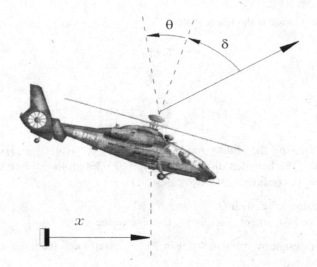

Fig. 2.20 Helicopter pitch angle, θ, control

the initial condition $x(0) = [1, 2, 3]^T$, obtain the appropriate value of K to ensure a reaching time $t_r = 2.5$ s and confirm your design by simulation.

Exercise 2.7. Consider the Chua's circuit of Example 2.7 with the output $y = x_1$. Propose a control law based on sliding modes in order to achieved the output tracking of $y_r = 5 \sin(2t)$. Confirm your design by simulation.

Exercise 2.8. The equations of motion of the high-performance helicopter in Fig. 2.20 are given by:

$$\begin{aligned}
\ddot{\theta} &= -a_1\dot{\theta} - b_1\dot{x} + n\delta \\
\ddot{x} &= g\theta - a_2\dot{\theta} - b_2\dot{x} + g\delta
\end{aligned}$$

where x is the translation in the horizontal direction. Design a sliding mode controller in terms of the rotor thrust angle δ that forces the pitch angle θ to asymptotically follow the reference profile $\theta_d = \frac{\pi}{6} + \frac{\pi}{12} \sin(t)$. Considering the initial conditions $\theta(0) = 0.2$ rad, $\dot{\theta} = 0$ rad/s, $x(0) = 0$ m, and $\dot{x}(0) = 50$ m/s, confirm your design by simulations of system with the parameters $a_1 = 0.415$, $a_2 = 0.0198$, $b_1 = 0.0111$, $b_2 = 1.43$, $n = 6.27$, and $g = 9.81$.

Exercise 2.9. Consider the satellite in Example 2.6 and design an ISM control $u = u_0 + u_1$. Preserve the linear control $u_0 = u_l$ as in Example 2.6 in order to stabilize the equilibrium point $(0, 0, 0)^T$ for the ideal system (without nonlinear disturbance terms). The control law u_0 should be designed using the LQR minimization algorithm. Design the ISM control component u_1 as in Eq. (2.174) to compensate the matched disturbances. Consider Q and R as identity matrices

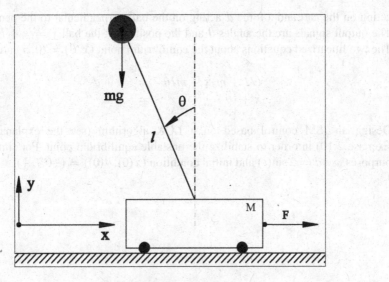

Fig. 2.21 Cart–pendulum

of appropriate dimensions. Simulate the system where for simulation purposes consider the initial condition $x(0) = [0.5, -1, 2]^T$ and the parameters $I_1 = 1 \, \text{kgm}^2$, $I_2 = 0.8 \, \text{kgm}^2$, and $I_3 = 0.4 \, \text{kgm}^2$.

Exercise 2.10. Consider the following linear system subject to external disturbances:

$$\dot{x} = Ax + Bu + \phi$$

where $A = \begin{bmatrix} -1 & 1 & 0 \\ 0 & -3 & 1 \\ -1 & 0 & -2 \end{bmatrix}$, $B = \begin{bmatrix} 0 & 0 \\ 1 & 0 \\ 0 & 1 \end{bmatrix}$ and $\phi = \begin{bmatrix} 3\sin(t) \\ 4\cos(t) \\ 4\cos(t) \end{bmatrix}$

(a) Identify if the disturbance vector ϕ is matched. If it is not, then project the disturbance into matched and unmatched terms.

(b) Design the control $u = u_0 + u_1$. The linear control u_0 is to be designed for nominal system (without disturbances) in order to stabilize the equilibrium point $(0, 0, 0)$ using the LQR algorithm. Consider Q and R identity matrices with appropriate dimensions. Design an ISM control component u_1 as in Eq. (2.174) to compensate the matched disturbances, and select the matrix $G = B^+ = (B^T B)^{-1} B^T$ that helps to accommodate the unmatched disturbance term. Confirm the effectiveness of the controller design via simulations. For simulation purposes consider the initial conditions $x(0) = (1, -2, 3)^T$.

Exercise 2.11. Consider the cart–pendulum system in Fig. 2.21 that consists of a cart of mass M that moves along the axis x, with a ball of mass m at the end of a rigid massless pendulum of length l. Shown as inputs are a horizontal force $F = u$

acting on the cart and a force d acting on the ball perpendicular to the pendulum. The output signals are the angles θ and the position of the ball $y = x + l \sin(\theta)$. The two linearized equations about the equilibrium point $(x, \theta) = (0, \pi)$ are

$$\begin{aligned} (M + m)\ddot{x} + ml\ddot{\theta} &= u + d \\ \ddot{x} + l\ddot{\theta} - g\theta &= \tfrac{1}{m}d \end{aligned}$$

Design an ISM control based on a LQR algorithm (see the explanation in Exercise 2.10) in order to stabilize the unstable equilibrium point. For simulation purposes use $d = 2\sin(t)$ and initial condition $(x(0), \theta(0)) = (-0.7, \frac{\pi}{3})$.

Chapter 3
Conventional Sliding Mode Observers

The purpose of an observer is to estimate the unmeasurable states of a system based only on the measured outputs and inputs. It is essentially a mathematical replica of the system, driven by the input of the system together with a signal representing the difference between the measured system and observer outputs. In the earliest observer, attributed to Luenberger, the difference between the output of the plant and the observer is fed back *linearly* into the observer. However, in the presence of unknown signals or uncertainty, a Luenberger observer is usually (a) unable to force the output estimation error to zero and (b) the observer states do not converge to the system states. A sliding mode observer, which feeds back the output estimation error via a nonlinear switching term, provides an attractive solution to this issue. Provided a bound on the magnitude of the disturbances is known, the sliding mode observer can force the output estimation error to converge to zero in *finite time*, while the observer states converge asymptotically to the system states. In addition, disturbances within the system can also be reconstructed.

3.1 Introduction

Consider initially a nominal linear system

$$\dot{x}(t) = Ax(t) + Bu(t) \qquad (3.1)$$

$$y(t) = Cx(t) \qquad (3.2)$$

where $A \in \mathbb{R}^{n \times n}$, $B \in \mathbb{R}^{n \times m}$, and $C \in \mathbb{R}^{p \times n}$. Without loss of generality assume that C has full row rank which means each of the measured outputs is independent. The objective is to obtain an estimate of the state $x(t)$ based only on knowledge of the quantities $y(t)$ and $u(t)$. An algebraic condition on the matrix pair (A, C)—the notion of observability—was proposed as a necessary and sufficient condition for

Y. Shtessel et al., *Sliding Mode Control and Observation*, Control Engineering, 105
DOI 10.1007/978-0-8176-4893-0_3, © Springer Science+Business Media New York 2014

state reconstruction (for details see Appendix C). For simplicity, the observability condition will be assumed to hold, although technically some of the developments only require the weaker restriction of detectability on the pair (A, C).

One way of viewing the approach of Luenberger is to think of the observer system as comprising a model of the plant together with a feedback term which corrects the estimates by injecting back the discrepancy between its output and the output of the system, through a designer-specified gain. In the simplest form of sliding mode observer, instead of feeding back the output error between the observer and the system in a linear fashion, the output error is fed back via a discontinuous switched signal.

3.2 A Simple Sliding Mode Observer

Consider a coordinate transformation $x \mapsto T_c x$ associated with the invertible matrix

$$T_c = \begin{bmatrix} N_c^T \\ C \end{bmatrix} \qquad (3.3)$$

where the submatrix $N_c \in \mathbb{R}^{n \times (n-p)}$ spans the null-space of C. By construction $\det(T_c) \neq 0$. Applying the change of coordinates $x \mapsto T_c x$, the triple (A, B, C) has the form

$$T_c A T_c^{-1} = \begin{bmatrix} A_{11} & A_{12} \\ A_{21} & A_{22} \end{bmatrix}, \quad T_c B = \begin{bmatrix} B_1 \\ B_2 \end{bmatrix}, \quad C T_c^{-1} = \begin{bmatrix} 0 & I_p \end{bmatrix} \quad (3.4)$$

where $A_{11} \in \mathbb{R}^{(n-p) \times (n-p)}$ and $B_1 \in \mathbb{R}^{(n-p) \times m}$. The structure imposed on the output distribution matrix in (3.4) is crucial to what follows. Assume without loss of generality the system (3.1) and (3.2) is already in the form of (3.4).

Utkin proposed an observer for (3.1) and (3.2) of the form

$$\dot{\hat{x}}(t) = A\hat{x}(t) + Bu(t) + G_n \nu \qquad (3.5)$$

$$\hat{y}(t) = C\hat{x}(t) \qquad (3.6)$$

where $(\hat{x}, \hat{y})$ are the estimates of (x, y) and ν is a discontinuous injection term. Define $e(t) := \hat{x}(t) - x(t)$ and $e_y(t) := \hat{y}(t) - y(t)$ as the state estimation and output estimation errors, respectively. The term ν is defined component-wise as

$$\nu_i = \rho \operatorname{sign}(e_{y,i}), \quad i = 1, 2, ..., p \qquad (3.7)$$

where ρ is a positive scalar and $e_{y,i}$ represents the ith component of e_y. The term ν is designed to be discontinuous with respect to the sliding surface $S = \{e : Ce = 0\}$ to force the trajectories of $e(t)$ onto S in finite time. Assume without loss of generality that the system is already in the coordinate associated with (3.4), then the gain G_n has the structure

$$G_n = \begin{bmatrix} L \\ -I_p \end{bmatrix} \tag{3.8}$$

where $L \in \mathbb{R}^{(n-p) \times p}$ represents the design freedom. It follows from the definition of $e(t)$ and Eqs. (3.1) and (3.5) that the error system is given by

$$\dot{e}(t) = Ae(t) + G_n \nu \tag{3.9}$$

From the structure of the output distribution matrix C in (3.4), the state estimation error can be partitioned as $e = \mathrm{col}(e_1, e_y)$ where $e_1 \in \mathbb{R}^{n-p}$. Consequently the error system from (3.9) can be written in the form

$$\dot{e}_1(t) = A_{11}e_1(t) + A_{12}e_y(t) + L\nu \tag{3.10}$$

$$\dot{e}_y(t) = A_{21}e_1(t) + A_{22}e_y(t) - \nu \tag{3.11}$$

Furthermore Eq. (3.11) can be written component-wise as

$$\dot{e}_{y,i}(t) = A_{21,i}e_1(t) + A_{22,i}e_y(t) - \rho \operatorname{sign}(e_{y,i}) \tag{3.12}$$

where $A_{21,i}$ and $A_{22,i}$ represent the ith rows of A_{21} and A_{22}, respectively. To develop conditions under which sliding will take place, the reachability condition will be tested. From (3.12)

$$\begin{aligned} e_{y,i}\dot{e}_{y,i} &= e_{y,i}(A_{21,i}e_1 + A_{22,i}e_y) - \rho|e_{y,i}| \\ &< -|e_{y,i}|(\rho - |(A_{21,i}e_1 + A_{22,i}e_y)|) \end{aligned}$$

Provided the scalar ρ is chosen large enough such that

$$\rho > |A_{21,i}e_1 + A_{22,i}e_y| + \eta \tag{3.13}$$

where the scalar $\eta \in \mathbb{R}_+$, then

$$e_{y,i}\dot{e}_{y,i} < -\eta|e_{y,i}| \tag{3.14}$$

This is the eta-reachability condition discussed in Chap. 2 and implies that $e_{y,i}$ will converge to zero in *finite time*. When every component of $e_y(t)$ has converged to zero, a sliding motion takes place on the surface $\mathcal{S}$.

Remark 3.1. Note that this is not a global result. For any given ρ, there will exist initial conditions of the observer (typically representing very poor estimates of the initial conditions of the plant) so that (3.13) is not satisfied.

During sliding, $e_y(t) = \dot{e}_y(t) = 0$, and the error system defined by (3.10) and (3.11) can be written in collapsed form as

$$\dot{e}_1(t) = A_{11}e_1(t) + L\nu_{eq} \tag{3.15}$$

$$0 = A_{21}e_1(t) - \nu_{eq} \tag{3.16}$$

where v_{eq} is the so-called *equivalent output error injection* that is required to maintain the sliding motion. This is the natural analogue of the equivalent control discussed in Chap. 2. Substituting for v_{eq} from (3.15) and (3.16) yields the following expression for the reduced-order sliding motion:

$$\dot{e}_1(t) = (A_{11} + LA_{21})e_1(t) \tag{3.17}$$

This represents the reduced-order motion (of order $n - p$) that governs the sliding mode dynamics.

It can be shown if (A, C) is observable, then (A_{11}, A_{21}) is also observable, and a matrix L can always be chosen to ensure that the reduced-order motion in (3.17) is stable.

Example 3.1. Consider a second-order state-space system described by (3.1) and (3.2) where

$$A = \begin{bmatrix} 0 & 1 \\ -2 & 0 \end{bmatrix}, \quad B = \begin{bmatrix} 0 \\ 1 \end{bmatrix}, \quad C = \begin{bmatrix} 1 & 1 \end{bmatrix} \tag{3.18}$$

which represents a simple harmonic oscillator. For simplicity, assume $u(t) = 0$. A suitable choice for the nonsingular matrix T_c from (3.3) is

$$T_c = \begin{bmatrix} 1 & -1 \\ 1 & 1 \end{bmatrix} \tag{3.19}$$

Following the change of coordinates $x \mapsto T_c x$, the system triple (A, B, C) becomes

$$T_c A T_c^{-1} = \begin{bmatrix} 0.5 & 1.5 \\ -1.5 & -0.5 \end{bmatrix}, \quad T_c B = \begin{bmatrix} -1 \\ 1 \end{bmatrix}, \quad C T_c^{-1} = \begin{bmatrix} 0 & 1 \end{bmatrix} \tag{3.20}$$

and $A_{11} = 0.5$ and $A_{21} = -1.5$. Suppose the nonlinear gain from (3.8) is chosen as $L = 3$. This results in the sliding motion being governed by $A_{11} + LA_{21} = -4$, which is stable. In the original coordinates of (3.18), the nonlinear gain can be calculated as

$$G_n = T_c^{-1} \begin{bmatrix} L \\ -1 \end{bmatrix} = \begin{bmatrix} 0.5 & 0.5 \\ -0.5 & 0.5 \end{bmatrix} \begin{bmatrix} 3 \\ -1 \end{bmatrix} = \begin{bmatrix} 1 \\ -2 \end{bmatrix} \tag{3.21}$$

The following simulation was performed with the system having initial conditions $x(0) = \text{col}(0.5, -0.8)$ and the observer having zero initial conditions.

Figure 3.1 shows the system states $x(t)$ and the estimates $\hat{x}(t)$. After approximately 1.5 s, excellent tracking of the states occurs. Figure 3.2 shows the output estimation error $e_y(t)$ and the state estimation errors $e(t)$. After approximately 0.66 s, $e_y(t)$ becomes zero and remains zero. This is indicative of a sliding motion taking place on S. Subsequently, the errors (Fig. 3.1) evolve governed the dynamics of the *reduced-order motion*. Figure 3.3 shows that during the sliding motion, the term v exhibits high-frequency switching. Figure 3.4 shows an approximation of the equivalent output error injection signal v_{eq} obtained from low-pass filtering v

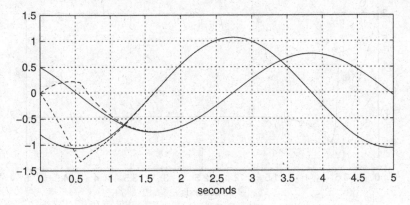

Fig. 3.1 System states $x(t)$ (*solid*) and the observer estimates $\hat{x}(t)$ (*dashed*)

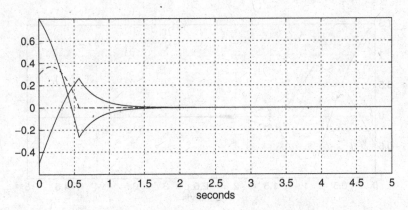

Fig. 3.2 The output estimation error $e_y(t)$ (*dashed*) and the components of the state estimation error $e(t)$ (*solid*)

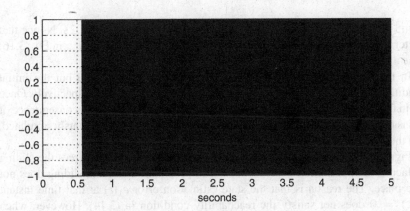

Fig. 3.3 The nonlinear injection switching term v

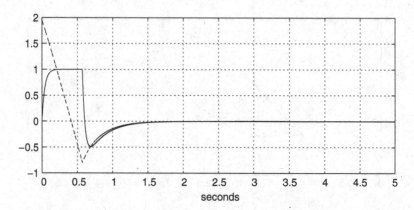

Fig. 3.4 The equivalent output error injection v_{eq} (*solid*) and $A_{21}e_1(t)$ (*dashed*)

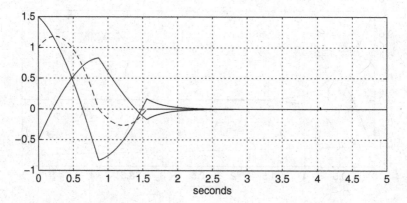

Fig. 3.5 The output estimation error $e_y(t)$ (*dashed*) and the components of the state estimation error $e(t)$ (*solid*)

from Fig. 3.3 through a first-order filter with time constant $\tau = 0.02\,\text{s}$. Notice that the term v_{eq} shows good agreement with the formal expression for v_{eq} in Eq. (3.16) once a sliding motion is taking place.

In the next set of simulations the same observer is employed but the initial conditions of the plant states have been changed to 0.5 and -1.5, respectively. Once again the initial conditions of the observer are set as zero. The direct consequence of this is that the initial conditions $e_1(0)$ and $e_y(0)$ are larger than previously simulated, and the response of the error system is qualitatively different.

Figure 3.5 shows that the output estimation error $e_y(t)$ now pierces the sliding surface S at approximately 0.87 s but does not remain there, and sliding does not take place. The reason is that the state estimation error $e_1(t)$ at that time instant $e_y(t) = 0$ does not satisfy the reachability condition in (3.14). However, when $e_y(t)$ becomes zero again at approximately 1.55 s, a sliding motion begins. At this

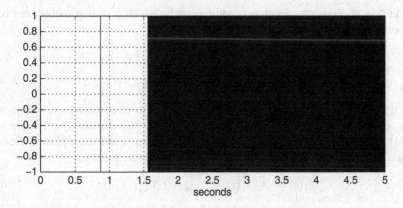

Fig. 3.6 The discontinuous term v with larger error initial conditions

point in time, the error $e_1(t)$ is much smaller, and the reachability condition (3.14) is satisfied. Figure 3.6 shows the injection term v in the case when the initial errors are large. It is clear that the onset of sliding occurs much later since the high-frequency sliding motion does not appear until 1.5 s.

3.3 Robustness Properties of Sliding Mode Observers

Now suppose the nominal linear system in Eq. (3.1) is replaced by the uncertain system

$$\dot{x}(t) = Ax(t) + Bu(t) + M\xi(t, x, u) \qquad (3.22)$$

where $\xi(t, x, u) \in \mathbb{R}^h$ is a disturbance and $M \in \mathbb{R}^{n \times h}$ is the associated distribution matrix. To exploit the robustness properties associated with sliding modes described in Chap. 2, suppose the gain G_n is designed such that it is matched to the disturbance distribution matrix, i.e., $M = G_n X$ for some $X \in \mathbb{R}^{p \times h}$. As a consequence in the coordinates of (3.4) and (3.8)

$$M = \begin{bmatrix} LX \\ -X \end{bmatrix} = \begin{bmatrix} L \\ -I \end{bmatrix} X \qquad (3.23)$$

and the error system in (3.10) and (3.11) becomes

$$\dot{e}_1(t) = A_{11}e_1(t) + A_{12}e_y(t) + Lv - LX\xi(t, x, u) \qquad (3.24)$$

$$\dot{e}_y(t) = A_{21}e_1(t) + A_{22}e_y(t) - v + X\xi(t, x, u) \qquad (3.25)$$

Again it is necessary to develop conditions under which a sliding motion can be enforced. To this end, from (3.25)

$$e_{y,i}\dot{e}_{y,i} = e_{y,i}(A_{21,i}e_1 + A_{22,i}e_y + X_i\xi) - \rho|e_{y,i}|$$
$$< -|e_{y,i}|(\rho - |A_{21,i}e_1 + A_{22,i}e_y + X_i\xi|)$$

and it is clear that if the gain of the switching term is large enough so that $\rho > |A_{21,i}e_1 + A_{22,i}e_y + X_i\xi| + \eta$ for a scalar $\eta > 0$ then the reachability condition in (3.14) is satisfied. An ideal sliding motion is then guaranteed to take place in finite time. During the sliding motion Eqs. (3.24) and (3.25) take the form

$$\dot{e}_1(t) = A_{11}e_1(t) + Lv_{eq} - LX\xi(t,x,u) \tag{3.26}$$
$$0 = A_{21}e_1(t) - v_{eq} + X\xi(t,x,u) \tag{3.27}$$

Substituting for v_{eq} from (3.27) and (3.26) yields

$$\dot{e}_1(t) = (A_{11} + LA_{21})e_1(t) \tag{3.28}$$

which is independent of the disturbance $\xi(t,x,u)$.

Remark 3.2. Notice that for the existence of an ideal sliding motion, the matching condition (3.23) is not required; a large enough ρ is sufficient to induce a sliding motion. The matching condition is only required for the reduced-order motion (3.28) to be independent of $\xi(t,x,u)$.

Example 3.2. Consider system (3.18), subject to uncertainty entering via the distribution matrix

$$M = \begin{bmatrix} 1 \\ -2 \end{bmatrix}$$

In the simulations which follow $\xi(t,x,u) = 0.2\sin(x_1(t))$ but this information is not available to the observer. Notice from (3.21) that the gain matrix $G_n = M$ and the so-called matching conditions in Eq. (3.23) are satisfied with $X = 1$.

Figures 3.7 and 3.8 show a sliding motion occurring after 0.66 s. Also notice that the state estimation errors evolve according to a first-order decay. Crucially the evolution is unaffected by the disturbance. This disturbance rejection property is a major advantage of sliding mode observers compared to traditional linear Luenberger observers. It can be seen from Fig. 3.9 that although the effect of the disturbance $\xi(t,x,u)$ is not present in the state estimation errors, it appears directly in the signal v_{eq}. In fact once the reduced-order motion $e_1(t)$ has become sufficiently small (at about 1.5 s), the signal v_{eq} exactly "reproduces" the disturbance $\xi(t,x,u)$ (with a small delay due to the low-pass filter used to obtain v_{eq}). This is a powerful result because the term v was not designed with any a priori knowledge about $\xi(t,x,u)$, except that it is bounded.

Figure 3.3 shows the term v consists of high-frequency switching once sliding is established. In the observer this does not present the sort of problem it does for

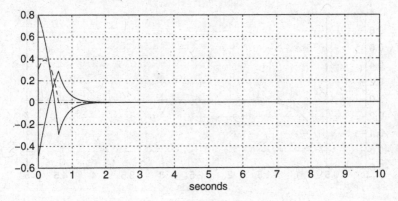

Fig. 3.7 The output estimation error $e_y(t)$ (*dashed*) and the components of the state estimation error $e(t)$ (*solid*)

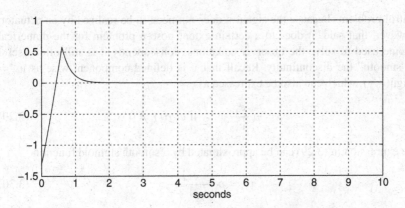

Fig. 3.8 The error vector associated with the sliding motion $e_1(t)$

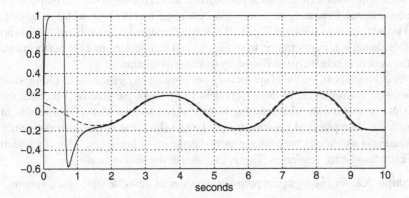

Fig. 3.9 The equivalent output error injection ν_{eq} (*solid*) and the disturbance $\xi(t, x, u)$ (*dashed*)

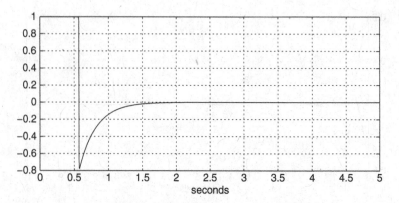

Fig. 3.10 The output error injection term v after being smoothed

control problems because the signal v does not need to be realized by an actuator. However, that said, it does to a certain extent pose a problem for the numerical scheme used to solve the observer equations; consequently it is still often useful to "smooth" the discontinuity. Recall that v is defined component-wise as $v_i = \rho \, \text{sign}(e_{y,i})$, which can also be expressed as

$$v_i = \rho \frac{e_{y,i}}{|e_{y,i}|} \qquad \text{if } e_{y,i}(t) \neq 0 \tag{3.29}$$

The expression in (3.29) can be approximated by a smooth sigmoid function

$$v_i = \rho \frac{e_{y,i}}{|e_{y,i}| + \delta} \tag{3.30}$$

where δ is a small positive scalar. As in the control problem this results in a trade-off between maintaining close to ideal performance and achieving a smooth output error injection signal. Repeating the simulation, now using v as in (3.30) with $\delta = 0.0001$, yields the responses shown in Fig. 3.10. Clearly the signal is smooth and there is no visible chattering. Furthermore, from Fig. 3.11, it can be seen that the performance of the system is relatively unaffected by this approximation.

As explained earlier, the observer does not provide any guarantee of global state estimation convergence. In the observer in (3.5)–(3.7), the size of ρ dictates the size of the domain in which sliding is guaranteed to take place. This results in a trade-off. For practical reasons, a very large value of ρ is not desirable (since chattering is amplified), but a large ρ increases the set of initial conditions for which the estimation error converges. This is explored in the next example.

Example 3.3. For pedagogical purposes consider an unstable state-space system

$$A = \begin{bmatrix} -2 & -3 \\ 1 & 3 \end{bmatrix}, \quad B = \begin{bmatrix} 0 \\ 1 \end{bmatrix}, \quad C = \begin{bmatrix} 0 & 1 \end{bmatrix} \tag{3.31}$$

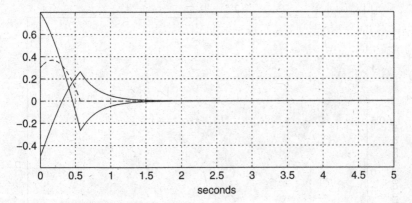

Fig. 3.11 The output estimation error $e_y(t)$ (*dashed*) and the components of the state estimation error $e(t)$ (*solid*)

Notice that in this example no change of coordinates is needed because the output distribution matrix C is already in the required structure of (3.4). Specifying $L = 0$ gives linear (reduced-order) first-order dynamics with a pole at -2. In the simulations which follow $\rho = 1$. Also a full-order state feedback $u(t) = Fx(t)$ controller is used where $F = \begin{bmatrix} 1 & -1 \end{bmatrix}$ so that $\text{eig}(A + BF) = \{\pm 1.4142i\}$. The reason for this choice of closed-loop eigenvalues is that the states will be oscillatory, and the tracking of the states can be more readily observed. In Fig. 3.12, the "shaded" area is the region in which the initial conditions $(e_1(0), e_y(0))$ must lie for a sliding motion to occur. Outside this set of points the observer fails to converge. The shaded region is sometimes referred to as the *sliding patch*. The size of the shaded area can of course be enlarged by increasing the value of ρ, but this is undesirable since increasing ρ tends to lead to more severe chattering.

A more elegant way to enlarge the sliding patch is to add a linear output error feedback term to the observer. Specifically Eq. (3.5) can be modified to take the form

$$\dot{\hat{x}}(t) = A\hat{x}(t) - G_l e_y(t) + Bu(t) + G_n \nu \qquad (3.32)$$

where $G_l \in \mathbb{R}^{n \times p}$. An appropriate choice of the gain G_l will enlarge the sliding patch. From Eqs. (3.1), (3.2), and (3.6), the state estimation error associated with the observer in (3.32) is

$$\dot{e}(t) = (A - G_l C)e(t) + G_n \nu \qquad (3.33)$$

The error system in (3.33) is nonlinear and so a good approach to try to establish global asymptotic stability is to consider Lyapunov-based methods. Consider the quadratic form

$$V = e^T P e \qquad (3.34)$$

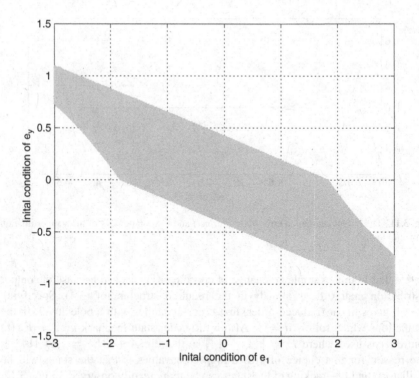

Fig. 3.12 The sliding region

where $P \in \mathbb{R}^{n \times n}$ is a symmetric positive definite matrix, as a candidate Lyapunov function. Differentiating (3.34) with respect to time yields

$$\dot{V} = \dot{e}^T P e + e^T P \dot{e}$$
$$= e^T (P(A - G_l C) + (A - G_l C)^T P)e + 2e^T P G_n v \qquad (3.35)$$

If P, G_l, and G_n can be chosen such that the expression in (3.35) is negative, then the error system in (3.33) is (globally) quadratically stable.

Example 3.4. For the system in (3.31), choosing

$$G_l = \begin{bmatrix} -3 \\ 6 \end{bmatrix} \quad \text{and} \quad G_n = \begin{bmatrix} 0 \\ 1 \end{bmatrix} \qquad (3.36)$$

results in a closed-loop error system

$$\dot{e}_1(t) = -2e_1(t) \qquad (3.37)$$
$$\dot{e}_y(t) = e_1(t) - 3e_y(t) - \text{sign}(e_y) \qquad (3.38)$$

Consider a positive definite quadratic function as in (3.34) where $e = (e_1, e_y)$ and

$$P = \begin{bmatrix} \frac{1}{4} & 0 \\ 0 & 1 \end{bmatrix} \tag{3.39}$$

Differentiating with respect to time yields

$$\begin{aligned}
\dot{V} &= \frac{1}{2} e_1 \dot{e}_1 + 2 e_y \dot{e}_y \\
&= \frac{1}{2} e_1 (-2 e_1) + 2 e_y (e_1 - 3 e_y - \text{sign}(e_y)) \\
&= -e_1^2 - 6 e_y^2 + 2 e_1 e_y - 2|e_y| \\
&= -(e_1 - e_y)^2 - 5 e_y^2 - 2|e_y|
\end{aligned}$$

If $(e_1, e_y) \neq 0$, then $\dot{V} < 0$ and global stability of the error system is proven. Furthermore it can be shown that a sliding motion takes place in finite time. Once the error $e(t)$ becomes sufficiently small, the reachability condition in (3.14) is satisfied and a sliding motion takes place. From the arguments above, since $e \to 0$, $e_1 \to 0$ as $t \to \infty$. In particular at some finite time t_0, $|e_1(t)| < \frac{1}{2}$ for all $t \geq t_0$. Now form a typical reachability test

$$\begin{aligned}
e_y \dot{e}_y &= e_y (e_1 - 3 e_y - \text{sign}(e_y)) \\
&\leq |e_1||e_y| - 3 e_y^2 - |e_y| \\
&\leq |e_1||e_y| - |e_y| \tag{3.40}
\end{aligned}$$

and therefore for all $t > t_0$,

$$e_y \dot{e}_y \leq -\frac{1}{2}|e_y| \tag{3.41}$$

which implies $e_y \to 0$ in finite time and a sliding mode takes place.

This example shows that the introduction of a linear output error injection term can be beneficial.

The problem of robust state estimation for systems with bounded matched uncertainty will now be explored. Consider the following uncertain system

$$\dot{x}(t) = Ax(t) + Bu(t) + Bf(t, y, u) \tag{3.42}$$

$$y(t) = Cx(t) \tag{3.43}$$

where $f : \mathbb{R}_+ \times \mathbb{R}^p \times \mathbb{R}^m \mapsto \mathbb{R}^m$ represents lumped uncertainty or nonlinearities. The function is assumed to be unknown but bounded so that

$$\|f(t, y, u)\| \leq \rho(t, y, u) \tag{3.44}$$

where $\rho(\cdot)$ is known. Consider an observer of the form

$$\dot{z}(t) = Az(t) + Bu(t) - GCe(t) - P^{-1}C^T F^T v \tag{3.45}$$

where $e = z - x$. The symmetric positive definite matrix $P \in \mathbb{R}^{n \times n}$ and the gain matrix G are assumed to satisfy

$$PA_0 + A_0^T P < 0 \tag{3.46}$$

where $A_0 := A - GC$, and the structural constraint

$$PB = (FC)^T \tag{3.47}$$

for some $F \in \mathbb{R}^{m \times p}$. The discontinuous scaled unit-vector term

$$v = \rho(t, y, u) \frac{FCe(t)}{\|FCe(t)\|} \tag{3.48}$$

and $e(t) = z(t) - x(t)$.

Under these circumstances the quadratic form given by $V(e) = e^T Pe$ can be shown to guarantee quadratic stability. Furthermore an ideal sliding motion takes place on

$$\mathcal{S}_F = \{e \in \mathbb{R}^n : FCe = 0\}$$

in finite time.

Remark 3.3. It should be noted that if $p > m$ then sliding on $\mathcal{S}_F$ is not the same as sliding on $Ce(t) = 0$.

Example 3.5. Consider the equations of motion for a pendulum system written as

$$\ddot{\phi}(t) = -\frac{g}{l} \sin(\phi(t))$$

where g is the gravitational constant and l is the length of the pendulum. These can be rewritten in state-space form as

$$\dot{x}(t) = \begin{bmatrix} 0 & 1 \\ 0 & 0 \end{bmatrix} x(t) + \begin{bmatrix} 0 \\ 1 \end{bmatrix} \xi(t, x) \tag{3.49}$$

where $x_1 = \phi$, $x_2 = \dot{\phi}$, and $\xi(t, x_1, x_2) = -\frac{g}{l} \sin(\phi)$. Somewhat artificially choose as an output measurement $y(t) = Cx(t)$ where

$$C = \begin{bmatrix} 1 & 1 \end{bmatrix} \tag{3.50}$$

i.e., the sum of position and velocity. Also assume the term $\xi(t, x_1, x_2)$ is unknown—possibly because the length of the pendulum is imprecisely known. The aim is both to estimate $x(t)$ and reconstruct $\xi(t, x)$ from $y(t)$.

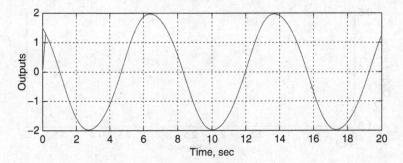

Fig. 3.13 Plant output $y(t)$ and the observer estimate $\hat{y}(t)$

Choosing

$$G_l = \begin{bmatrix} 1 \\ 1 \end{bmatrix}$$

means the eigenvalues of $A - G_l C$ are $\{-1, -1\}$ and furthermore

$$P(A - G_l C) + (A - G_l C)^T P < 0$$

if

$$P = \begin{bmatrix} 2 & 1 \\ 1 & 1 \end{bmatrix}$$

Notice also for this choice of P, the structural equation $PB = C^T$ is satisfied. The sliding mode observer is

$$\dot{z}(t) = \begin{bmatrix} 0 & 1 \\ 0 & 0 \end{bmatrix} z(t) - \begin{bmatrix} 1 \\ 1 \end{bmatrix} e_y(t) - \begin{bmatrix} 0 \\ 1 \end{bmatrix} 2\,\mathrm{sign}(e_y) \qquad (3.51)$$

where

$$e_y(t) = Cz(t) - y(t)$$

is the output estimation error.

When the initial conditions of the true states and observer states are deliberately set to different values the output of the observer tracks the output of the plant in finite time as shown in Fig. 3.13.

Figure 3.14 shows the same information as a plot of e_y. A sliding motion takes place after 0.2 s. The finite time response is a characteristic of sliding modes.

Figure 3.15 shows the states and the state estimation errors. Although the output estimation error converges to zero in finite time, the state estimation error is asymptotic. However, asymptotic convergence has been achieved despite the plant/model mismatch resulting from the term $\xi(t, x_1, x_2)$ which is not used in the sliding mode observer.

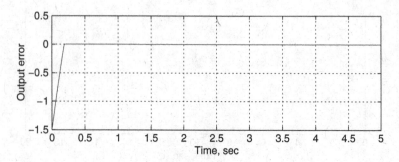

Fig. 3.14 The output estimation error e_y

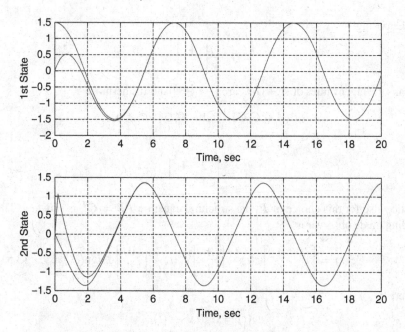

Fig. 3.15 System states $x(t)$ (*solid*) and the observer estimates $\hat{x}(t)$ (*dashed*)

Figure 3.16 shows a low-pass-filtered version of the injection signal $v = -2\text{sign}(e_y)$ from the observer. It clearly replicates the "unknown signal" $\xi(t, x_1, x_2)$ and compensates for it to allow perfect (asymptotic) convergence of the state estimates.

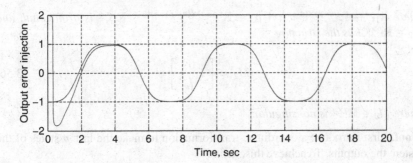

Fig. 3.16 The injection signal and the unknown input

3.4   A Generic Conventional Sliding Mode Observer

Consider the following uncertain dynamical system:

$$\dot{x}(t) = Ax(t) + Bu(t) + Mf(t, y, u) \tag{3.52}$$

$$y(t) = Cx(t) \tag{3.53}$$

where $x \in \mathbb{R}^n$ are the states, $y \in \mathbb{R}^p$ are the measurable outputs, and $u \in \mathbb{R}^m$ are the measurable inputs. The signal $f \in \mathbb{R}^q$ is the lumped uncertainty acting upon the system where $q < p$. It is unknown but assumed to be bounded so that

$$\|f(t, t, u)\| \leq \alpha(t, y, u) \tag{3.54}$$

where $\alpha(\cdot)$ is known. Assume without loss of generality that the matrices C and M are full rank. The objective is to reconstruct both the states and the unknown input $f(t)$ based only on the measured signals $u(t), y(t)$.

Two lemmas will now be presented which underpin the rest of this chapter. They will describe a canonical form which will help facilitate understanding of the problem and provide a framework for solving the problem.

Lemma 3.1. *Let the triple* (A, M, C) *represent a linear system with* $p > q$ *and suppose* $rank(CM) = rank(M) = q$, *then there exists a change of coordinates* $x \mapsto T_o x$ *for the system* (3.52) *and* (3.53) *such that in the new coordinates the triple* (A, M, C) *of the transformed system has the following structure:*

$$A = \begin{bmatrix} A_{11} & A_{12} \\ \begin{matrix} A_{211} \\ A_{212} \end{matrix} & A_{22} \end{bmatrix}, \quad M = \begin{bmatrix} 0 \\ M_2 \end{bmatrix}, \quad C = \begin{bmatrix} 0 & T \end{bmatrix} \tag{3.55}$$

where $A_{11} \in \mathbb{R}^{(n-p)\times(n-p)}$, $A_{211} \in \mathbb{R}^{(p-q)\times(n-p)}$, $T \in \mathbb{R}^{p\times p}$ is orthogonal, and $M_2 \in \mathbb{R}^{p\times q}$ has the structure

$$M_2 = \begin{bmatrix} 0 \\ M_o \end{bmatrix} \tag{3.56}$$

where $M_o \in \mathbb{R}^{q\times q}$ is nonsingular.

Proof. First introduce a coordinate transformation to make the last p states of the system the outputs. To achieve this, define

$$T_c = \begin{bmatrix} N_c^T \\ C \end{bmatrix} \tag{3.57}$$

where $N_c \in \mathbb{R}^{n\times(n-p)}$ is such that its columns span the null space of C. The coordinate transformation $x \mapsto T_c x$ is nonsingular by construction and in the new coordinate system

$$C = \begin{bmatrix} 0 & I_p \end{bmatrix}$$

Suppose in the new coordinate system

$$M = \begin{bmatrix} M_1 \\ M_2 \end{bmatrix} \begin{matrix} \updownarrow n-p \\ \updownarrow p \end{matrix}$$

in which M_1 and M_2 have no particular structure other than the fact that $\text{rank}(M_2)=q$. This follows because $CM = M_2$ and so by assumption $\text{rank}(M_2)=q$. Hence the left pseudo-inverse $M_2^\dagger$ is well defined. Also there exists an orthogonal matrix $T \in \mathbb{R}^{p\times p}$ such that

$$T^T M_2 = \begin{bmatrix} 0 \\ M_o \end{bmatrix} \tag{3.58}$$

where $M_o \in \mathbb{R}^{q\times q}$ is nonsingular. Such a matrix can be found in QR factorization. Consequently, the coordinate transformation $x \mapsto T_b x$ where

$$T_b = \begin{bmatrix} I_{n-p} & -M_1 M_2^\dagger \\ 0 & T^T \end{bmatrix} \tag{3.59}$$

is nonsingular, and in the new coordinates the triple (A, M, C) is in the form

$$A = \begin{bmatrix} A_{11} & A_{12} \\ A_{21} & A_{22} \end{bmatrix}, \quad M = \begin{bmatrix} 0 \\ M_o \end{bmatrix}, \quad C = \begin{bmatrix} 0 & T \end{bmatrix} \tag{3.60}$$

where $A_{11} \in \mathbb{R}^{(n-q)\times(n-q)}$ and the remaining subblocks in the system matrix are partitioned accordingly. The triple in (3.60) has exactly the structure claimed in the lemma statement. □

Lemma 3.2. *The pair* (A_{11}, A_{211}) *is detectable if and only if the invariant zeros of* (A, M, C) *are stable.*

Proof. From the PBH rank test,[1] the unobservable modes of (A_{11}, A_{211}) are given by the values of s that make the following matrix pencil lose rank:

$$P_{obs}(s) = \begin{bmatrix} sI - A_{11} \\ A_{211} \end{bmatrix}$$

The zeros of (A, M, C) are given by the values of s that make the Rosenbrock matrix $R_{obs}(s)$ lose rank, where

$$R_{obs}(s) = \begin{bmatrix} sI - A & -M \\ C & 0 \end{bmatrix} = \left[\begin{array}{ccc|c} sI - A_{11} & -A_{12} & & 0 \\ -A_{211} & \star & & 0 \\ -A_{212} & \star & & -M_o \\ \hline 0 & T & & 0 \end{array} \right]$$

and the $\star$ represents elements that do not play a role in the subsequent analysis. Since M_o and T are both square and invertible, $R_{obs}(s)$ loses rank if and only if $P_{obs}(s)$ loses rank. Clearly the unobservable modes of (A_{11}, A_{211}) are the invariant zeros of (A, M, C), and hence the proof is complete. □

Assume that for the system in (3.52) and (3.53):

A1. $\text{rank}(CM) = \text{rank}(M)$.
A2. The invariant zeros (if any) of (A, M, C) are stable.

The canonical form associated with Lemma 3.1 will be used as a basis for the solution to the problem of estimating both the states and unknown inputs.
   The following observer will be considered:

$$\dot{\hat{x}}(t) = A\hat{x}(t) + Bu(t) - G_l e_y(t) + G_n v \qquad (3.61)$$

$$\hat{y}(t) = C\hat{x}(t) \qquad (3.62)$$

where $e_y(t) := \hat{y}(t) - y(t)$ is the output estimation error. The design freedom is associated with the two gains $G_l \in \mathbb{R}^{n \times p}$ and $G_n \in \mathbb{R}^{n \times p}$ which are design matrices to be determined. The vector v is defined by

$$v(t) = -\rho(t, y, u)\frac{e_y}{\|e_y\|} \qquad \text{if } e_y(t) \neq 0 \qquad (3.63)$$

where $\rho(\cdot)$ is a positive scalar function dependent on the magnitude of the unknown input signal $f(t)$. Condition A1 means that the canonical form in (3.55) can be

---

[1]For details see appendix C.

attained, and hence without loss of generality assume that the triple (A, M, C) has the form given in (3.55). In the coordinates of (3.55), let

$$G_n = \begin{bmatrix} -L \\ I_p \end{bmatrix} T^T P_o^{-1} \tag{3.64}$$

where $L = \begin{bmatrix} L^o & 0 \end{bmatrix}$ and $L^o \in \mathbb{R}^{(n-p) \times (p-q)}$. The symmetric positive definite matrix $P_o \in \mathbb{R}^{p \times p}$ and the gain L^o are design matrices. Defining $A_o = A - G_l C$ and the state estimation error as $e(t) := \hat{x}(t) - x(t)$. From (3.52) and (3.61) and (3.53) and (3.62), the following error system can be obtained:

$$\dot{e}(t) = A_o e(t) + G_n v - M f(t, \hat{x} - e, y) \tag{3.65}$$

Proposition 3.1. *If there exists a matrix G_l and a Lyapunov matrix P of the form*

$$P = \begin{bmatrix} P_1 & P_1 L \\ L^T P_1 & T^T P_o T + L^T P_1 L \end{bmatrix} > 0 \tag{3.66}$$

where $P_1 \in \mathbb{R}^{(n-p) \times (n-p)}$, which satisfies

$$P A_o + A_o^T P < 0 \tag{3.67}$$

and $\rho(t, y, u) \geq \|P_o C M\| \alpha(t, y, u) + \eta_o$ where $\eta_o > 0$, then the state estimation error $e(t)$ is asymptotically stable.

Proof. Consider as a candidate Lyapunov function

$$V = e^T P e \tag{3.68}$$

where P is given in (3.66). Differentiating (3.68) with respect to time yields

$$\dot{V} = \dot{e}^T P e + e^T P \dot{e}$$
$$= e^T (P A_o + A_o^T P) e + 2 e^T P G_n v - 2 e^T P M f$$

From the definitions of P, G_n, and M in (3.66), (3.64), and (3.55), respectively, it can be verified that

$$P G_n = C^T \tag{3.69}$$

and

$$P M = C^T P_o C M \tag{3.70}$$

Using (3.67) and (3.69)–(3.70), $\dot{V}$ becomes

$$\dot{V} \leq 2 e^T C^T v - 2 e^T C^T P_o C M f$$

From the definition of v in (3.63) and using the bound of f in (3.54)

$$\dot{V} < -2\rho\|e_y\| - 2e_y^T P_o C M f$$
$$\leq -2\|e_y\|(\rho - \|P_o C M\|\alpha)$$
$$\leq -2\eta_o\|e_y\|$$
$$< 0 \text{ for } e \neq 0$$

which proves the state estimation error is quadratically stable.   □

Using this result, it can be further shown that a sliding motion can be achieved in finite time.

Corollary 3.1. *A stable sliding motion takes place on the surface*

$$S = \{e : Ce = 0\} \tag{3.71}$$

in finite time and the sliding motion is governed by $A_{11} + L^o A_{211}$.

Proof. To prove a sliding motion is attained, firstly apply a change of coordinates $x \mapsto T_L x$ where

$$T_L = \begin{bmatrix} I_{n-p} & L \\ 0 & T \end{bmatrix} \tag{3.72}$$

such that the triple (A, M, C) in (3.55) is transformed to be

$$\bar{A} = \begin{bmatrix} \bar{A}_{11} & \bar{A}_{12} \\ \bar{A}_{21} & \bar{A}_{22} \end{bmatrix}, \quad \bar{M} = \begin{bmatrix} 0 \\ \bar{M}_2 \end{bmatrix}, \quad \bar{C} = \begin{bmatrix} 0 & I_p \end{bmatrix} \tag{3.73}$$

where $\bar{A}_{11} = A_{11} + L^o A_{211}$. This follows from the structure of L in (3.64). Using (3.73), the error system (3.65) can be partitioned as

$$\dot{e}_1(t) = \bar{A}_{11}e_1(t) + (\bar{A}_{12} - \bar{G}_{l,1})e_y(t) \tag{3.74}$$

$$\dot{e}_y(t) = \bar{A}_{21}e_1(t) + (\bar{A}_{22} - \bar{G}_{l,2})e_y(t) + P_o^{-1}v - \bar{M}_2 f(t) \tag{3.75}$$

where

$$\begin{bmatrix} \bar{G}_{l,1} \\ \bar{G}_{l,2} \end{bmatrix} = T_L G_l \tag{3.76}$$

Introduce a candidate Lyapunov function for the subsystem (3.75) as

$$V_s = e_y^T P_o e_y$$

Differentiating V_s with respect to time and using (3.75) yield

$$\dot{V}_s = e_y^T (P_o(\bar{A}_{22} - \bar{G}_{l,2}) + (\bar{A}_{22} - \bar{G}_{l,2})^T P_o)e_y + 2e_y^T P_o \bar{A}_{21}e_1 + 2e_y^T v - 2e_y^T P_o \bar{M}_2 f$$

It can be shown that

$$(T_L^T)^{-1} P T_L^{-1} = \begin{bmatrix} P_1 & 0 \\ 0 & P_o \end{bmatrix} \tag{3.77}$$

and

$$T_L A_o T_L^{-1} = \begin{bmatrix} \bar{A}_{11} & \bar{A}_{12} - \bar{G}_{l,1} \\ \bar{A}_{21} & \bar{A}_{22} - \bar{G}_{l,2} \end{bmatrix} \tag{3.78}$$

By direct calculation it follows that $(T_L^T)^{-1}(PA_o + A_o^T P)T_L^{-1}$ can be expanded as

$$\begin{bmatrix} P_1 \bar{A}_{11} + \bar{A}_{11}^T P & \star \\ \star & P_o(\bar{A}_{22} - \bar{G}_{l,2}) + (\bar{A}_{22} - \bar{G}_{l,2})^T P_o \end{bmatrix} < 0 \tag{3.79}$$

where the $\star$ represents elements that do not play any significant role in the analysis. Since (3.79) is symmetric, $P_o(\bar{A}_{22} - \bar{G}_{l,2}) + (\bar{A}_{22} - \bar{G}_{l,2})^T P_o < 0$ and hence

$$\dot{V}_s < 2e_y^T P_o \bar{A}_{21} e_1 + 2e_y^T v - 2e_y^T P_o \bar{M}_2 f$$

$$\leq 2\|e_y\| \|P_o \bar{A}_{21} e_1\| - 2\eta_o \|e_y\|$$

$$= 2\|e_y\|(\|P_o \bar{A}_{21} e_1\| - \eta_o) \tag{3.80}$$

Notice that

$$\|e_y\|^2 = (\sqrt{P_o} e_y)^T P_o^{-1}(\sqrt{P_o} e_y) \geq \lambda_{min}(P_o^{-1})\|\sqrt{P_o} e_y\|^2 = \lambda_{min}(P_o^{-1}) V_s \tag{3.81}$$

Define η as a scalar satisfying $0 < \eta < \eta_o$. Since from Proposition 3.1 the state estimation error is quadratically stable, in finite time $e_1(t)$ enters the domain

$$\Omega_\eta = \{e_1 : \|P_o \bar{A}_{21} e_1\| < \eta_o - \eta\}$$

and remains there. Inside the domain Ω_η inequality (3.80) becomes

$$\frac{dV_s}{dt} < -2\eta\|e_y\| < -2\eta \sqrt{\lambda_{min}(P_o^{-1})} \sqrt{V_s}$$

This proves that a sliding motion takes place on $\mathcal{S}$ in finite time.

When a sliding motion has been achieved, $e_y(t) = \dot{e}_y(t) = 0$ and from (3.74) the remaining dynamics $e_1(t)$ are governed by $\bar{A}_{11} = A_{11} + L^o A_{211}$. Since from inequality (3.79) $P_1 \bar{A}_{11} + \bar{A}_{11}^T P_1 < 0$ and $P_1 > 0$, the matrix $\bar{A}_{11}$ is stable. This completes the proof. □

Remark 3.4. The following observations can be made:

• One possible choice of linear gain in (3.61) is

$$G_l = \begin{bmatrix} A_{12} - A_{11}L + LA_{22}^s \\ A_{22} - LA_{21} - A_{22}^s \end{bmatrix}$$

where $A_{22}^s \in \mathbb{R}^{p \times p}$ is any Hurwitz stable matrix. For a given L which makes $A_{11} + LA_{21}$ stable, this guarantees the existence of a matrix P of the form in (3.66) such that $P(A - G_l C) + (A - G_l C)^T P < 0$. This choice is far from unique and optimization methods can be applied to synthesize G_l according to some criteria.

- During the sliding motion when $\dot{e}_y = e_y = 0$, from (3.75)

$$0 = \bar{A}_{21} e_1(t) + P_o^{-1} v_{eq} - \bar{M}_2 f(t, \hat{x} - e, y) \tag{3.82}$$

Furthermore since the autonomous dynamical system in (3.74) is asymptotically stable, $e_1(t) \to 0$ as $t \to \infty$. Therefore

$$P_o^{-1} v_{eq} - \bar{M}_2 f(t, \hat{x} - e, y) \to 0 \tag{3.83}$$

and the equivalent injection compensates for the unknown uncertain term $f(t, \hat{x} - e, y)$. Consequently not only does the state $\hat{x}$ asymptotically tend to $x(t)$ but also the scaling of the equivalent injection

$$\bar{M}_2^\dagger P_o^{-1} v_{eq}(t) \to f(t, \hat{x} - e, y)$$

where $\bar{M}_2^\dagger$ is any left pseudo-inverse of $\bar{M}_2$.

Example 3.6. Consider the fourth-order system

$$A = \begin{bmatrix} -3.9354 & 0 & 0 & -14.7110 \\ 0 & 0 & 0 & 1.0000 \\ 1.0000 & 14.9206 & 0 & 1.6695 \\ 0.7287 & 0 & 0 & -2.1963 \end{bmatrix} \quad B = \begin{bmatrix} 0 \\ 0 \\ 0 \\ 0.8116 \end{bmatrix} \tag{3.84}$$

$$C = \begin{bmatrix} 0 & 0 & 1 & 0 \\ 0 & 0 & 0 & 1 \end{bmatrix} \tag{3.85}$$

This represents a linearization of the rigid body dynamics of a passenger vehicle (Fig. 2.10). The first state is an average of the lateral velocity v and yaw rate r; the second state represents Ψ, the vehicle orientation; the third state, Y, is the lateral deviation from the intended lane position, and the fourth state, r, is the yaw rate.

It is easy to identify the following subcomponents:

$$A_{11} = \begin{bmatrix} -3.9354 & 0 \\ 0 & 0 \end{bmatrix} \quad A_{211} = \begin{bmatrix} 1.0000 & 14.9206 \end{bmatrix} \tag{3.86}$$

It can be verified that (A_{11}, A_{211}) is observable and so a matrix L^o can be found such that $A_{11} + L^o A_{211}$ is stable. Choosing

$$L^o = \begin{bmatrix} -0.0318 \\ -0.1362 \end{bmatrix}$$

ensures the eigenvalues of $A_{11} + L^o A_{211}$ are at $\{-2, -4\}$. A particular choice of the Luenberger gain in the sliding mode observer is

$$G_l = \begin{bmatrix} -0.0297 & -14.7111 \\ 0.4087 & 1.0000 \\ 5.0646 & 1.6695 \\ 0.0232 & 0.8037 \end{bmatrix}$$

which is obtained by choosing $A_{22}^s = -3I_2$.

3.5 A Sliding Mode Observer for Nonlinear Systems

In this section, a class of nonlinear uncertain systems is considered which might be termed semi-linear. Consider a system described by

$$\dot{x} = Ax + \phi(x, u) + Df(y, u, t) \tag{3.87}$$

$$y = Cx \tag{3.88}$$

where $x \in \mathbb{R}^n$, $u \in \mathbb{R}^m$, and $y \in \mathbb{R}^p$ are the state variables, inputs, and outputs, respectively. The matrices $A \in \mathbb{R}^{n \times n}$, $D \in \mathbb{R}^{n \times q}$, and $C \in \mathbb{R}^{p \times n}$ ($q \leq p < n$) are constant, with D and C both full rank. The known nonlinear term $\phi(x, u)$ is assumed to be Lipschitz with respect to x for all $u \in \mathcal{U}$ (here $\mathcal{U}$ is an admissible control set): i.e., there exists a positive scalar $\mathcal{L}_\phi$ such that

$$\|\phi(x, u) - \phi(\hat{x}, u)\| \leq \mathcal{L}_\phi \|x - \hat{x}\| \tag{3.89}$$

The gain $\mathcal{L}_\phi$ can be thought of as the steepest possible gradient that $\phi(\cdot)$ achieves with variations in x. The unknown function $f(y, u, t) \in \mathbb{R}^q$ is assumed to satisfy

$$\|f(y, u, t)\| \leq \rho(y, u, t) \tag{3.90}$$

where the function $\rho(y, u, t)$ is known.

The system in (3.87) and (3.88) might represent a nonlinear system subject to faults captured by the unknown input signal $f(y, u, t)$. In particular if $D = B$, then the faults are associated with the actuators of the system—hence the direct dependence of the signal $f(\cdot)$ on the control signal $u(t)$.

The following assumptions will be imposed on system (3.87) and (3.88):
Assumption 1. $\text{rank}(CD) = \text{rank}(D)$.
Assumption 2. All the invariant zeros of the matrix triple (A, D, C) lie in the left half plane.

From Lemma 3.1, it can be assumed without loss of generality that system (3.87) and (3.88) already has the form

$$\dot{x}_1 = A_1 x_1 + A_{12} x_2 + \phi_1(x, u) \tag{3.91}$$

$$\dot{x}_2 = A_{21} x_1 + A_{22} x_2 + \phi_2(x, u) + D_2 f(y, u, t) \tag{3.92}$$

$$y = C_2 x_2 \tag{3.93}$$

where $x = \text{col}(x_1, x_2)$ with $x_1 \in \mathbb{R}^{n-p}$, and $\phi_1(x, u)$ and $\phi_2(x, u)$ are the first $n - p$ and the last p components of $\phi(x, u)$.

Now a robust sliding mode observer will be proposed using the system structure characteristics shown in Sect. 3.4. First introduce a coordinate transformation $z = Tx$ where

$$T := \begin{bmatrix} I_{n-p} & L \\ 0 & I_p \end{bmatrix} \tag{3.94}$$

where L has the structure

$$L = \begin{bmatrix} L^o & 0 \end{bmatrix} \tag{3.95}$$

and $L^o \in \mathbb{R}^{(n-p) \times (p-q)}$. Then, it follows that in the new z coordinates, system (3.91)–(3.93) has the following form:

$$\dot{z}_1 = (A_{11} + L A_{21}) z_1 + (A_{12} + L A_{22} - (A_{11} + L A_{21}) L) z_2$$
$$+ \begin{bmatrix} I_{n-p} & L \end{bmatrix} \phi(T^{-1} z, u) \tag{3.96}$$

$$\dot{z}_2 = A_{21} z_1 + (A_{22} - A_{21} L) z_2 + \phi_2(T^{-1} z, u) + D_2 f(y, u, t) \tag{3.97}$$

$$y = C_2 z_2 \tag{3.98}$$

where $z := \text{col}(z_1, z_2)$ with $z_1 \in \mathbb{R}^{n-p}$. Notice that (3.96) is independent of the unknown function $f(\cdot)$. This very specific structure (3.96) occurs because $L D_2 = 0$. For system (3.96)–(3.98), consider a dynamical system (the observer) given by

$$\dot{\hat{z}}_1 = (A_{11} + L A_{21}) \hat{z}_1 + (A_{12} + L A_{22} - (A_{11} + L A_{21}) L) C_2^{-1} y$$
$$+ \begin{bmatrix} I_{n-p} & L \end{bmatrix} \phi(T^{-1} \hat{z}, u) \tag{3.99}$$

$$\dot{\hat{z}}_2 = A_{21} \hat{z}_1 + (A_{22} - A_{21} L) \hat{z}_2 - K(y - C_2 \hat{z}_2) + \phi_2(T^{-1} \hat{z}, u) + v \tag{3.100}$$

$$\hat{y} = C_2 \hat{z}_2 \tag{3.101}$$

where $\hat{z} := \text{col}(\hat{z}_1, C_2^{-1} y)$ and $\hat{y}$ is the output of the dynamical system. Note that $\hat{z}$ *does not represent the state estimate* $\text{col}(\hat{z}_1, \hat{z}_2)$. It is merely used as a convenient notation in the developments which follow. Also note that the unknown input $f(\cdot)$ does not appear in (3.99) and (3.100) but the known nonlinearity $\phi(\cdot)$ does—except its arguments depend on $\hat{z}$ and not z. The gain matrix K is chosen so that the matrix $C_2(A_{22} - A_{21} L) C_2^{-1} + C_2 K$ is symmetric negative definite. This is always possible since C_2 is nonsingular and an explicit formula is given later. Define the output estimation error as $e_y = y - \hat{y}$, and the injection signal v as

$$v := k(\cdot)C_2^{-1}\frac{e_y}{\|e_y\|} \qquad (3.102)$$

where $k(\cdot)$ is a positive scalar function to be determined.

Let $e_1 = z_1 - \hat{z}_1$ and $e_y = y - \hat{y} = C_2(z_2 - \hat{z}_2)$. Then from (3.96) to (3.98) and (3.99) to (3.101), the state estimation error dynamical system is described by

$$\dot{e}_1 = (A_{11} + LA_{21})e_1 + \left[I_{n-p}\ L \right]\left(\phi(T^{-1}z, u) - \phi(T^{-1}\hat{z}, u)\right) \qquad (3.103)$$

$$\dot{e}_y = C_2 A_{21} e_1 + \left(C_2(A_{22} - A_{21}L)C_2^{-1} + C_2 K\right)e_y + C_2 D_2 f(t) - C_2 v$$

$$+ C_2\left(\phi_2(T^{-1}z, u) - \phi_2(T^{-1}\hat{z}, u)\right) \qquad (3.104)$$

where $\hat{z} = \mathrm{col}(\hat{z}_1, C_2^{-1}y)$ and v is defined by (3.102). From Lemma 3.1 and Assumptions 1–2, a matrix L of the form in (3.95) can be found to make $A_{11} + LA_{21}$ stable. As argued earlier, a gain matrix $K \in \mathbb{R}^{p \times p}$ can be chosen as

$$K = -(A_{22} - A_{21}L)C_2^{-1} - C_2^{-1}A_s \qquad (3.105)$$

where A_s is symmetric positive definite to ensure $C_2(A_{22} - A_{21}L)C_2^{-1} + C_2 K$ is negative definite, and hence the nominal linear system matrix of the state estimation error dynamical system (3.103) and (3.104) given by

$$\begin{bmatrix} A_{11} + LA_{21} & 0 \\ C_2 A_{21} & C_2(A_{22} - A_{21}L)C_2^{-1} + C_2 K \end{bmatrix} \qquad (3.106)$$

is stable.

From (3.94) and (3.98), it follows that

$$T^{-1}z - T^{-1}\hat{z} = \begin{bmatrix} I_{n-p} & -L \\ 0 & I_p \end{bmatrix}\begin{bmatrix} z_1 - \hat{z}_1 \\ z_2 - C_2^{-1}y \end{bmatrix} = \begin{bmatrix} e_1 \\ 0 \end{bmatrix}$$

$$\Rightarrow \|T^{-1}z - T^{-1}\hat{z}\| = \|e_1\| \qquad (3.107)$$

For system (3.103) and (3.104), consider a sliding surface

$$S = \{(e_1, e_y) \mid e_y = 0\} \qquad (3.108)$$

Then, the following can be proved:

Proposition 3.2. *Under Assumptions 1–2, the sliding motion of system (3.103) and (3.104), associated with the surface (3.108), is asymptotically stable if the matrix inequality*

$$\bar{A}^T \bar{P}^T + \bar{P}\bar{A} + \frac{1}{\varepsilon}\bar{P}\bar{P}^T + \varepsilon\left(\mathcal{L}_\phi\right)^2 I_{n-p} + \alpha P < 0 \qquad (3.109)$$

where ε and α are positive constants, $\mathcal{L}_\phi$ is the Lipschitz constant for $\phi(x, u)$ with respect to x, and L has the structure in (3.95), is solvable for $\bar{P}$ where

$$\bar{P} := P \begin{bmatrix} I_{n-p} & L \end{bmatrix} \quad \text{and} \quad \bar{A} := \begin{bmatrix} A_{11} \\ A_{21} \end{bmatrix} \tag{3.110}$$

with $P > 0$.

Proof. The analysis above has shown that (3.103) represents the sliding dynamics when restricted to the sliding surface (3.108). Therefore, it is only required to prove that (3.103) is asymptotically stable. Consider a candidate Lyapunov function $V = e_1^T P e_1$. The time derivative of V along the trajectories of system (3.103) is given by

$$\dot{V} |_{(3.103)} = e_1^T \left((A_{11} + L A_{21})^T P + P(A_{11} + L A_{21}) \right) e_1$$
$$+ 2 e_1^T P \begin{bmatrix} I_{n-p} & L \end{bmatrix} \left(\phi(T^{-1}z, u) - \phi(T^{-1}\hat{z}, u) \right)$$
$$= e_1^T \left(\bar{A}^T \bar{P}^T + \bar{P} \bar{A} \right) e_1 + 2 \left(\bar{P}^T e_1 \right)^T \left(\phi(T^{-1}z, u) - \phi(T^{-1}\hat{z}, u) \right)$$

From the well-known inequality[2] that $2 X^T Y \leq \varepsilon X^T X + \frac{1}{\varepsilon} Y^T Y$ for any scalar $\varepsilon > 0$, it follows that

$$\dot{V} |_{(3.103)} \leq e_1^T \left(\bar{P} \bar{A} + \bar{A}^T \bar{P}^T \right) e_1 + \varepsilon e_1^T \bar{P} \bar{P}^T e_1$$
$$+ \frac{1}{\varepsilon} \left(\phi(T^{-1}z, u) - \phi(T^{-1}\hat{z}, u) \right)^T \left(\phi(T^{-1}z, u) - \phi(T^{-1}\hat{z}, u) \right)$$

From (3.107),

$$\| \phi(T^{-1}z, u) - \phi(T^{-1}\hat{z}, u) \| \leq \mathcal{L}_\phi \| e_1 \| \tag{3.111}$$

Consequently

$$\dot{V} |_{(3.103)} \leq e_1^T \left(\bar{P} \bar{A} + \bar{A}^T \bar{P}^T \right) e_1 + \varepsilon e_1^T \bar{P} \bar{P}^T e_1 + \frac{1}{\varepsilon} \left(\mathcal{L}_\phi \right)^2 \| e_1 \|^2$$
$$= e_1^T \left(\bar{P} \bar{A} + \bar{A}^T \bar{P}^T + \varepsilon \bar{P} \bar{P}^T + \frac{1}{\varepsilon} \left(\mathcal{L}_\phi \right)^2 I \right) e_1$$
$$\leq -\alpha e_1^T P e_1 = -\alpha V \tag{3.112}$$

where (3.109) has been used to obtain the last inequality, and the proof is complete.   □

Remark 3.5. Note that inequality (3.109) can be transformed into the following Linear Matrix Inequality (LMI) problem: for a given scalar $\alpha > 0$, find matrices P and Y and a scalar ε such that

---

[2]This is sometimes known as Young's Inequality.

$$\begin{bmatrix} \Theta(P,Y) + \alpha P + \varepsilon\left(\mathcal{L}_\phi\right)^2 & P & Y \\ P & -\varepsilon I_{n-p} & 0 \\ Y^T & 0 & -\varepsilon I_p \end{bmatrix} < 0 \qquad (3.113)$$

where $\Theta(P,Y) := PA_1 + A_1^T P + YA_{21} + A_{21}^{TY^T}$ and the decision variable $Y := PL$ with $P > 0$. This problem can be solved by LMI techniques. If $\mathcal{L}_\phi$ is known, then for a given α, the problem of finding P, Y, and ε to satisfy (3.113) is a standard LMI feasibility problem. Alternatively, an optimization problem can be posed which is to find P, Y, and ε which maximizes $\mathcal{L}_\phi$ in (3.113). This is a convex eigenvalue optimization problem and can be solved using standard LMI algorithms.

Since $\dot{V}(t) \leq -\alpha V(t)$ in (3.112), it follows that there exists a positive scalar M such that

$$\|e_1(t)\| \leq M \|e_1(0)\| \exp\{-\alpha t/2\} \qquad (3.114)$$

where a choice is $M := \sqrt{\frac{\lambda_{\max}(P)}{\lambda_{\min}(P)}}$. Based on inequality (3.114), introduce a dynamic system given by

$$\dot{\hat{w}}(t) = -\tfrac{1}{2}\alpha\hat{w}(t) \qquad (3.115)$$

For any value $e_1(0)$, choose $\hat{w}(0)$ such that $M \|e_1(0)\| \leq \hat{w}(0)$. Then, it is easy to see that the available solution $\hat{w}(t)$ to Eq. (3.115) is an upper bound on the size of the corresponding state estimation error $e_1(t)$; specifically $\|e_1(t)\| \leq \hat{w}(t)$ for all $t \geq 0$.

Proposition 3.3 has shown that the sliding mode associated with the sliding surface $\mathcal{S}$ given in (3.108) is stable if the matrix inequality (3.109) is solvable. The objective is now to determine the scalar gain function $k(\cdot)$ in (3.102) such that the system can be driven to the surface $\mathcal{S}$ in finite time and a sliding motion can be maintained.

Proposition 3.3. *Under Assumptions 1–2, system (3.103) and (3.104) is driven to the sliding surface (3.108) in finite time and remains there if the gain $k(\cdot)$ in (3.102) is chosen to satisfy*

$$k(t,u,y,\hat{z}) \geq (\|C_2 A_{21}\| + \|C_2\|\mathcal{L}_\phi)\hat{w}(t) + \|C_2 D_2\|\rho(y,u,t) + \eta \qquad (3.116)$$

where η is a positive constant and $\hat{w}$ is the solution to the differential Eq. (3.115).

Proof. Let $\tilde{V}(e_y) = e_y^T e_y$. From the expression for the output estimation error in (3.104), it follows that

$$\dot{\tilde{V}} = e_y^T \left(C_2(A_{22} - A_{21}L)C_2^{-1} + C_2 K + \left(C_2(A_{22} - A_{21}L)C_2^{-1} + C_2 K\right)^T\right)e_y +$$
$$+ 2e_y^T \left(C_2 A_{21} e_1 + C_2\big(\phi_2(T^{-1}z) - \phi_2(T^{-1}\hat{z}_y)\big) + C_2 D_2 f(t) - C_2 v\right) \qquad (3.117)$$

Since, by design $C_2(A_{22} - A_{21}L)C_2^{-1} + C_2K$ is symmetric negative definite, it follows that

$$\left(C_2(A_{22} - A_{21}L)C_2^{-1} + C_2K\right)^T + C_2(A_{22} - A_{21}L)C_2^{-1} + C_2K < 0 \qquad (3.118)$$

By applying (3.90) and (3.118) to (3.117), it follows from (3.107) that

$$\dot{V} \leq 2\|e_y\| \left(\|C_2A_{21}\| + \|C_2\|\mathcal{L}_\phi\right)\|e_1\| + \|C_2D_2\|\rho(y, u, t)) - 2e_y^T C_2 v \qquad (3.119)$$

From the arguments above, $\|e_1(t)\| \leq \hat{w}(t)$, and substituting the v given in (3.102) into (3.119) yields

$$\dot{V} = 2\|e_y\|((\|C_2A_{21}\| + \|C_2\|\mathcal{L}_\phi)\hat{w} - 2k(\cdot))\|e_y\| \qquad (3.120)$$

From (3.116) and (3.120) it follows that $\dot{V} \leq -2\eta\|e_y\| \leq -2\eta\tilde{V}^{\frac{1}{2}}$. This shows that the reachability condition is satisfied. It follows that $\tilde{V} = 0$ in finite time and consequently a sliding motion is achieved and maintained after some finite time $t_s > 0$. Hence the proof is complete. $\qquad\qquad\square$

3.6 Fault Detection: A Simulation Example

Consider a single-link flexible joint robot system, where the system nonlinearities come from the joint flexibility modeled as a stiffened torsional spring, and the gravitational force. The dynamical model for the robot can be described by

$$\dot{\theta}_1 = \omega_1 \qquad (3.121)$$

$$\dot{\omega}_1 = \frac{1}{J_1}(\kappa_1(\theta_2 - \theta_1) + \kappa_2(\theta_2 - \theta_1)^3) - \frac{B_v}{J_1}\omega_1 + \frac{K_\tau}{J_1}u \qquad (3.122)$$

$$\dot{\theta}_2 = \omega_2 \qquad (3.123)$$

$$\dot{\omega}_2 = -\frac{1}{J_2}(\kappa_1(\theta_2 - \theta_1) + \kappa_2(\theta_2 - \theta_1)^3) - \frac{m_l gh}{J_2}\sin(\theta_2) \qquad (3.124)$$

where θ_1 and ω_1 are the motor position and velocity, respectively; θ_2 and ω_2 are the link position and velocity, respectively; J_1 is the inertia of the DC motor; J_2 is the inertia of the link; $2h$ is the length of the link, while m_l represents its mass; B_v is the viscous friction; κ_1 and κ_2 both are positive constants; and K_τ is the amplifier gain. The domain considered here is $\{(\theta_1, \omega_1, \theta_2, \omega_2) \mid |\theta_2 - \theta_1| < 2.8, |\omega_1| \leq 50\}$. It is assumed that the motor position, motor velocity and the sum of link velocity, and link position are measured. Suppose that a fault occurs in the input channel in the robot system. Therefore the fault distribution matrix D will be equal to the input distribution matrix. Suitable values for the parameters are: $J_1 = 3.7 \times 10^{-3}\text{kgm}^2$, $J_2 = 9.3 \times 10^{-3}\text{kgm}^2$, $h = 1.5 \times 10^{-1}\text{m}$, $m = 0.21\text{kg}$, $B_v = 4.6 \times 10^{-2}\text{m}$, $\kappa_1 = \kappa_2 = 1.8 \times 10^{-1}\text{Nm/rad}$, and $K_\tau = 8 \times 10^{-2}\text{Nm/V}$. Let $x = \text{col}(x_1, x_2, x_3, x_4) :=$

$(\theta_1, \omega_1, \theta_2, \omega_2)$. Then, the robot system can be described in the form (3.87) and (3.88) where

$$\phi(x, u) = \begin{bmatrix} 0 \\ 0.0194(x_3 - x_1)^3 + 21.6216u \\ 0 \\ 0.0486(x_3 - x_1)^3 - 83.4324\sin(x_3) \end{bmatrix}$$

A suitable transformation is given by $z = Tx$ with T defined as

$$T = \begin{bmatrix} -1.4142 & 0 & 0 & 0 \\ 0 & 0 & 1 & 0 \\ -1 & -1 & 0 & 0 \\ 0 & 0 & 0 & -0.01 \end{bmatrix}$$

It follows that

$$\left[\begin{array}{c|c} A_{11} & A_{12} \\ \hline A_{21} & A_{22} \end{array} \right] = \left[\begin{array}{c|ccc} -1 & 0 & 1.4142 & 0 \\ \hline 0 & 0 & 0 & -1 \\ 8.0496 & -0.0486 & -11.4324 & 0 \\ 0.0137 & 0.0194 & 0 & 0 \end{array} \right] \tag{3.125}$$

$$D_2 = \begin{bmatrix} 0 \\ \hline D_{22} \end{bmatrix} = \begin{bmatrix} 0 \\ \hline -21.6216 \\ 0 \end{bmatrix} \tag{3.126}$$

$$C_2 = \begin{bmatrix} 0 & -1 & 0 \\ 1 & 0 & 0 \\ 0 & 0 & -1 \end{bmatrix} \tag{3.127}$$

and

$$\phi(T^{-1}z, u) = \begin{bmatrix} 0 \\ 0 \\ -21.6216u - 0.0486(z_2 + 0.7071z_1)^3 \\ 0.0002(z_2 + 0.7071z_1)^3 + 0.3319\sin(z_2) \end{bmatrix}$$

Let $\alpha = 0.5$. From the LMI synthesis, the optimal value of the Lipschitz gain $\mathcal{L}_\phi = 0.75$ when $L = \begin{bmatrix} 0 & 0 & 0 \end{bmatrix}$, $\varepsilon = 2$, and $P = 1.5$ and the conditions of Proposition 3.2 are satisfied. Finally, by choosing

$$K = \begin{bmatrix} 0 & 1.1 & 1 \\ 10.2324 & -0.0486 & 0 \\ 0 & 0.0194 & -1 \end{bmatrix}$$

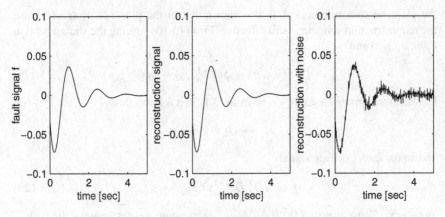

Fig. 3.17 Fault estimation

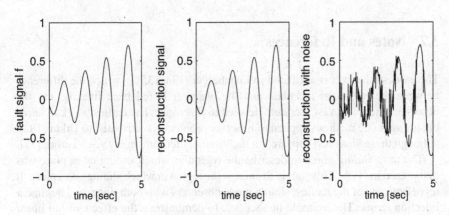

Fig. 3.18 Fault estimation

it follows that

$$C_2(A_{22} - A_{21}L)C_2^{-1} + C_2K = \begin{bmatrix} -1.2 & 0 & 0 \\ 0 & -1.1 & 0 \\ 0 & 0 & -1 \end{bmatrix}$$

and thus (3.118) is true.

For simulation purposes, a linear state feedback controller has been introduced to stabilize the system. In the first case, the fault signal is $f(y, u, t) = 0.5 \sin(u)$ which does not affect the stability of the system. The associated simulation is shown in Fig. 3.17. In the second case, the fault signal is $f(y, u, t) = \sin(u)$ which destroys the system stability. The corresponding simulation is shown in Fig. 3.18. The simulations show that the signal $\tilde{f}$ can reconstruct the fault perfectly, even if the fault destroys closed-loop stability. However, in the second simulation, the reconstruction properties will eventually be lost over time as the states of the plant become unbounded. It also shows that in the presence of sensor noise the reconstruction scheme is still effective. From (3.104) during the sliding motion $e_y = \dot{e}_y = 0$ and

$$v_{eq} = D_2 f(y, u, t) + \phi(T^{-1}z, u) - \phi(T^{-1}\hat{z}, u) \qquad (3.128)$$

Furthermore since $\hat{z} \to z$ as $t \to \infty$ from (3.128) it follows that

$$v_{eq} \to D_2 f(y, u, t)$$

and so the fault estimate signal

$$\hat{f} := (D_2^T D_2)^{-1} D_2 v_{eq} \to f \qquad (3.129)$$

as $t \to \infty$. Consequently $(D_2^T D_2)^{-1} D_2 v_{eq}$ is an asymptotic estimate of the fault.

3.7 Notes and References

The earliest observer is attributed to Luenberger [134, 135] in which the difference between the output of the plant and the observer is fed back linearly into the observer. There is a vast literature devoted to this topic. The earliest work in terms of sliding mode methods applied to observer problems is attributed to Utkin [182], although these ideas had appeared in the Russian literature many years earlier [37].

The term *sliding patch* to describe the region in which sliding takes place was first coined in [172]. This sought to enlarge the region in which sliding takes place. It represents one of the earliest sliding mode observers with both linear and nonlinear injection terms. The example in Sect. 3.2, to demonstrate the effects of the linear and nonlinear terms on the observer performance, is taken from [7].

The problem of robust state estimation for systems with bounded matched uncertainty was first explored by Walcott and Żak [188]. In terms of the design of Walcott and Żak, a system theoretic interpretation of the constraints in (3.46) and (3.47) by Steinberg and Corless [175] is that the transfer function matrix $G(s) = FC(sI - A_0)^{-1}B$ is strictly positive real. The problem of synthesizing $P, G,$ and F (and incorporating some sort of design element) is nontrivial. The structural requirements (3.46) and (3.47) of Walcott and Żak were shown in [70] to be solvable if and only if:

- Rank$(CB) = m$.
- All the invariant zeros of (A, B, C) have negative real parts.

Under these circumstances a (semi-)analytic expression for the solution to the observer design problem is given in terms of a gain matrix $L \in \mathbb{R}^{(n-p)\times(p-m)}$ and a stable matrix $A_{22}^s \in \mathbb{R}^{p\times p}$ that defines G_l (see [70]). The approach in this chapter builds on the work of Edwards and Spurgeon [65] which builds on the ideas of Walcott and Żak [188]. An interesting comparison between sliding mode observers and other nonlinear techniques appears in [189]. A well-cited tutorial on sliding mode observers is presented in [62]. A recent tutorial on this material appears in [174]. The observer from Sect. 3.5 to deal with Lipschitz nonlinear systems is given in Yan and Edwards [193]. The single-link flexible joint robot system is taken from [78]. A comparison of these observers with so-called unknown input observers appears in [69].

The optimization problem in Sect. 3.5 is a convex eigenvalue optimization problem and can be solved using standard LMI algorithms [99] (largely based on interior point methods). A now classical account of the different applications of convex optimization to control problems is given by Boyd et al. [38].

One of the earliest paper to apply sliding mode observers to fault detection and isolation problems is Edwards et al. [70]. These ideas have been expanded upon to incorporate directly robust reconstruction [178]. Subsequently the relative degree one restrictions in this chapter has been removed using a cascade of conventional sliding mode observers [179]. An overview of applications of these ideas to fault detection and fault tolerant control problems appears in [7].

3.8 Exercises

Exercise 3.1. Show that for the system

$$\dot{x}(t) = Ax(t) \tag{3.130}$$
$$y(t) = Cx(t) \tag{3.131}$$

where

$$A = \begin{bmatrix} 0 & 1 \\ 0 & 0 \end{bmatrix} \quad C = \begin{bmatrix} 1 & 0 \end{bmatrix} \tag{3.132}$$

choosing the gain

$$G_n = \begin{bmatrix} -1 \\ 1 \end{bmatrix} \tag{3.133}$$

in the observer from (3.5) and (3.6) yields a stable sliding motion.

Exercise 3.2. Consider the system

$$\dot{x}(t) = Ax(t) \tag{3.134}$$
$$y(t) = Cx(t) \tag{3.135}$$

where

$$A = \begin{bmatrix} 0 & 1 \\ 0 & 0 \end{bmatrix} \quad C = \begin{bmatrix} 1 & 1 \end{bmatrix} \tag{3.136}$$

Find the range of values of the scalar gain γ for which the gain vector

$$G_n = \begin{bmatrix} \gamma \\ 1 \end{bmatrix} \tag{3.137}$$

in the observer from (3.5) and (3.6) yields a stable sliding motion.

Exercise 3.3. For the triple integrator

$$\dddot{y}(t) = u(t) \tag{3.138}$$

written in state-space form, with the states chosen as $\mathrm{col}(y, \dot{y}, \ddot{y})$, design an observer of the form (3.5) and (3.6).

Exercise 3.4. Consider the system

$$\dot{x}(t) = Ax(t) + Bu(t) + D\xi(t, x) \tag{3.139}$$

$$y(t) = Cx(t) \tag{3.140}$$

where $\xi(t, x)$ is an unknown but bounded disturbance and

$$A = \begin{bmatrix} 0 & 1 \\ 0 & 0 \end{bmatrix} \quad B = D = \begin{bmatrix} 0 \\ 1 \end{bmatrix} \quad C = \begin{bmatrix} 1 & \alpha \end{bmatrix} \tag{3.141}$$

where α is a scalar. Find the range of value of α for which it is possible to design an observer of the form (3.61) and (3.62).

Exercise 3.5. Consider a simple quarter-car vehicle model written in state space form choosing as states v (forward velocity) and ω (wheel angular velocity)

$$\dot{x}(t) = Ax(t) + Bu(t) + Df(t, x) \tag{3.142}$$

where $x = \mathrm{col}(v, \omega)$ and

$$A = \begin{bmatrix} -\sigma_v g & 0 \\ 0 & -\frac{\sigma_\omega}{J} \end{bmatrix} \quad B = \begin{bmatrix} 0 \\ -\frac{K_b}{J} \end{bmatrix} \quad D = \begin{bmatrix} \frac{4}{m} \\ -\frac{r}{J} \end{bmatrix} \tag{3.143}$$

In the above, $f(t, x)$ represents the unmeasured breaking force applied to the wheel. The positive scalars $\sigma_v, \sigma_w, J, K_b, r, m, g$, and r are physical parameters associated with the vehicle. It is assumed that only angular wheel speed ω is measured, and in this case the output distribution matrix is given as

$$C = \begin{bmatrix} 0 & 1 \end{bmatrix} \tag{3.144}$$

The following model-based nonlinear observer is proposed

$$\dot{\hat{x}}(t) = A\hat{x}(t) + Bu(t) + G_l e_y + Dv \tag{3.145}$$

where $\hat{x} = \text{col}(\hat{v}, \hat{\omega})$, the design gain $G_l \in \mathbb{R}^2$, $e_y = \omega - \hat{\omega}$, and $v = -k\,\text{sign}(Fe_y)$ where k is a scalar gain, F is a scalar which will be defined later, and $\hat{v}$ and $\hat{\omega}$ are estimates of v and ω, respectively. The design problem is to compute G_l, F, and k so that the estimated angular velocity $\hat{\omega}$ is such that $e_y = \omega - \hat{\omega} \equiv 0$ in finite time: i.e., a sliding motion is achieved in finite time.

Here the gain G_l is proposed as

$$G_l := \begin{bmatrix} \frac{4J\sigma_v g}{rm} - \frac{4J}{rm}\alpha \\ -\frac{\sigma_\omega}{J} + \alpha \end{bmatrix} \tag{3.146}$$

where α is a positive scalar.

1. Show that the eigenvalues of $(A - G_lC) = \{-\sigma_v g, -\alpha\}$ which implies $(A - G_lC)$ is stable by design.
2. Consider as a Lyapunov function $V = e^T P e$ where

$$P := \begin{bmatrix} P_1 & \frac{4J}{rm}P_1 \\ \frac{4J}{rm}P_1 & P_2 + \frac{16J^2}{r^2m^2}P_1 \end{bmatrix} \tag{3.147}$$

and P_1 and P_2 are positive scalars. Define

$$Q := -\left(P(A - G_lC) + (A - G_lC)^T P\right) \tag{3.148}$$

Show that Q is symmetric positive definite.
3. Show that $PD = FC^T$ for the scalar $F := -2\frac{\alpha r}{J}P_2$.

Use this information to demonstrate that (3.145) induces a sliding motion on the hyperplane $S = \{e : Ce = 0\}$ in finite time.

Exercise 3.6. A more complex model of the same system, but now including a LuGre friction model, is

$$\dot{x}_p(t) = A_p x(t) + B_p u(t) + D_p \theta x_1(t) f(x_3) \tag{3.149}$$

where the control signal $u(t) = P_b(t)$ and

$$A_p := \begin{bmatrix} 0 & 0 & 1 \\ g\sigma_0 & -g\sigma_v & g(\sigma_1 + \sigma_2) \\ q\sigma_0 & g\sigma_v & q(\sigma_1 + \sigma_2) \end{bmatrix} \qquad B_p := \begin{bmatrix} 0 \\ 0 \\ -\frac{rk_b}{J} \end{bmatrix} \tag{3.150}$$

The distribution matrix through which the nonlinear terms operate is

$$D_p := \begin{bmatrix} -1 \\ -g\sigma_1 \\ -q\sigma_1 \end{bmatrix} \tag{3.151}$$

In these matrices, the aggregate parameter $q := -(g + \frac{F_n r^2}{J})$. Since it is assumed that only angular wheel speed ω is measured, the output distribution matrix

$$C_p = \begin{bmatrix} 0 & \frac{1}{r} & \frac{1}{r} \end{bmatrix} \tag{3.152}$$

Consider an observer of the form

$$\dot{\hat{x}}_p(t) = A_p \hat{x}_p(t) + B_p u(t) + G_p e_y + D_p \nu \tag{3.153}$$

where $G_p \in \mathbb{R}^3$ and $\nu = -k_p \text{sign}(e_y)$ where k_p is a scalar gain. The main objective is to synthesize an observer to generate an estimated angular velocity $\hat{\omega} = C\hat{x}$ such that $e_y = \omega - \hat{\omega} \equiv 0$ in finite time despite the nonlinear friction terms which have been ignored in (3.153). It is shown in [148] that if

$$\bar{g}_1 = r + (g + \frac{\beta}{\sigma_1}) \frac{J}{F_n r} \tag{3.154}$$

$$\bar{g}_2 = g(\beta + g\bar{\sigma} - \frac{\sigma_0}{\sigma_1}) \frac{J}{F_n r} + gr(\sigma_1 + \sigma_2) \tag{3.155}$$

$$\bar{g}_3 = (\sigma_1 + \sigma_2)q - \frac{\sigma_w}{J} + \frac{\sigma_0}{\sigma_1} - \beta \tag{3.156}$$

where $\bar{\sigma} = \sigma_v + \sigma_1 + \sigma_2$ and β is a negative scalar, then

$$G_p = \begin{bmatrix} \bar{g}_1 \\ \bar{g}_2 \\ r\bar{g}_3 - \bar{g}_2 \end{bmatrix} \tag{3.157}$$

is an appropriate choice of gain in (3.153). Define $e_p = x_p - \hat{x}_p$ as the state estimation error, where $x_p = \text{col}(x_1, x_2, x_3)$. The dynamics of the error system can be obtained from (3.149) and (3.153) as

$$\dot{e}_p = (A_p - G_p C_p)e_p + D_p \left(\theta f(x_3) x_1 - k_p \, \text{sign}(e_y) \right) \tag{3.158}$$

Show the reduced-order sliding motion is governed by the *linear system*

$$\begin{bmatrix} \dot{e}_1 \\ \dot{e}_2 \end{bmatrix} = \begin{bmatrix} -\frac{\sigma_0}{\sigma_1} & \frac{\sigma_2}{\sigma_1} \\ 0 & -\sigma_v g \end{bmatrix} \begin{bmatrix} e_1 \\ e_2 \end{bmatrix} \tag{3.159}$$

and so the sliding motion is stable

Exercise 3.7. Consider the following Lur'e-type representation for Chua's circuit:

$$\dot{x}_i = A x_i + D \xi_i(y_i) + B u_i \tag{3.160}$$

$$y_i = C x_i \tag{3.161}$$

where

$$A = \begin{bmatrix} -al_1 & a & 0 \\ 1 & -1 & 1 \\ 0 & -b & 0 \end{bmatrix}, \quad D = \begin{bmatrix} -a(l_0 - l_1) \\ 0 \\ 0 \end{bmatrix} \tag{3.162}$$

$$C = \begin{bmatrix} 1 & 0 & 0 \\ 0 & 1 & 0 \end{bmatrix} \tag{3.163}$$

The nonlinearity is $\xi_i(y_i) = \frac{1}{2}(|x_{i1} + c| - |x_{i1} - c|)$, which has a sector bound $[0, 1]$. The chosen values of the parameters are $a = 9, b = 14.286, c = 1, l_0 = -1/7$, and $l_1 = 2/7$ in order to obtain double-scroll attractor behavior.

Design a sliding mode observer for this system.

Chapter 4
Second-Order Sliding Mode Controllers and Differentiators

As we have seen, classic sliding modes provide robust and high-accuracy solutions for a wide range of control problems under uncertainty conditions. However, two main restrictions remain. First, the constraint to be held at zero in conventional sliding modes has to be of relative degree 1, which means that the control needs to explicitly appear in the first time derivative of the constraint. Thus, one has to search for an appropriate constraint. Second, high-frequency control switching may easily cause unacceptable practical complications (chattering effect), if the control has any physical sense.

Suppose that the problem is to keep the sliding variable s at zero, while the control appears only in $\ddot{s}$. Usually the constraint function $\sigma = s + \dot{s}$ is chosen. By construction, $\dot{\sigma} = \dot{s} + \ddot{s}$ contains the control, and σ can be kept at zero in a classic sliding mode (Chap. 2). As a result s tends asymptotically to zero. Keeping it at exact zero is not possible. One also needs to calculate $\dot{s}$ to realize this scheme. Both of these goals, exact robust differentiation and exactly keeping $s = 0$, can be accomplished by the second-order sliding mode technique to be developed in this chapter.

Suppose that the problem is to keep s at zero, while the control appears already in $\dot{s}$. This problem is easily solved by means of conventional sliding modes (Chap. 2). But often the chattering effect makes the solution unacceptable. A possible solution is to consider the control derivative as a new virtual control. Then the above reasoning can be applied, and using a second-order sliding mode technique, the task can be accomplished exactly, and in finite time, by means of *continuous* control. As a consequence it can be expected that the chattering effect is significantly attenuated.

4.1 Introduction

Consider a simple control system involving target pointing by means of a pendulum (with the angle coordinate measured from $q = \pi/6$) given by

$$\dot{x} = -\sin(x + q) + u, \quad q = \pi/6 \tag{4.1}$$

Y. Shtessel et al., *Sliding Mode Control and Observation*, Control Engineering,
DOI 10.1007/978-0-8176-4893-0_4, © Springer Science+Business Media New York 2014

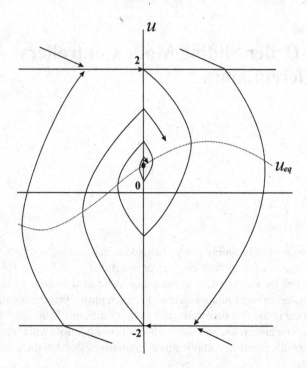

Fig. 4.1 Asymptotically stable second-order sliding mode at $x = 0, u = \sin(q)$

The targeting problem is reformulated as the stabilization of Eq. (4.1) at $x = 0$. It is easily solved by means of the standard relay controller

$$u = -2\,\mathrm{sign}(x) \qquad (4.2)$$

As we have already seen, this controller produces considerable chattering. One of the natural ways to avoid chattering is to introduce dynamical regularization gradually switching the control

$$\dot{u} = \begin{cases} -u & \text{if } |u| > 2 \\ -\alpha\,\mathrm{sign}(x) & \text{if } |u| \leq 2 \end{cases} \qquad (4.3)$$

where $u(0) = u_0$.

Let x and u be the new coordinates (Fig. 4.1), and suppose α is sufficiently large. Obviously, we get $\dot{x} > 0$ with $u > u_{eq} = \sin(x + \pi/6)$ and $\dot{x} < 0$ with $u < \sin(x + q)$. Each trajectory starting from the point $(0, u_0)$, with $u_0 > \sin(q) = \sin(\pi/6) = 0.5$, revolves around the point $(0, 0.5)$. The closer is the initial point to $(0, 0.5)$, the closer is the trajectory to $(0, 0.5)$. Thus the point $(0, 0.5)$ is the *limit of trajectories*. Furthermore it can even be shown that solutions asymptotically converge to this point. From the theory of ordinary differential equations one learns that a limit trajectory also has to be a solution trajectory. In other words,

$x = 0, u = 0.5$ is a constant solution of the system (4.1), (4.3). Moreover, obviously the solution should still be $x = 0, u = \sin(q)$, with q slowly changing. The point $x = 0, u = 0.5$ does not represent a solution in any classical sense, but is a solution in the sense of Filippov.

The point $x = 0, u = \sin(q) = 0.5$ satisfies the conditions

$$x = 0, \dot{x} = 0 \tag{4.4}$$

Such a constant solution would be an ideal solution for the stated control problem, provided it can indeed be considered as a solution of Eqs. (4.1), (4.3). Motions satisfying (4.4) are said to be in *the second-order sliding mode* or *2-sliding mode*. The point $x = 0, u = 0.5$ is the *2-sliding manifold*. In this chapter we will learn how to establish such modes and to ensure their *finite-time stability*. But first we need to redefine the very notion of the solution for the case of differential equations with discontinuous right-hand sides.

Definition 4.1. Consider a discontinuous differential equation $\dot{x} = f(x)$ (Filippov differential inclusion $\dot{x} \in F(x)$) with a smooth output function $\sigma = \sigma(x)$, and let it be understood in the Filippov sense. Then, provided that:

1. σ and the total time derivative $\dot{\sigma} = \sigma'_x(x) f(x)$ are continuous functions of x
2. The set

$$\sigma = \dot{\sigma} = 0 \tag{4.5}$$

   is a nonempty integral set
3. The Filippov set of admissible velocities at the set defined by Eq. (4.5) contains more than one vector

the motion on the set (4.5) is said to exist in a 2-sliding (second-order sliding) mode (Fig. 4.2), and the set (4.5) is called a 2-sliding set. The nonautonomous case is reduced to the considered one by introducing the fictitious equation $\dot{t} = 1$.

Note that the third requirement means that set (4.5) is a discontinuity set of the equation, and it is introduced here only to exclude extraneous cases of integral manifolds of continuous differential equations. That condition is illustrated by the two limit velocity vectors at the 2-sliding point M in Fig. 4.2. Also note that the extension of the above definitions by the introduction of the fictitious equation $\dot{t} = 1$ actually makes time similar to other coordinates. This approach is different from the standard definition by Filippov, it is simpler, and it provides for more solutions.

The conventional sliding mode described in Chap. 2 is called first order (σ is continuous, and $\dot{\sigma}$ is discontinuous). The general definition of the sliding mode order is very similar and is introduced in Chap. 6.

Remark 4.1. The notion of the sliding order appears to be connected with the notion of relative degree.

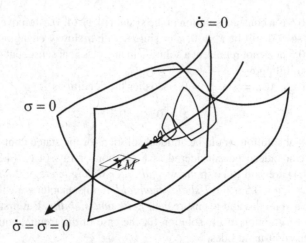

Fig. 4.2 2-sliding mode

Consider a general system linear in the control given by

$$\dot{x} = a(t, x) + b(t, x)u \tag{4.6}$$

$x \in \mathbb{R}^n$, with the output

$$\sigma = \sigma(t, x) \tag{4.7}$$

The functions a, b, σ are assumed to have all the necessary derivatives. In this chapter we consider the simplest case when $\sigma, u \in \mathbb{R}$. The total time derivative of σ is defined as

$$\dot{\sigma} = \sigma'_t + \sigma'_x a + \sigma'_x bu$$

Suppose that $\sigma'_x b \equiv 0$. Then calculating the second total derivative yields

$$\ddot{\sigma} = \sigma''_{tt} + 2\sigma''_{tx} a + \sigma'_x a'_t + [\sigma''_{xx}(a + bu)]a + \sigma'_x[a'_x(a + bu)]$$

Thus,

$$\ddot{\sigma} = h(t, x) + g(t, x)u, \ g(t, x) = (\sigma''_{xx} b)a + \sigma'_x(a'_x b) \tag{4.8}$$

where h is another appropriately defined function. Hence, the relative degree equals 1 if $\sigma'_x b \neq 0$, and it equals 2 , if $\sigma'_x b \equiv 0$ and $(\sigma''_{xx} b)a + \sigma'_x(a'_x b) \neq 0$.

Suppose that the system relative degree exists, and the control function u is defined by some discontinuous feedback. Then with relative degree 1 the function $\dot{\sigma}$ is discontinuous, while σ of course is continuous. On the other hand with relative degree 2 and discontinuous u get that $\sigma, \dot{\sigma}$ are continuous functions, while $\ddot{\sigma}$ is discontinuous. Therefore we come to conclusion that the conventional sliding (1-sliding) mode can only be achieved with relative degree 1, while the second-order sliding (2-sliding) mode requires relative degree 2 with respect to discontinuous control.

4.2   2-Sliding Mode Controllers

Once more consider a dynamic system of the form

$$\dot{x} = a(t,x) + b(t,x)u, \ \sigma = \sigma(t,x) \tag{4.9}$$

where $x \in \mathbb{R}^n$, $u \in \mathbb{R}$ is control, σ is the only measured output, and the smooth functions a, b, σ (and the dimension n) are unknown. The task is to make the output σ vanish in finite time and to keep $\sigma \equiv 0$ by means of discontinuous globally bounded feedback control. The system trajectories are supposed to be infinitely extendible in time for any bounded input. The system is understood in the Filippov sense.

Assume that the measured output $\sigma(t,x)$ is twice differentiable with respect to time and the condition $\sigma'_x b \equiv 0$ and $(\sigma''_{xx}b)a + \sigma'_x(a'_x b) \neq 0$ hold. Then calculating the second total time derivative $\ddot{\sigma}$ along the trajectories of Eq. (4.9), under the conditions outlined above in Eq. (4.8), we obtain

$$\ddot{\sigma} = h(t,x) + g(t,x)u$$

where the functions $h = \ddot{\sigma}|_{u=0}, g = \frac{\partial}{\partial u}\ddot{\sigma} \neq 0$ are some unknown smooth functions. Suppose that the inequalities

$$0 < K_m \leq g \leq K_M, \ |h| \leq C \tag{4.10}$$

hold globally for some $K_m, K_M, C > 0$. Note that, at least locally, Eq. (4.10) is satisfied for any smooth system (4.8) with the well-defined relative degree 2.

Obviously, no continuous feedback controller of the form $u = \varphi(\sigma, \dot{\sigma})$ can solve the stated problem. Indeed, such a control ensuring $\sigma \equiv 0$ has to satisfy the equality $\ddot{\sigma} \equiv 0$ as well, which means that $\varphi(0,0) = -h(t,x)/g(t,x)$, whenever $\sigma = \dot{\sigma} = 0$ holds. The uncertainty in the problem prevents it, since the controller will not be effective for the simple autonomous linear system $\ddot{\sigma} = c + ku$, $K_m \leq k \leq K_M$, $|c| \leq C$, with $\varphi(0,0) \neq -c/k$. In other words, due to the uncertainty, the 2-sliding mode $\sigma = \dot{\sigma} = 0$ needs to be established.

Assume now that Eq. (4.10) holds globally. Then Eqs. (4.8), (4.10) imply the differential inclusion

$$\ddot{\sigma} \in [-C, C] + [K_m, K_M]u \tag{4.11}$$

Most 2-sliding controllers may be considered as controllers for Eq. (4.11) steering $\sigma, \dot{\sigma}$ to 0 in (preferably) finite time. Since the inclusion (4.11) does not "remember" the original system (4.9), such controllers are obviously robust with respect to any perturbations preserving (4.10).

Hence, the problem is to find a feedback

$$u = \varphi(\sigma, \dot{\sigma}) \tag{4.12}$$

such that all the trajectories of Eqs. (4.11), (4.12) converge in finite time to the origin $\sigma = \dot{\sigma} = 0$ of the phase plane $\sigma, \dot{\sigma}$. We will now consider a number of the simplest and most popular controllers solving this problem.

4.2.1 Twisting Controller

The twisting controller described below is historically the first 2-sliding controller which was proposed. It is defined by the formula

$$u = -(r_1 \text{sign}(\sigma) + r_2 \text{sign}(\dot{\sigma})), \quad r_1 > r_2 > 0 \tag{4.13}$$

Theorem 4.1. *Let r_1 and r_2 satisfy the conditions*

$$(r_1 + r_2)K_m - C > (r_1 - r_2)K_M + C, \ (r_1 - r_2)K_m > C \tag{4.14}$$

The controller in Eq. (4.13) guarantees the appearance of a 2-sliding mode $\sigma = \dot{\sigma} = 0$ attracting the trajectories of the sliding variable dynamics (4.11) in finite time.

Proof. It is easy to see that every trajectory of the system crosses the axis $\sigma = 0$ in finite time. Indeed, due to Eqs. (4.13), (4.14) $\text{sign}(\sigma) \, \text{sign}(\ddot{\sigma}) < 0$ and with $\text{sign}(\sigma)$ being constant for a long time, $\sigma\dot{\sigma} < 0$ is established, while the absolute value of $\dot{\sigma}$ tends to infinity. It follows from Eq. (4.14) that with $\sigma \neq 0$

$$-[K_M(r_1 + r_2) + C] \leq \ddot{\sigma}\,\text{sign}(\sigma) \leq -[K_M(r_1 + r_2) - C] < 0 \text{ with } \sigma\dot{\sigma} > 0$$
$$-[K_M(r_1 - r_2) + C] \leq \ddot{\sigma}\,\text{sign}(\sigma) \leq -[K_M(r_1 - r_2) - C] < 0 \text{ with } \sigma\dot{\sigma} < 0$$
$$\tag{4.15}$$

According to the Filippov definitions, the values taken on a set of the measure 0 (in particular on any curve) do not matter. Let $\dot{\sigma}_0, \sigma_M, \dot{\sigma}_M$ (Fig. 4.3.) be the trajectory of differential equation

$$\ddot{\sigma} = \begin{cases} -[K_m(r_1 + r_2) - C]\text{sign}(\sigma) & \text{with } \dot{\sigma}\sigma > 0 \\ -[K_M(r_1 - r_2) + C]\text{sign}(\sigma) & \text{with } \dot{\sigma}\sigma \leq 0, \end{cases} \tag{4.16}$$

with the same initial conditions. Assume now for simplicity that the initial values are $\sigma = 0, \dot{\sigma} = \dot{\sigma}_0 > 0$ at $t = 0$. Thus, the trajectory enters the half-plane $\dot{\sigma} > 0$. Simple calculation shows that with $\sigma > 0$ the solution of Eq. (4.16) is determined by the equalities

$$\sigma = \sigma_M - \frac{\dot{\sigma}^2}{2[K_m(r_1+r_2)-C]} \text{ with } \dot{\sigma} > 0$$

$$\sigma = \sigma_M - \frac{\dot{\sigma}^2}{2[K_M(r_1-r_2)+C]} \text{ with } \dot{\sigma} \leq 0 \tag{4.17}$$

where σ_M is determined from the equation

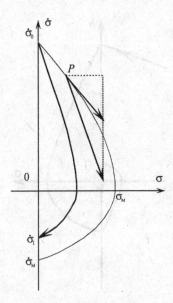

Fig. 4.3 Construction of a majorant trajectory for the twisting controller

$$2[K_M(r_1 + r_2) - C]\sigma_M = \dot{\sigma}_0^2 \tag{4.18}$$

Consider any point $P(\sigma_P, \dot{\sigma}_P)$ of this curve (Fig. 4.3). The velocity of Eqs. (4.11), (4.13) at this point has coordinates $(\dot{\sigma}_P, \ddot{\sigma}_P)$. Hence, the horizontal component of the velocity depends only on the point itself. Since the vertical component satisfies the inequalities (4.15), the velocity of Eqs. (4.11), (4.13) always "looks" into the region bounded by the axis $\sigma = 0$ and curve (4.17). That curve is called the majorant. Let the trajectory of Eqs. (4.11), (4.13) next intersect the axis $\sigma = 0$ at the point $\dot{\sigma}_1$. Then, obviously, $|\dot{\sigma}_1| \le |\dot{\sigma}_M|$ and

$$|\dot{\sigma}_1| / |\dot{\sigma}_0| \le [K_M(r_1 - r_2) + C] / [K_M(r_1 + r_2) - C]^{1/2} = q < 1 \tag{4.19}$$

Extending the trajectory into the half-plane $\sigma < 0$, after similar reasoning, guarantees that the successive crossings of the axis $\sigma = 0$ satisfy the inequality

$$|\dot{\sigma}_{i+1}| / |\dot{\sigma}_i| \le q < 1 \tag{4.20}$$

as shown in (Fig. 4.3). Therefore, the algorithm obviously converges. Next the convergence time is to be estimated. The real trajectory consists of an infinite number of segments belonging to the half-planes $\sigma \ge 0$ and $\sigma \le 0$ (Fig. 4.4). On each of these segments $\dot{\sigma}$ changes monotonously according to Eq. (4.15). The total variance of the function $\dot{\sigma}(t)$ is

$$Var\,(\dot{\sigma}(\cdot)) = |\dot{\sigma}_{i+1}| \le |\dot{\sigma}_0|\,(1 + q + q^2 + \ldots) = \frac{|\dot{\sigma}_0|}{1 - q} \tag{4.21}$$

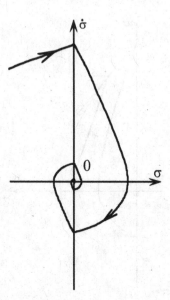

Fig. 4.4 Twisting controller trajectory

and the total convergence time is estimated as

$$T \leq \sum \frac{|\dot{\sigma}_i|}{[K_m(r_1 - r_2) - C]} \leq \frac{|\dot{\sigma}_0|}{(1 - q)[K_m(r_1 - r_2) - C]} \qquad (4.22)$$

The proof of the theorem is complete. $\square$

Remark 4.2. Note that considering the successive intersections of the trajectory with the σ axis, a similar inequality can be obtained:

$$|\sigma_{i+1}| / |\sigma_i| \leq [K_M(r_1 - r_2) + C] / [K_M(r_1 + r_2) - C]^{1/2} = q^2 < 1 \qquad (4.23)$$

which also can be used for the proof. The same majorant curves are used, taken in the half-plane $\dot{\sigma} \leq 0$ or $\dot{\sigma} \geq 0$.

Remark 4.3. In practice the parameters are *never* assigned according to inequalities (4.14). Usually the real system is not exactly known, the model itself is not really adequate, and the estimations of parameters K_M, K_m, C are much larger than the actual values (often 100 times larger!). The larger the controller parameters, the more sensitive is the controller to any switching imperfections and measurement noises. Thus, a pragmatic way is to adjust the controller parameters via computer simulations. (In fact this is true with respect to all controllers described in this chapter.)

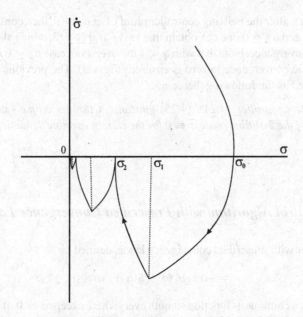

Fig. 4.5 Suboptimal controller trajectory convergence (the case of $q < 1$)

4.2.2   Suboptimal Algorithm

The so-called suboptimal controller is given by

$$u = -r_1 \text{sign}\left(\sigma - \sigma^*/2\right) + r_2 \text{sign}\left(\sigma^*\right), \quad r_1 > r_2 > 0, \tag{4.24}$$

where

$$r_1 - r_2 > \frac{C}{K_m}, \quad r_1 + r_2 > \frac{4C + K_M(r_1 - r_2)}{3K_m}, \tag{4.25}$$

and σ^* is the value of σ detected at the last time when $\dot\sigma$ was equal to 0. The initial value of σ^* is equal to 0. Any computer implementation of this controller requires successive measurements of $\dot\sigma$ or σ. Usually, the detection $\dot\sigma = 0$ occurs when the difference between successive measurements of $\Delta\sigma$ changes sign. The idea of the controller is directly derived from time-optimal control of a double integrator. A trajectory of the suboptimal controller is shown in the coordinates $\sigma, \dot\sigma$ in Fig. 4.5.

In the figure σ_0, σ_2 are two successive points of the intersection with the axis $\dot\sigma = 0$ and $\sigma_1 = \sigma_0/2$. Similar to the proof of the twisting controller inequality (4.19), this implies that

$$|\sigma_1 - \sigma_2| / |\sigma_0 - \sigma_1| \le [K_M(r_1 - r_2) + C] / [K_M(r_1 + r_2) - C] = q^2 < 2$$

$$\tag{4.26}$$

(see the remark after the twisting controller proof). Let $\sigma_0 > 0$, then considering the cases $\sigma_2 > 0$ and $\sigma_2 \leq 0$ one can obtain that $|\sigma_2| / |\sigma_0| \leq 1/2$, which also provides finite-time convergence. Note that with $q < 1$ the overshoot case $\sigma_2 \leq 0$ is excluded and monotonic convergence to zero is ensured (Fig. 4.5). The previous results can be summarized as the following theorem:

Theorem 4.2. *Controller* (4.24), (4.25) *guarantees the finite-time establishment and keeping of the 2-sliding mode $\sigma \equiv 0$ for the sliding variable dynamics satisfying* (4.11).

$\square$

4.2.3   Control Algorithm with Prescribed Convergence Law

The controller with prescribed convergence law is defined as

$$u = -\alpha \, \text{sign} \, (\dot{\sigma} + \xi(\sigma)), \; \alpha > 0 \qquad (4.27)$$

where $\xi(\sigma)$ is a continuous function smooth everywhere except $\sigma = 0$. It is assumed that all solutions of the differential equation $\dot{\sigma} + \xi(\sigma) = 0$ converge to 0 in finite time. The idea is to keep $\dot{\sigma} + \xi(\sigma) = 0$ in the 1-sliding mode.
Choosing $\xi(\sigma) = \beta |\sigma|^{1/2} \, \text{sign} \, \sigma$, $\beta > 0$, in Eq. (4.27) yields the controller

$$u = -\alpha \, \text{sign}(\dot{\sigma} + \beta |\sigma|^{1/2} \, \text{sign} \, \sigma) \qquad (4.28)$$

The following result can be proved:

Theorem 4.3. *Controller* (4.28) *guarantees the establishment and maintenance of a 2-sliding mode $\sigma \equiv 0$ for the sliding variable dynamics given by Eq.* (4.11), *in finite time.*

Proof. Differentiating the function $\Sigma = \dot{\sigma} + \beta \, |\sigma|^{1/2} \, \text{sign} \, (\sigma)$ along the trajectory yields

$$\dot{\Sigma} \in [-C, C] - \alpha \, [K_m, K_M] \, \text{sign} \, (\Sigma) + \tfrac{1}{2} \beta \dot{\sigma} \, |\sigma|^{-1/2} \qquad (4.29)$$

Checking the condition $\dot{\Sigma} \, \text{sign}(\Sigma) < const < 0$ in a vicinity of each point on the curve $\Sigma = 0$, using $\dot{\sigma} = -\beta \, |\sigma|^{1/2} \, \text{sign} \, (\sigma)$, implies that the 1-sliding-mode existence condition holds at each point except at the origin, if $\alpha K_m - C > \beta^2/2$.

The trajectories of the inclusion inevitably hit the curve $\Sigma = 0$ due to geometrical reasons. Indeed, each trajectory, starting with $\Sigma > 0$, terminates sooner or later at the semi-axis $\sigma = 0, \dot{\sigma} < 0$, if $u = -\alpha \cdot \text{sign}(\Sigma)$ keeps its constant value $-\alpha$ (Fig. 4.6). Thus, on the way it inevitably hits the curve $\Sigma = 0$. The same is true for the trajectory starting with $\Sigma < 0$. From that moment the trajectory slides along the curve $\Sigma = 0$ towards the origin and reaches it in finite time. Obviously, each trajectory starting from a disk centered at the origin comes to the origin in a finite time, the convergence time being uniformly bounded in the disk.

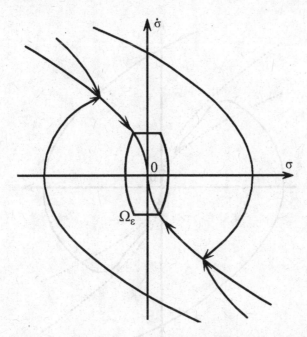

Fig. 4.6 Trajectories of the controller with the prescribed convergence law

Consider the region Ω_ε confined by the lines $\dot\sigma = \pm\varepsilon$ and the trajectories of the differential equations $\ddot\sigma = -C + K_m\alpha$ with initial conditions $\sigma = \varepsilon^2/\beta^2$, $\dot\sigma = \varepsilon$, and $\ddot\sigma = C - K_m\alpha$ with initial conditions $\sigma = -\varepsilon^2/\beta^2$, $\dot\sigma = \varepsilon$ (Fig. 4.6). No trajectory starting from the origin can leave Ω_ε. Since ε can be taken arbitrarily small, the trajectory cannot leave the origin. This completes the proof. $\qquad\square$

4.2.4 Quasi-Continuous Control Algorithm

An important class of controllers comprises the recently proposed so-called *quasi-continuous* controllers, featuring control continuous everywhere except the 2-sliding manifold $\sigma = \dot\sigma = 0$ itself. Since the 2-sliding condition requires the simultaneous fulfillment of two exact equalities, in the presence of any small noises and disturbances, the general-case trajectory does not ever hit the 2-sliding set. Hence, in practice the condition $\sigma = \dot\sigma = 0$ is never fulfilled, and the control remains continuous function of time, all the time. The larger the noises and switching imperfections, the worse the accuracy and the slower the changing rate of u. As a result, chattering is significantly reduced. The following is a 2-sliding controller with such features:

$$u = -\alpha \frac{\dot\sigma + \beta|\sigma|^{1/2}\mathrm{sign}(\sigma)}{|\dot\sigma| + \beta|\sigma|^{1/2}} \qquad (4.30)$$

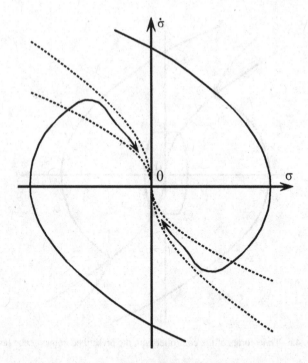

Fig. 4.7 Trajectories of the quasi-continuous controller

This control is continuous everywhere except the origin and it vanishes on the parabola $\dot{\sigma} + \beta \, |\sigma|^{1/2} \, \mathrm{sign}(\sigma) = 0$. For sufficiently large α, there are numbers $\rho_1, \rho_2 : 0 < \rho_1 < \beta < \rho_2$ such that all the trajectories enter the region between the curves $\dot{\sigma} + \rho_i \, |\sigma|^{1/2} \, \mathrm{sign}(\sigma) = 0$, $i = 1, 2$ and cannot leave it (Fig. 4.7).

Theorem 4.4. *Let*

$$\alpha, \beta > 0, \ \alpha K_m - C > 0 \tag{4.31}$$

and suppose the inequality

$$\alpha K_m - C - 2\alpha K_m \frac{\beta}{\rho + \beta} - \frac{1}{2}\rho^2 > 0 \tag{4.32}$$

holds for some positive $\rho > \beta$ (it is always true for a sufficiently large α), then the controller (4.30) guarantees the establishment of a stable 2-sliding mode $\sigma \equiv 0$ for the sliding variable dynamics given by Eq. (4.11), in finite time.

Remark 4.4. The conditions of the theorem can be solved for α, but the resulting expressions are redundantly cumbersome.

Proof. Denote $\rho = -\dot{\sigma} / |\sigma|^{1/2}$. Due to the symmetry of the problem, it is enough to consider the case of $\sigma > 0$ and $-\infty < \rho < \infty$. Calculations show that $u = \alpha (\rho - \beta) / (|\rho| + \beta)$ and

$$\dot{\rho} \in \left([-C, C] - [K_m, K_M] \alpha \frac{\rho - \beta}{|\rho| + \beta} + \frac{1}{2} \rho^2 \text{sign}(\sigma) \right) |\sigma|^{-1/2} \qquad (4.33)$$

With a negative or small positive ρ, the rotation velocity $\dot{\rho}$ is always positive due to Eq. (4.31). Thus there is a positive $\rho_1 < \beta$ such that the trajectories enter the region $\rho > \rho_1$. It is now necessary to show that there is a $\rho_2 > \beta$ such that in some vicinity of $\rho = \rho_2$ the inequality $\dot{\rho} < 0$ holds. This is exactly condition (4.32). Thus, conditions (4.31), (4.32) provide for the establishment and keeping of the inequality $\rho_1 < \rho < \rho_2$ and the proof of the theorem is complete. $\square$

4.2.5   Accuracy of 2-Sliding Mode Controllers

Consider the cases of noisy and/or discrete measurements with respect to the sampling interval τ. We will see in Chap. 6 that the discrete-sampling versions based on the Euler scheme provide an accuracy level of $\sigma = O(\tau^2)$, $\dot{\sigma} = O(\tau)$ in the absence of noise. Noisy measurements lead to the accuracy $\sigma = O(\varepsilon)$, $\dot{\sigma} = O(\varepsilon^{1/2})$, if the maximal errors of σ and $\dot{\sigma}$ and the sampling are of the order of ε and $\varepsilon^{1/2}$, respectively, and the maximal sampling interval τ is of the order $\varepsilon^{1/2}$. Note that this result does not require any practical dependence between τ and noise magnitudes. Indeed, in practice there are always specific values of noise magnitudes and sampling intervals, which can always be considered as a sample of an infinite family (in a nonunique way). Moreover, one can always reduce either the noise magnitudes or the sampling interval, preserving the same upper accuracy estimation.

4.3   Control of Relative Degree One Systems

All the controllers described this far require real-time measurements of $\dot{\sigma}$ or at least of sign($\dot{\sigma}$). In other words, in order to guarantee $\sigma = \dot{\sigma} = 0$, both σ and $\dot{\sigma}$ measurements are needed. This is reasonable but, nevertheless, not inevitable. The following controller can be used instead of the conventional (first-order) sliding mode using the same available information.

4.3.1   Super-Twisting Controller

Consider once more the dynamical system (4.9) of relative degree 1 and suppose that

$$\dot{\sigma} = h(t, x) + g(t, x)u \qquad (4.34)$$

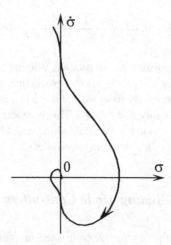

Fig. 4.8 Trajectory of the super-twisting controller

Furthermore assume that for some positive constants C, K_M, K_m, U_M, q

$$|\dot{h}| + U_M|\dot{g}| \leq C, \quad 0 \leq K_m \leq g(t, x) \leq K_M, \quad |h/g| < qU_M, \quad 0 < q < 1$$

$$(4.35)$$

hold and define

$$u = -\lambda|\sigma|^{1/2}\text{sign}(\sigma) + u_1, \quad \dot{u}_1 = \begin{cases} -u, & |u| > U_M \\ -\alpha\text{sign}(\sigma), & |u| \leq U_M \end{cases} \quad (4.36)$$

Then the following result is obtained.

Theorem 4.5. *With $K_m\alpha > C$ and λ sufficiently large, the controller* (4.36) *guarantees the appearance of a 2-sliding mode $\sigma = \dot{\sigma} = 0$ in system* (4.34), *which attracts the trajectories in finite time. The control u enters in finite time the segment $[-U_M, U_M]$ and stays there. It never leaves the segment, if the initial value is inside at the beginning.*

Remark 4.5. Note that the controller does not need measurements of $\dot{\sigma}$.

The controller given in Eq. (4.36) is called the super-twisting controller. The corresponding phase portrait is shown in Fig. 4.8. A sufficient (*very crude!*) condition for validity of the theorem is

$$\lambda > \sqrt{\frac{2}{(K_m\alpha - C)}\frac{(K_m\alpha + C)K_M(1 + q)}{K_m^2(1 - q)}} \quad (4.37)$$

Proof. Computing $\dot{u}$ with $|u| > U_M$ yields $\dot{u} = -\frac{1}{2}\lambda\dot{\sigma}|\sigma|^{-1/2} - u$. It follows from Eqs. (4.34), (4.35) that $\dot{\sigma}u > 0$ with $|u| > U_M$ and thus, $\dot{u}u < 0$, and u moves

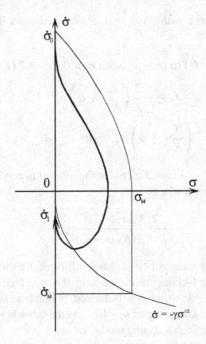

Fig. 4.9 A majoring curve for the super-twisting controller

towards the segment $|u| \le U_M$. Therefore $|u| \le U_M$ is established in finite time, for $|\dot{u}| > U_M$ when $|u| > U_M$. Note that a 1-sliding mode with $u = -U_m \text{sign}(\sigma)$ could exist during time intervals of constant $\text{sign}(\sigma)$ (e.g., see Fig. 4.1). The following equation is satisfied with $|u| < U_M$, $\sigma \ne 0$:

$$\ddot{\sigma} = \dot{h} + \dot{g}u - g\frac{1}{2}\lambda\frac{\dot{\sigma}}{|\sigma|^{1/2}} - g\,\alpha\,\text{sign}(\sigma)$$

The trivial identity $\frac{d}{dt}|\sigma| = \dot{\sigma}\,\text{sign}(\sigma)$ is used here. Note that once more, the values taken on sets of measure 0 are not accounted for; thus the differentiation is performed with $\text{sign}(\sigma) = const$. The latter equation may be rewritten as

$$\ddot{\sigma} \in [-C, C] - [K_m, K_M]\left(\frac{1}{2}\lambda\frac{|\dot{\sigma}|}{|\sigma|^{1/2}} + \alpha\,\text{sign}(\sigma)\right) \qquad (4.38)$$

This inclusion does not 'remember' anything about the original system. Then similarly to the proof of Theorem 4.1, with $\sigma > 0$, $\dot{\sigma} > 0$, the real trajectory is confined by the axes $\sigma = 0, \dot{\sigma} = 0$ and the trajectory of the equation $\ddot{\sigma} = -(K_m\alpha - C)$. Let σ_M be the intersection of this curve with axis $\dot{\sigma} = 0$. Obviously, $2(K_m\alpha - C)\sigma_M = \dot{\sigma}_0^2$ (Fig. 4.9). It is easy to see from Fig. 4.9 that

$$\sigma > 0, \quad \dot{\sigma} > 0, \quad \frac{1}{2}\lambda\frac{|\dot{\sigma}|}{|\sigma|^{1/2}} > \frac{C}{K_m} + \alpha \;\Rightarrow\; \ddot{\sigma} > 0$$

Thus, the majoring curve with $\sigma > 0$ is constructed from the following curves (Fig. 4.9):

$$\dot{\sigma}^2 = 2\,(K_m\alpha - C)\,(\sigma_M - \sigma) \text{ with } \dot{\sigma} > 0,\ \dot{\sigma}_0^2 = 2\,(K_m\alpha - C)\,\sigma_M$$

$$\sigma = \sigma_M \text{ with } 0 \geq \dot{\sigma} \geq -\frac{2}{\lambda}\left(\frac{C}{K_m} + \alpha\right)\sigma^{1/2}$$

$$\dot{\sigma} = \dot{\sigma}_M = -\frac{2}{\lambda}\left(\frac{C}{K_m} + \alpha\right)\sigma_M^{1/2} \text{ with } 0 \leq \sigma \leq \sigma_M$$

The condition $|\dot{\sigma}_M/\dot{\sigma}_0| < 1$ is sufficient for the algorithm convergence while $|u| < U_M$. That condition is rewritten as

$$\frac{2(K_m\alpha + C)^2}{\lambda^2 K_m^2(K_m\alpha - C)} < 1$$

Unfortunately, the latter inequality is still not sufficient, for this consideration does not include the possible 1-sliding mode keeping of $u = \pm U_M$. It is easy to see that such a mode is not possible with $\sigma\dot{\sigma} > 0$. Indeed, in that case $u\dot{\sigma}$ stays negative and does not allow any sign switching of $u - U_m$. On the other hand, from Eqs. (4.34), (4.35) and $|u| \leq U_M$, in such a sliding mode

$$K_m(1 - q)U_M \leq |\dot{\sigma}| = g\,|h/g + u| \leq K_M(1 + q)U_M$$

Thus, $\dot{\sigma}_0 \leq K_M(1 + q)U_M$, and the condition

$$\left|\frac{\dot{\sigma}_M}{\dot{\sigma}_0}\right| < \frac{K_m(1 - q)U_M}{K_M(1 + q)U_M} = \frac{K_m(1 - q)}{K_M(1 + q)}$$

is sufficient to avoid keeping $u = \pm U_M$ in sliding mode. The resulting condition above coincides with Eq. (4.37).

It is now required to prove the finite-time convergence. It is enough to consider only a sufficiently small vicinity of the origin, where $|u| < U_M$ is guaranteed. Consider an auxiliary variable $\xi = h(t, x) + g(t, x)u_1$. Obviously, $\xi = \dot{\sigma}$ at the moments when $\sigma = 0$, and $u_1 \rightarrow -h/g$ as $t \rightarrow \infty$. Thus, $\xi = g(h/g + u_1)$ tends to zero. Starting from the moment when $|u_1| < U_M$ holds, its derivative $\dot{\xi} = \dot{h} + \dot{g}u_1 - g\,\alpha\,\mathrm{sign}(x)$ satisfies the inequalities

$$0 < K_m\alpha - C \leq -\dot{\xi}\,\mathrm{sign}(\sigma) \leq K_M\alpha + C$$

As in the proof of Theorem 4.1, the total variation of ξ is equal to $\sum |\dot{\sigma}_i|$, is bounded by a geometric series, and therefore converges. The total convergence time $T \leq \sum |\dot{\sigma}_i|/(K_m\alpha - C)$ and the proof of the theorem is complete.   □

Note that the accuracy estimations formulated at the end of Sect. 4.2 remain valid for sufficiently small noises and/or sampling intervals. This robustness feature leads to the application of the controller in observation and identification. One of the most important applications is considered in the next Subsection.

4.3.2 First-Order Differentiator

The super-twisting controller is used for systems of relative degree 1. In other words it can be used instead of a standard 1-sliding-mode controller in order to avoid chattering. However for relative degree 2 systems a 2-sliding controller, like a twisting one, is needed to stabilize system (4.6) in finite time. In order to avoid the use of $\dot{\sigma}$ measurements, a differentiator (observer) is needed. Popular linear high-gain observers cannot fulfill this task because they only provide asymptotic stabilization at an equilibrium state. The differentiator needed here has to feature robust exact differentiation with finite-time convergence in the absence of the measurement noise.

Let the input signal $f(t)$ be a function defined on $[0, \infty)$ consisting of a bounded Lebesgue-measurable noise with unknown features and an unknown base signal $f_0(t)$ with the first derivative having a known global Lipschitz constant $L > 0$. The problem is to find real-time robust estimations of $f_0(t)$ and $\dot{f}_0(t)$ which are exact in the absence of measurement noise.

Consider the auxiliary system $\dot{z}_0 = v$, where v is a control input. Let $\sigma_0 = z_0 - f_0(t)$ and let the task be to keep $\sigma_0 = 0$ in a 2-sliding mode. In that case $\sigma_0 = \dot{\sigma}_0 = 0$, which means that $z_0 = f_0(t)$ and $\dot{f}_0 = v$. The system can be rewritten as

$$\dot{\sigma}_0 = -\dot{f}_0(t) + v, \quad |\ddot{f}_0| \le L$$

The function $\dot{f}_0$ can be not smooth, but its derivative $\ddot{f}_0$ exists almost everywhere due to the Lipschitz property of $\dot{f}_0$. A modification of the super-twisting controller

$$v = -\lambda_1 |\sigma_0|^{1/2} \text{sign}(\sigma_0) + z_1$$
$$\dot{z}_1 = -\lambda_2 \text{sign}(\sigma_0)$$

is applied here. The modification is needed, for neither $\dot{f}_0(t)$ nor v is bounded. The resulting form of the differentiator is

$$\dot{z}_0 = v = -\lambda_1 |z_0 - f(t)|^{1/2} \text{sign}(z_0 - f(t)) + z_1$$
$$\dot{z}_1 = -\lambda_0 \text{sign}(z_0 - f(t)) \tag{4.39}$$

where both v and z_1 can be taken as the differentiator outputs.

Theorem 4.6. *In the absence of noise for any $\lambda_0 > L$ for every sufficiently large λ_1, both v and z_1 converge in finite time to $\dot{f}_0(t)$, while z_0 converges to $f_0(t)$.*

The proof of the theorem is actually contained in the proof of Theorem 4.5. Sufficient crude convergence conditions are

$$\lambda_0 > L, \quad \frac{2(\lambda_0 + L)^2}{\lambda_1^2(\lambda_0 - L)} < 1 \tag{4.40}$$

Theorem 4.7. *Let the input noise satisfy the inequality $|f(t) - f_0(t)| \leq \varepsilon$. Then the following inequalities are established in finite time for some positive constants μ_1, μ_2, μ_3, depending exclusively on the parameters of the differentiator and L:*

$$|z_0 - f_0(t)| \leq \mu_1 \varepsilon, \quad |z_1 - \dot{f}_0(t)| \leq \mu_2 \varepsilon^{1/2}, \quad |v - \dot{f}_0(t)| \leq \mu_3 \varepsilon^{1/2}$$

Moreover, these asymptotics cannot be improved.

Sketch of the proof. Let $\sigma_0 = z_0 - f_0(t)$, $\sigma_1 = z_1 - \dot{f}_0(t)$, then

$$\dot{\sigma}_1 = -\ddot{f}_0(t) - \lambda_0 \text{sign}(\sigma_0) \in [-L, L] - \lambda_0 \text{sign}(\sigma_0),$$

and the differentiator equations in the absence of the input noise may be replaced by the inclusion

$$\begin{aligned}
\dot{\sigma}_0 &= -\lambda_1 |\sigma_0|^{1/2} \text{sign}(\sigma_0) + \sigma_1 \\
\dot{\sigma}_1 &\in -[\lambda_0 - L, \lambda_0 + L] \text{sign}(\sigma_0)
\end{aligned} \tag{4.41}$$

Its solutions converge to the origin $\sigma_0 = 0$, $\sigma_1 = 0$ in finite time. With $\varepsilon \neq 0$ inclusion (4.41) turns into

$$\begin{aligned}
\dot{\sigma}_0 &\in -\lambda_1 |\sigma_0 + [-\varepsilon, \varepsilon]|^{1/2} \text{sign}(\sigma_0 + [-\varepsilon, \varepsilon]) + \sigma_1 \\
\dot{\sigma}_1 &\in -[\lambda_0 - L, \lambda_0 + L] \text{sign}(\sigma_0 + [-\varepsilon, \varepsilon])
\end{aligned}$$

For small $\varepsilon = \varepsilon_0$, the trajectories are concentrated in a small set $\sigma_0 \leq \kappa_0$, $\sigma_1 \leq \kappa_1$ and stay there forever. Apply a combined transformation of coordinates, time, and parameters:

$$G_v \quad : (\sigma_0, \sigma_1, t, \varepsilon_0) \mapsto (v^2 \sigma_0, v\sigma_1, vt, v^2 \varepsilon_0)$$

Then it is easy to see that the trajectories of inclusion (4.40) are transferred into the trajectories of the same inclusion, but with different noise magnitude $\varepsilon = v^2 \varepsilon_0$. Now define $v = \sqrt{\varepsilon/\varepsilon_0}$ and get that the new attracting invariant set satisfies the inequalities $\sigma_0 \leq v^2 \kappa_0 = (\kappa_0/\varepsilon_0)\varepsilon$, $\xi \leq v\kappa_1 = (\kappa_1/\sqrt{\varepsilon_0})\varepsilon$. $\qquad \square$

Theorem 4.8. *Let parameters $\lambda_1 = \Lambda_1$, $\lambda_0 = \Lambda_0$ of the differentiator in Eqs. (4.39), (4.40) guarantee exact differentiation with $L = 1$. Then parameters $\lambda_1 = \Lambda_1 L^{1/2}$, $\lambda_0 = \Lambda_0 L$ are valid for any $L > 0$ and guarantee the accuracy level*

$$|z_0 - f_0(t)| \leq \mu_1 \varepsilon, \quad |z_1 - \dot{f}_0(t)| \leq \mu_2 L^{1/2} \varepsilon^{1/2}, \quad |v - \dot{f}_0(t)| \leq \mu_3 L^{1/2} \varepsilon^{1/2}$$

for some positive constants μ_1, μ_2, μ_3.

Proof. Denote $\tilde{f} = f/L$, then the following differentiator provides for the exact differentiation of $\tilde{f}(t)$:

$$\begin{aligned}
\dot{\tilde{z}}_0 &= -\Lambda_1 |\tilde{z}_0 - \tilde{f}(t)|^{1/2} \text{sign}(\tilde{z}_0 - \tilde{f}(t)) + \tilde{z}_1 \\
\dot{\tilde{z}}_1 &= -\Lambda_0 \text{sign}(\tilde{z}_1 - \tilde{f}(t))
\end{aligned}$$

By multiplying by L and defining $z_0 = L\tilde{z}_0$, $z_1 = L\tilde{z}_1$, the statement of the theorem is proven.   □

The parameter choices $\lambda_1 = 1.5L^{1/2}$, $\lambda_0 = 1.1L$ and $\lambda_1 = L^{1/2}$, $\lambda_0 = 2L$ are valid, even though they do not satisfy (4.40). The first one of these choices seems to be a good compromise providing a reasonably fast convergence and high accuracy.

Remark 4.6. Note that while v is noisy in the presence of the input noise, z_1 is a Lipschitzian signal, but small input noises lead to a small phase delay of z_1.

Example 4.1. Suppose that $t_0 = 0$, the initial values of the internal variable $z_0(0)$ and the "measured" input signal $f(0)$ coincide, and the initial value of the output signal z_1 is zero. The simulation was carried out using the Euler method with measurement and integration steps equaling 10^{-4}.

The proposed differentiator (4.39), (4.40) was compared with a simple linear differentiator described by the transfer function $\frac{s}{(0.1s+1)^2}$. Such a differentiator is actually a combination of the ideal differentiator and a low-pass filter. The differentiator parameters were chosen as $\lambda_1 = 6$, $\lambda_0 = 8$. The output signals $f(t) = \sin(t) + 5t$, $f(t) = \sin(t) + 5t + 0.01\cos(10t)$, and $f(t) = \sin(t) + 5t + 0.001\cos(30t)$ together with the ideal derivatives $\dot{f}_0(t)$ are shown in Fig. 4.10. The linear differentiator is seen not to differentiate exactly. At the same time it is highly insensitive to any signals with frequency above 30. The proposed differentiator handles properly any input signal f with $|\ddot{f}| \leq 7$ regardless the signal spectrum.

4.4 Differentiator-Based Output-Feedback 2-SM Control

We are now able to construct a robust output-feedback 2-sliding mode (2-SM) controller for the system with relative degree 2. Recall that the system is described by the equation and conditions

$$\dot{x} = a(t, x) + b(t, x)u$$

$$0 < K_m \leq \frac{\partial}{\partial u}\ddot{\sigma} \leq K_M, \ |\ddot{\sigma}| \leq C$$

The control is to solve the stabilization problem in finite time, only using measurements of σ. The robust exact differentiation of σ is always possible due to the boundedness of $\ddot{\sigma} \in [-C, C] + [K_m, K_M u]$ with bounded control u. Combining any above 2-sliding controller $u = -U(\sigma, \dot{\sigma})$ and the differentiator achieves

$$\begin{aligned}
&u = -U(\sigma, z_1) \\
&\dot{z} = -\lambda_1 |z - \sigma|^{1/2}\text{sign}(z - \sigma) + z_1, \\
&\dot{z}_1 = -\lambda_2 \text{sign}(z - \sigma), \ \ \lambda_1 = 1.5L^{1/2}, \ \lambda_2 = 1.1L
\end{aligned} \qquad (4.42)$$

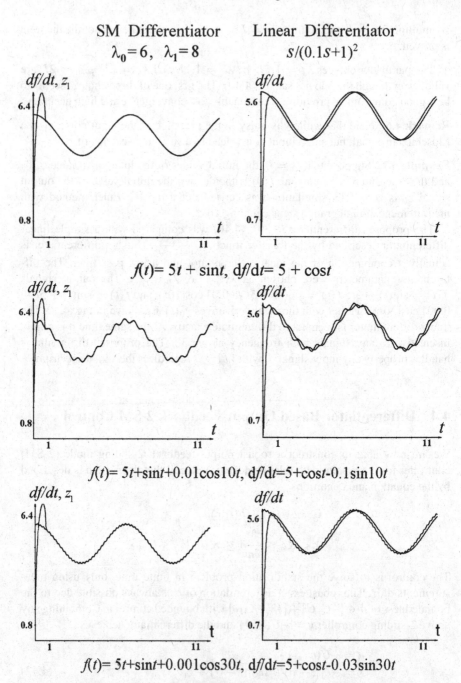

Fig. 4.10 Comparison of the 2-sliding mode-based differentiator and a linear filter

Any value $L > C + K_M \sup |U|$ can be used here. As a consequence of Theorem 4.1 and Theorem 4.6 the controller provides exact stabilization and finite-time convergence. It can be proven that in the presence of a bounded Lebesgue-measurable noise with the maximal magnitude ε, the steady-state accuracies $\sup |\sigma|$ and $\sup |\dot\sigma|$ are proportional to ε and $\sqrt{\varepsilon}$, respectively. Note that in practice the differentiator parameter L is often taken conservatively large to provide for the better closed-loop performance in the presence of noises.

Example 4.2. Consider the dynamic system

$$\ddot{x} = \sin(14.12t) + (1.5 + 0.5\cos(21t))u$$

$$\sigma = x,$$

with $C = 1$, $K_m = 1$, $K_M = 2$ and the output-feedback control

$$u = -5\,\mathrm{sign}\,z_0 - 3\,\mathrm{sign}\,z_1$$
$$\dot{z}_0 = -7|z_0 - x|^{1/2}\mathrm{sign}(z_0 - x) + z_1$$
$$\dot{z}_1 = -18\mathrm{sign}(z_0 - x)$$

At the time instant $t = 0$ the initial values $z_0(0) = x(0)$, $z_1 = 0$ were taken. The trajectory in the plane $x\dot{x}$ and the mutual graph of x, $\dot{x}$, and z_1 are shown in Fig. 4.11a,b, respectively. The graph of z_0 is not shown, since one cannot distinguish it from x. Convergence in the presence of a high-frequency noise with magnitude 0.01 is shown in Fig. 4.11c,d, respectively. The resulting steady-state accuracies are $|x| \leq 0.041$ and $|\dot{x}| \leq 0.79$.

4.5 Chattering Attenuation

A problem with conventional (first-order) sliding mode control is attenuation of the chattering effect. However 2-sliding mode control provides effective tools for the reduction or even practical elimination of the chattering, without compromising the benefits of the standard sliding mode. Recalling the problem statement from Sect. 4.3.1, let the relative degree of the system (4.5) be 1, and instead of Eqs. (4.8), (4.10) assume

$$\dot\sigma = h(t,x) + g(t,x)u, \ 0 < K_m \leq g \leq K_M, \ |h| \leq C \qquad (4.43)$$

where the functions g, h are some unknown smooth functions. Let also the control $u = -k\,\mathrm{sign}\,(\sigma)$ solve the problem of establishing and keeping $\sigma \equiv 0$. In particular, assume that

$$kK_m - C > 0 \qquad (4.44)$$

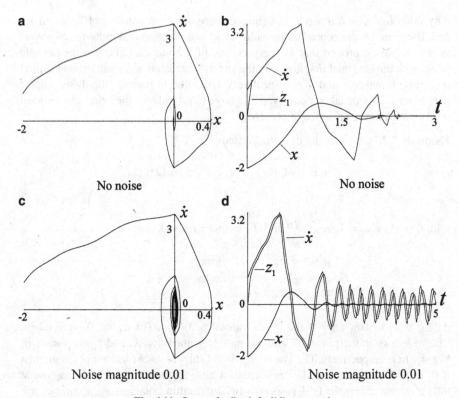

Fig. 4.11   Output-feedback 2-sliding control

Consider $\dot{u}$ as a new virtual control, in order to overcome the chattering. Differentiating (4.43) yields

$$\ddot{\sigma} = h_1(t, x, u) + g(t, x)\dot{u}$$

$$h_1 = h'_t + h'_x (a + bu) + (g'_t + g'_x (a + bu)) u$$

Assume that the function $h_1(t, x, u)$ is bounded so that

$$\sup_{|u| \leq k_1} |h_1(t, x, u)| = C_1 \tag{4.45}$$

Any previously discussed controller $\dot{u} = U(\alpha, \sigma, \dot{\sigma})$ can be used here in order to overcome the chattering and improve the sliding accuracy of the standard sliding mode. Indeed, define

$$\dot{u} = \begin{cases} -u, |u| > k \\ U(\alpha, \sigma, \dot{\sigma}), |u| \leq k \end{cases} \tag{4.46}$$

Theorem 4.9. *Let U be any one of the 2-sliding controllers considered in Sect. 4.2, and suppose the controller parameters are properly chosen in accordance with the corresponding convergence conditions. Then for a sufficiently large parameter α, the controller (4.46) guarantees the establishment of the finite-time stable 2-sliding mode on $\sigma = \dot{\sigma} \equiv 0$.*

Proof. It follows from Eqs. (4.43), (4.44) that the inequality $|\dot{\sigma}| < k K_m - C$ implies $|u| \leq k$. Thus, within the set $|\dot{\sigma}| < k K_m - C$, the system is driven by the controller $\dot{u} = U(\alpha, \sigma, \dot{\sigma})$. Controller (4.46) keeps $|u| \leq k$, and on certain time intervals $u \equiv k$ or $u \equiv -k$ is kept in a 1-sliding mode and the proof of the theorem is complete. $\square$

Lemma 4.1. *Any trajectory of the system (4.43), (4.46) hits in finite time the manifold $\sigma = 0$ or enters the set $\sigma \dot{\sigma} < 0$, $|u| \leq k$.*

Proof. Indeed, suppose that σ does not change its sign. Obviously, the inequality $|u| \leq k$ is established in finite time. If the condition $\sigma \dot{\sigma} < 0$ is attained, the statement of the lemma is true. Suppose that $\sigma \dot{\sigma} \geq 0$ holds, then, according to (4.46), u moves towards $u = -k \operatorname{sign}(\sigma)$ with $|\dot{u}| \geq \min(\alpha, k)$, both if $|u| > k$. or $|u| \leq k$. The conclusion that $u = -k \operatorname{sign}(\sigma)$ can be established only with $\sigma \dot{\sigma} < 0$ proves the lemma. $\square$

Lemma 4.2. *With sufficiently large α any trajectory of the system (4.43), (4.46) hits in finite time the manifold $\sigma = 0$.*

Proof. Denote by S the set defined by the inequalities $|\dot{\sigma}| < k K_m - C, \sigma \dot{\sigma} < 0$. There is a specific set Θ for each controller, adjacent to the axis $\sigma = 0$ and lying in the strip S, such that any trajectory entering it either converges in finite time to $\sigma = \dot{\sigma} = 0$ or hits the axis $\sigma = 0$; also no trajectory can enter S outside of Θ. For example, Θ is defined by the inequalities $\left(\dot{\sigma} + \lambda |\sigma|^{1/2} \operatorname{sign}(\sigma) \right) \sigma \leq 0$ and $|\dot{\sigma}| < k K_m - C$ for the controller in Eq. (4.21). Any trajectory starting in S either leaves it in finite time or enters Θ. Thus, there are two options: at some moment on a trajectory that stays out of S, which means that $|\dot{\sigma}| \geq k K_m - C, \sigma \dot{\sigma} < 0$, or it enters Θ. In both cases the trajectory hits $\sigma = 0$. The lemma is proven. $\square$

The following lemma is obviously true for any convergent 2-sliding controller.

Lemma 4.3. *There is a vicinity Ω of the origin within the strip $|\dot{\sigma}| < k K_m - C$, which is invariant with respect to the controller $\dot{u} = U(\alpha, \sigma, \dot{\sigma})$.*

Proof. Consider the auxiliary problem when Eq. (4.45) holds independently of the control value and the corresponding differential inclusion. Since all trajectories starting in a closed disk centered at the origin converge to the origin in finite time, the set, which comprises these transient trajectory segments, is an invariant compact for the controller $\dot{u} = U(\alpha, \sigma, \dot{\sigma})$.

All the proposed controllers produce the closed system (4.11) which is invariant with respect to the transformation

$$G_\kappa : (t, \sigma, \dot{\sigma}) \longmapsto \left(\kappa t, \kappa^2 \sigma, \kappa \dot{\sigma} \right)$$

Applying now this transformation, the set can be retracted into the strip $|\dot\sigma| < kK_m - C$, where Eq. (4.45) is really kept, and the proof of the lemma is complete. $\square$

Lemma 4.4. *For a sufficiently large α, any trajectory starting on the manifold $\sigma = 0$ with $|u| \le k$ enters the invariant set Ω.*

Proof. Any trajectory starting with $\sigma = 0$ and $\dot\sigma \ne 0$ inevitably enters the region $\sigma\dot\sigma > 0$, $|u| < k$. Within this region $\dot u = -\alpha\,\mathrm{sign}\,(\sigma)$ holds. Hence, the control u moves towards the value $-k\,\mathrm{sign}\,(\sigma)$, and on the way the trajectory hits the set $\dot\sigma = 0$, which still features $|u| < k$. From Eq. (4.43), $|u| \le k$ implies the global bound $|\dot\sigma| \le kK_M + C$. That restriction is true also at the initial point on the axis $\sigma = 0$. Simple calculations show that the inequality $|\sigma| \le \frac{1}{2}\,(kK_M + C)^2 / (\alpha K_1 - C_1)$ holds at the moment when $\dot\sigma$ vanishes. With sufficiently large α that point inevitably belongs to Ω.

Once the trajectory enters Ω, it continues to converge to the 2-sliding mode according to the corresponding 2-sliding dynamics considered in Sect. 4.2. This proves convergence to the 2-sliding mode. In the presence of small noises and sampling intervals, the resulting motion will take place in a small vicinity of the 2-sliding mode $\sigma = \dot\sigma = 0$. Thus, if this motion does not leave Ω, the studied 2-sliding dynamics is still in charge, and the corresponding accuracy estimations remain true. The proof of the lemma is now complete. $\square$

4.6 Case Study: Pendulum Control

Consider a variable-length pendulum control problem where all the motions are restricted to some vertical plane. A load of some known mass m is moving along the pendulum rod (Fig. 4.12).

Its distance from the origin O equals $R(t)$ and is not measured. There is no friction. An engine transmits a torque w that is considered as the control input. The task is to force the angular coordinate x of the rod to follow some profile $x_c(t)$ given in current time. The system is described by the differential equation

$$\ddot x = -2\frac{\dot R}{R}\dot x - g\frac{1}{R}\sin(x) + \frac{1}{mR^2}w \qquad (4.47)$$

where $g = 9.81 m/s^2$ is the gravitational constant and the mass m is taken as $m = 1kg$. Let $0 < R_m \le R \le R_M$; also assume that $\dot R, \ddot R, \dot x_c, \ddot x_c$ are bounded and $\sigma = x - x_c$ is available. The initial conditions are $x(0) = \dot x(0) = 0$. The following functions R and x_c are considered in the simulation:

$$R = 1 + 0.25\sin(4t) + 0.5\cos(t)$$

$$x_c = 0.5\sin(0.5t) + 0.5\cos(t)$$

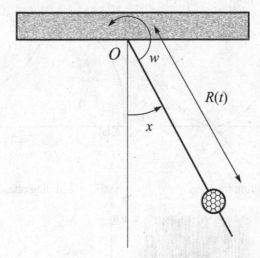

Fig. 4.12 Variable-length pendulum

4.6.1 Discontinuous Control

The relative degree of the system is 2. Here condition (4.10) holds only locally, since $\ddot{\sigma}|_{u=0}$ depends on $\dot{x}$ and is not uniformly bounded. Thus, the controllers are effective only in a bounded vicinity of the origin $x = \dot{x} = w = 0$. The appropriate discontinuous controller, Eq. (4.42) based on a quasi-continuous controller, has the form

$$w = -10 \frac{z_1 + |\sigma|^{1/2}\text{sign}(\sigma)}{|z_1| + |\sigma|^{1/2}}, \quad \sigma = x - x_c \tag{4.48}$$

$$\dot{z}_0 = -10.61 |z_0 - \sigma|^{1/2} \text{sign}(z_0 - \sigma) + z_1 \tag{4.49}$$

$$\dot{z}_1 = -55 \, \text{sign}(z_0 - \sigma) \tag{4.50}$$

where z_0, z_1 are real-time estimations of $\sigma, \dot{\sigma}$, respectively. The differentiator (4.49), (4.50) is exact for the input signal σ, with a second time derivative not exceeding 50 in absolute value.

The initial conditions $x(0) = \dot{x}(0) = 0$ have been taken as $z_0(0) = x(0) - x_c(0) = -0.5$, $z_1(0) = 0$. The sampling time step τ and the integration step have been chosen as 0.0001.

2-sliding tracking performance and trajectory tracking in the absence of noise, are shown in Fig. 4.13a, b, respectively. The corresponding achieved accuracies are $|\sigma| = |x - x_c| \leq 5.4 \times 10^{-6}$, $|\dot{x} - \dot{x}_c| \leq 1.0 \times 10^{-2}$ with $\tau = 0.0001$. The control signal associated with Eq. (4.48) is shown in Fig. 4.13c. It is seen from the graph that the control remains continuous until a 2-sliding mode $\sigma = \dot{\sigma} = 0$ takes place. The differentiator convergence is demonstrated in Fig. 4.13d.

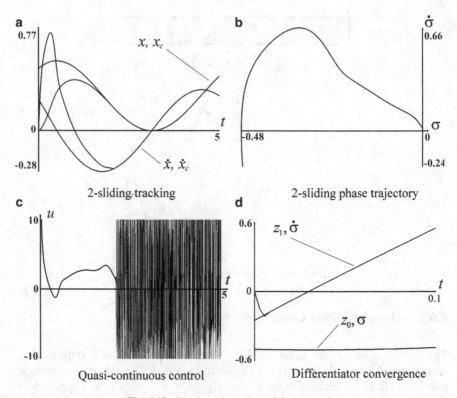

Fig. 4.13 Quasi-continuous pendulum control

The tracking results obtained from using Eqs. (4.48), (4.49), (4.50) and the differentiator performance in the presence of noise with the magnitude 0.01 are demonstrated in Fig. 4.14a, b, respectively. The tracking accuracy is $|\sigma| = |x - x_c| \leq 0.036$ (the noise is a periodic non-smooth function with nonzero average). The performance does not significantly change, when the frequency of the noise varies from $101/s$ to $1000001/s$.

Any other 2-sliding controller could also be implemented. Consider a twisting controller

$$w = -10\,\text{sign}(z_0) - 5\,\text{sign}(z_1) \tag{4.51}$$

The trajectory of the twisting controller (4.49)–(4.51) in the coordinates $x - x_c$ and $\dot{x} - \dot{x}_c$, in the absence of noise, is shown in Fig. 4.15b. The corresponding accuracy is $|x - x_c| \leq 6.7 \times 10^{-6}$, $|\dot{x} - \dot{x}_c| \leq 0.01$.

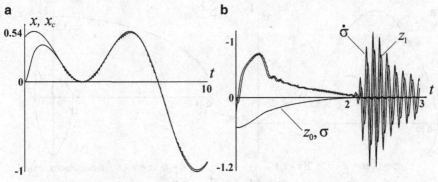

Fig. 4.14 Performance of the quasi-continuous controller with noisy measurements

4.6.2 Chattering Attenuation

In the case when torque chattering is unacceptable, $u = \dot{w}$ is considered as a new control. Define

$$\sigma = (\dot{x} - \dot{x}_c) + 2(x - x_c)$$

Again, the relative degree of the system with respect to the new input w is equal 2. Also condition (4.10) holds only locally, and thus the controllers are effective only in a bounded vicinity of the origin $x = \dot{x} = w = 0$. Their global application requires the standard method described in Sect. 4.5, which is not implemented here for simplicity.

The applied output-feedback controller is of the form Eq. (4.42) and is based on the twisting controller (4.13):

$$\dot{w} = u = -15 \operatorname{sign}(z_0) - 10 \operatorname{sign}(z_1) \tag{4.52}$$

$$\dot{z}_0 = -35 |z_0 - \sigma|^{1/2} \operatorname{sign}(z_0 - \sigma) + z_1 \tag{4.53}$$

$$\dot{z}_1 = -70 \operatorname{sign}(z_0 - \sigma), \ \sigma = (\dot{x} - \dot{x}_c) + 2(x - x_c) \tag{4.54}$$

Here the angular velocity $\dot{x}$ is assumed to be directly measured.[1]

The initial values $x(0) = \dot{x}(0) = 0$ are taken in the simulations. The value $w(0) = 0$ is taken for controller (4.52)–(4.54), and the sampling step $\tau = 0.0001$. The trajectory in the coordinates $x - x_c$ and $\dot{x} - \dot{x}_c$, in the absence of noise, is shown in Fig. 4.15a. The accuracy $|x - x_c| \leq 1.6 \times 10^{-6}$, $|\dot{x} - \dot{x}_c| \leq 1.8 \times 10^{-5}$ has been achieved. The trajectories in the presence of noise with magnitude 0.02 in

---

[1]Otherwise, a 3-sliding controller can be applied together with a second-order differentiator (see chap. 6) producing both $\dot{x} - \dot{x}_c$ and $\ddot{x} - \ddot{x}_c$.

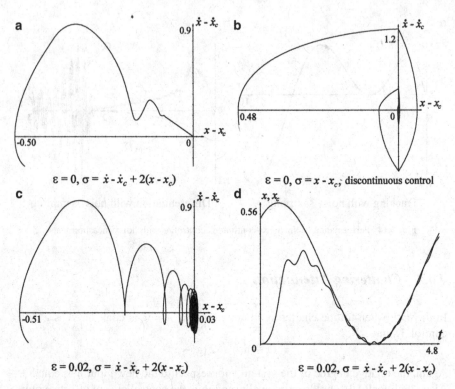

Fig. 4.15 Pendulum output-feedback twisting control, a,c,d: $\sigma = (\dot{x} - \dot{x}_c) + 2(x - x_c)$, b: $\sigma = x - x_c$

the σ-measurements are shown in Fig. 4.15c, and the tracking results are shown in Fig. 4.15d. The tracking accuracy $|x - x_c| \leq 0.018$, $|\dot{x} - \dot{x}_c| \leq 0.16$ is achieved. The performance does not differ when the frequency of the noise changes from $101/s$ to $100001/s$.

4.7 Variable-Gain Super-Twisting Control

An extension of the standard super-twisting algorithm for the conventional two-step SM control design procedure that provides exact compensation of smooth uncertainties/disturbances bounded together with their derivatives by *known functions* is considered in this section.

4.7.1 Problem Statement

Consider a linear time-invariant system (LTI) with a matching nonlinear perturbation

$$\dot{x} = Ax + B(u + \xi(x,t)) \tag{4.55}$$

where $x \in \mathbb{R}^n$ is the state vector, $u \in \mathbb{R}^m$ is the control input, the A and B are constant matrices of appropriate dimensions, and ξ is an absolutely continuous uncertainty/disturbance in the system (4.55). As in Chap. 2 the system in Eq. (4.55) is first transformed into regular form. The following properties are assumed:

(A1) Rank $B = m$.
(A2) The pair (A, B) is controllable.
(A3) The function ξ together with its gradient is bounded by known continuous functions almost everywhere.

Under assumptions (A1) and (A2), after the linear state transformation,

$$\begin{pmatrix} z_1 \\ z_2 \end{pmatrix} = Tx, \ T = \begin{bmatrix} B^\perp \\ B^+ \end{bmatrix}, \ B^+ = (B^T B)^{-1} B^T, \ B^\perp B = 0 \tag{4.56}$$

system (4.55) has the regular form

$$\begin{aligned} \dot{z}_1 &= A_{11} z_1 + A_{12} z_2 \\ \dot{z}_2 &= A_{21} z_1 + A_{22} z_2 + u + \tilde{\xi}(z_1, z_2, t) \end{aligned} \tag{4.57}$$

where $z_1 \in \mathbb{R}^{n-m}$ and $z_2 \in \mathbb{R}^m$. The structure of the system allows us, without loss of generality, to restrict ourselves to the single input case $(m = 1)$. The results are easily extended to the multi-input case. The sliding surface is chosen to have the form

$$\sigma = z_2 - K z_1 = 0 \tag{4.58}$$

As a consequence, when the motion is restricted to the manifold, the reduced-order model

$$\dot{z}_1 = (A_{11} + A_{12} K) z_1 \tag{4.59}$$

has the required performance. Since the pair (A_{11}, A_{12}) is controllable, the matrix K can be designed using any linear control design method for system (4.59); see, for example, Chap. 2.
Using (z_1, σ) as state variables and applying the controller

$$u = -(A_{21} + A_{22} K - K(A_{11} + A_{12} K)) z_1 - (A_{22} - K A_{12}) \sigma + v \tag{4.60}$$

system (4.57) takes the form

$$\dot{z}_1 = (A_{11} + A_{12} K) z_1 + A_{12} \sigma \tag{4.61}$$

$$\dot{\sigma} = v + \tilde{\xi}(z_1, \sigma + K z_1, t) \tag{4.62}$$

When the perturbation is bounded by a known function $\varrho(x)$

$$|\xi(x,t)| \le \varrho(x) \tag{4.63}$$

a (first-order) sliding mode can be enforced by a variable-gain controller

$$v = -(\varrho(x) + \varrho_0)\,\text{sign}(\sigma) \tag{4.64}$$

with $\varrho_0 > 0$. Alternatively, unit vector controllers can also be used for this purpose (see Chap. 2). The main disadvantage of these controllers is that they produce *chattering*, which grows with the uncertainty bound $\varrho(x)$.

Here a Lyapunov-based design is employed.

4.7.2   The Variable-Gain Super-Twisting Algorithm

The variable-gain super-twisting algorithm (VGSTA) proposed here is given by

$$v = -k_1(t,x)\,\phi_1(\sigma) - \int_0^t k_2(t,x)\,\phi_2(\sigma)\,dt \tag{4.65}$$

where

$$\phi_1(\sigma) = |\sigma|^{\frac{1}{2}}\,\text{sign}(\sigma) + k_3\sigma$$
$$\phi_2(\sigma) = \tfrac{1}{2}\,\text{sign}(\sigma) + \tfrac{3}{2}k_3|\sigma|^{\frac{1}{2}}\,\text{sign}(\sigma) + k_3^2\sigma, \quad k_3 > 0$$

When $k_3 = 0$ and the gains k_1 and k_2 are constant, we recover the standard super-twisting algorithm. The additional term $k_3 > 0$ allows us to deal with perturbations growing linearly in s, i.e., outside of the sliding surface, and the variable gains k_1 and k_2 make it possible to render the sliding surface insensitive to perturbations growing with bounds given by known functions. Note that the uncertainty/disturbance can always be written as

$$\tilde{\xi}(z_1,\sigma + Kz_1,t) = \underbrace{\left[\tilde{\xi}(z_1,\sigma + Kz_1,t) - \tilde{\xi}(z_1,Kz_1,t)\right]}_{g_1(z_1,\sigma,t)} + \underbrace{\tilde{\xi}(z_1,Kz_1,t)}_{g_2(z_1,t)}$$

where $g_1(z_1,\sigma,t) = 0$, when $\sigma = 0$. It follows from assumption (A3) that the uncertainty/disturbance $\xi(x,t)$ is bounded almost everywhere:

$$\begin{aligned} |g_1(z_1,\sigma,t)| &\le \varrho_1(t,x)\,|\phi_1(\sigma)| \\ \left|\tfrac{d}{dt}g_2(z_1,t)\right| &\le \varrho_2(t,x)\,|\phi_2(\sigma)| \end{aligned} \tag{4.66}$$

where $\varrho_1(t,x) \ge 0$, $\varrho_2(t,x) \ge 0$ are known continuous functions.

System (4.62) driven by the VGSTA (4.65) can be written as

$$
\begin{aligned}
\dot{z}_1 &= (A_{11} + A_{12}K)z_1 + A_{12}\sigma \\
\dot{\sigma} &= -k_1(t,x)\phi_1(\sigma) + z + g_1(z_1,\sigma,t) \\
\dot{z}_0 &= -k_2(t,x)\phi_2(\sigma) + \frac{d}{dt}g_2(z_1,t)
\end{aligned}
\tag{4.67}
$$

The algorithm is presented in the following theorem:

Theorem 4.10. *Suppose that for some known continuous functions $\varrho_1(t,x) \geq 0$, $\varrho_2(t,x) \geq 0$ the inequalities (4.66) are satisfied. Then for any initial condition $(z_1(0), \sigma(0), z_0(0))$ the sliding surface $\sigma = 0$ will be reached in finite time if the variable gains are selected as*

$$
\begin{aligned}
k_1(t,x) &= \delta + \tfrac{1}{\beta}\left(\tfrac{1}{4\epsilon}(2\epsilon\varrho_1 + \varrho_2)^2 + +2\epsilon\varrho_2 + \epsilon + (2\epsilon + \varrho_1)(\beta + 4\epsilon^2)\right) \\
k_2(t,x) &= \beta + 4\epsilon^2 + 2\epsilon k_1(t,x)
\end{aligned}
\tag{4.68}
$$

where $\beta > 0$, $\epsilon > 0$, $\delta > 0$ are arbitrary positive constants. The reaching time of the sliding surface can be estimated by

$$
T = \frac{2}{\gamma_2}\ln\left(\frac{\gamma_2}{\gamma_1}V^{\frac{1}{2}}(\sigma(0), z_0(0)) + 1\right)
\tag{4.69}
$$

where $V(\sigma, z_0) = \zeta^T P \zeta$, with $\zeta^T = \left[|\sigma|^{\frac{1}{2}}\operatorname{sign}(\sigma) + k_3\sigma, \quad z_0\right]$ and

$$
\gamma_1 = \frac{\epsilon\lambda_{\min}^{\frac{1}{2}}\{P\}}{\lambda_{\max}\{P\}}, \quad \gamma_2 = \frac{2\epsilon k_3}{\lambda_{\max}\{P\}}
\tag{4.70}
$$

Proof. We will show that the quadratic form

$$
V(\sigma, z_0) = \zeta^T P \zeta
\tag{4.71}
$$

where

$$
\zeta^T = \left[|\sigma|^{\frac{1}{2}}\operatorname{sign}(\sigma) + k_3\sigma, \quad z_0\right]
\tag{4.72}
$$

and

$$
P = \begin{bmatrix} p_1 & p_3 \\ p_3 & p_2 \end{bmatrix} = \begin{bmatrix} \beta + 4\epsilon^2, & -2\epsilon \\ -2\epsilon & 1 \end{bmatrix}
\tag{4.73}
$$

with arbitrary positive constants $\beta > 0$, $\epsilon > 0$, is a Lyapunov function for the subsystem (σ, z_0) of Eq. (4.67), showing finite-time convergence. Function (4.71) is positive definite, everywhere continuous, and differentiable everywhere except on the set $S = \{(\sigma, z_0) \in \mathbb{R}^2 \mid \sigma = 0\}$. The inequalities (4.66) can be rewritten as $g_1(z_1, \sigma, t) = \alpha_1(t,x)\phi_1(\sigma)$ and $\frac{d}{dt}g_2(z_1,t) = \alpha_2(t,x)\phi_2(\sigma)$ for some functions $|\alpha_1(t,x)| \le \varrho_1(t,x)$ and $|\alpha_2(t,x)| \le \varrho_2(t,x)$. Using these functions and noting that $\phi_2(\sigma) = \phi'_1(\sigma)\phi_1(\sigma)$ one can show that

$$\dot{\zeta} = \begin{bmatrix} \phi'_1(\sigma)\{-k_1(t,x)\phi_1(\sigma) + z_0 + g_1(x,t)\} \\ -k_2(t,x)\phi_2(\sigma) + \frac{d}{dt}g_2(x,t) \end{bmatrix}$$

$$= \phi'_1(\sigma) \begin{bmatrix} -(k_1(t,x) - \alpha_1(t,x)), & 1 \\ -(k_2(t,x) - \alpha_2(t,x)) & 0 \end{bmatrix} \zeta = \phi'_1(\sigma)\mathcal{A}(t,x)\zeta$$

for every point in $\mathbb{R}^2 \backslash \mathcal{S}$, where this derivative exists. Similarly one can calculate the derivative of $V(x)$ on the same set as

$$\dot{V}(\sigma, z_0) = \phi'_1(\sigma)\zeta^T \left(\mathcal{A}^T(t,x)P + P\mathcal{A}(t,x)\right)\zeta$$

$$= -\phi'_1(\sigma)\zeta^T Q(t,x)\zeta$$

where

$$Q(t,x) = \begin{bmatrix} 2(k_1(t,x) - \alpha_1)p_1 + 2(k_2(t,x) - \alpha_2)p_3 & \bigstar \\ (k_1(t,x) - \alpha_1)p_3 + (k_2(t,x) - \alpha_2)p_2 - p_1, & -2p_3 \end{bmatrix}$$

Selecting P as in Eq. (4.73) and the gains as in Eq. (4.68), we have

$$Q - 2\epsilon I = \begin{bmatrix} 2\beta k_1 + 4\epsilon(2\epsilon k_1 - k_2) - 2(\beta + 4\epsilon^2)\alpha_1 + 4\epsilon\alpha_2 - 2\epsilon & \bigstar \\ k_2 - 2\epsilon k_1 - (\beta + 4\epsilon^2) + 2\epsilon\alpha_1 - \alpha_2, & 2\epsilon \end{bmatrix}$$

$$= \begin{bmatrix} 2\beta k_1 - (\beta + 4\epsilon^2)(4\epsilon + 2\alpha_1) + 4\epsilon\alpha_2 - 2\epsilon & \bigstar \\ 2\epsilon\alpha_1 - \alpha_2, & 2\epsilon \end{bmatrix}$$

that is positive definite for every value of (t,x). This shows that

$$\dot{V} = -\phi'_1(\sigma)\zeta^T Q(t,x)\zeta \le -2\epsilon\phi'_1(\sigma)\zeta^T\zeta = -2\epsilon\left(\frac{1}{2|\sigma|^{\frac{1}{2}}} + k_3\right)\zeta^T\zeta$$

Since $\lambda_{\min}\{P\}\|\zeta\|_2^2 \le \zeta^T P\zeta \le \lambda_{\max}\{P\}\|\zeta\|_2^2$, where

$$\|\zeta\|_2^2 = \zeta_1^2 + \zeta_2^2 = |\sigma| + 2k_3|\sigma|^{\frac{3}{2}} + k_3^2\sigma^2 + z_0^2$$

is the Euclidean norm of ζ, and

$$|\zeta_1| \le \|\zeta\|_2 \le \frac{V^{\frac{1}{2}}(\sigma, z_0)}{\lambda_{\min}^{\frac{1}{2}}\{P\}}$$

we can conclude that

$$\dot{V} \leq -\gamma_1 V^{\frac{1}{2}} (\sigma, z_0) - \gamma_2 V (\sigma, z_0)$$

$$\gamma_1 = \frac{\epsilon \lambda_{\min}^{\frac{1}{2}} \{P\}}{\lambda_{\max} \{P\}}, \quad \gamma_2 = \frac{2\epsilon k_3}{\lambda_{\max} \{P\}} \tag{4.74}$$

Note that the trajectories cannot stay on the set $\mathcal{S} = \{(\sigma, z_0) \in \mathbb{R}^2 \mid \sigma = 0\}$. This means that V is a continuously decreasing function and we can conclude that the equilibrium point $(\sigma, z_0) = 0$ is reached in finite time from every initial condition.[2]

Since the solution of the differential equation

$$\dot{v} = -\gamma_1 v^{\frac{1}{2}} - \gamma_2 v, \quad v(0) \geq 0$$

is given by

$$v(t) = \exp\left(-\gamma_2 t\right) \left[v(0)^{\frac{1}{2}} + \frac{\gamma_1}{\gamma_2} \left(1 - \exp\left(\frac{\gamma_2}{2} t\right) \right) \right]^2$$

it follows that $(\sigma(t), z_0(t))$ converges to zero in finite time and reaches that value at most after a time given by Eq. (4.69). This concludes the proof of Theorem 4.10. $\qquad\qquad\qquad\qquad\qquad\qquad\qquad\qquad\qquad\qquad\qquad\qquad \square$

Remark 4.7. Theorem 4.10 proposes a methodology to design a sliding mode controller ensuring a sliding motion on the surface (4.58) substituting the discontinuous control law (4.64) by an absolutely continuous VGSTA (4.65). In this case the chattering level can be substantially reduced.

When $\rho(x) = const$, first-order sliding mode controllers (4.64) are able to compensate the bounded perturbations $\xi(x(t), t)$ measurable along the system trajectories. On the other hand the super-twisting algorithm with constant gains k_1 and k_2 is able to compensate for the Lipschitz continuous perturbations $\xi(x, t)$ along the system trajectories, but their absolute value cannot grow faster than a linear function of t, nor faster than linear with respect to $|\sigma(t)|^{\frac{1}{2}}$ along the system trajectories. Theorem 4.10 extends the VGSTA design for the class of perturbations (4.66).

---

[2]For details see Zubov's stability theorem [196].

4.8   Case Study: The Mass–Spring–Damper System

4.8.1   Model Description

The mass–spring–damper (MSD) system consists of two masses, three springs, one damper, and a DC motor in the configuration shown in Fig. 4.16. The system is the Educational Control Products (ECP) model 210a.

The dynamics of the system are given by the following set of ordinary differential equations:

$$m_2 \ddot{\chi}_2 + (\kappa_3 + \kappa_2)\chi_2 + c_1 \dot{\chi}_2 - \kappa_2 \chi_1 = 0 \tag{4.75}$$

$$m_1 \ddot{\chi}_1 + (\kappa_1 + \kappa_2)\chi_1 - \kappa_2 \chi_2 = F \tag{4.76}$$

where $\chi_1, \dot{\chi}_1, \ddot{\chi}_1, \chi_2, \dot{\chi}_2, \ddot{\chi}_2$ are the position, velocity, and acceleration of the masses 1 and 2, respectively. The term F is the force that the DC motor inputs into mass 1. The state vector is selected as $x_1 = \chi_1$, $x_2 = \dot{\chi}_1$, $x_3 = \chi_2$, and $x_4 = \dot{\chi}_2$, and the input $u = F$. The state space representation is

$$\dot{x}_1 = x_2 \tag{4.77}$$

$$\dot{x}_2 = -\frac{\kappa_1}{m_1}x_1 - \frac{\kappa_2}{m_1}x_1 + \frac{\kappa_2}{m_1}x_3 + \frac{1}{m_1}u \tag{4.78}$$

$$\dot{x}_3 = x_4 \tag{4.79}$$

$$\dot{x}_4 = -\frac{(\kappa_3 + \kappa_2)}{m_2}x_3 - \frac{c_1}{m_2}x_4 + \frac{\kappa_2}{m_2}x_1 \tag{4.80}$$

The nominal values are shown in Table 4.1.

It is possible to measure the positions x_1, x_3 through the encoders that are coupled to mass 1 and mass 2 respectively.

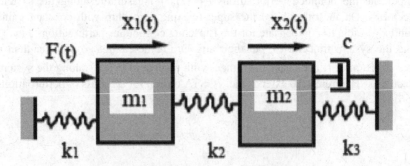

Fig. 4.16 The mass–spring–damper (MSD) system

Table 4.1 Model Nominal Values

Name	m_1	m_2	κ_1	κ_2	κ_3	c_1
Value	1.28	1.05	190	780	450	15
Units	$[kg]$	$[kg]$	$[N/m]$	$[N/m]$	$[N/m]$	$[N \cdot s/m]$

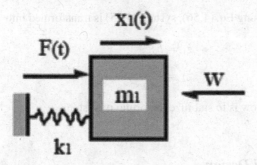

Fig. 4.17 The mass–spring system with disturbance

4.8.2   Problem Statement

To design the control we will consider just the mass m_1 and the spring κ_1 as part of the system and everything else is considered a disturbance as shown in Fig. 4.17. This configuration yields the state-space representation

$$\dot{x}_1 = x_2 \tag{4.81}$$

$$\dot{x}_2 = -\frac{\kappa_1}{m_1}x_1 + \frac{1}{m_1}(u + w) \tag{4.82}$$

where

$$w = \kappa_2(x_3 - x_1) \tag{4.83}$$

The goal of the control is to track the desired position $[x_d, 0]^T$ where x_d is constant. To work at the equilibrium point instead of the point $[x_d, 0]^T$ the following change of coordinates can be applied:

$$\begin{bmatrix} \bar{x}_1 \\ \bar{x}_2 \end{bmatrix} = \begin{bmatrix} x_1 - x_d \\ x_2 \end{bmatrix}$$

and using the control law $u = \kappa_1 x_d + u_1$ we can obtain the system

$$\dot{\bar{x}}_1 = \bar{x}_2 \tag{4.84}$$

$$\dot{\bar{x}}_2 = -\frac{\kappa_1}{m_1}\bar{x}_1 + \frac{1}{m_1}(u_1 + w) \tag{4.85}$$

such that when $\bar{x} = 0$ then $[x_1, x_2]^T = [x_d, 0]^T$. System (4.85) satisfies assumptions (A1) and (A2) and therefore it can be transformed to the regular form using Eq. (4.56) so that

$$\begin{bmatrix} z_1 \\ z_2 \end{bmatrix} = \begin{bmatrix} b & 0 \\ 0 & b^{-1} \end{bmatrix} \begin{bmatrix} \bar{x}_1 \\ \bar{x}_2 \end{bmatrix} \tag{4.86}$$

where $b = \frac{1}{m_1}$. Using Eq. (4.56), system (4.85) is transformed into

$$\dot{z}_1 = b^2 z_2 \qquad (4.87)$$

$$\dot{z}_2 = -\frac{\kappa_1}{b} z_1 + u_1 + w \qquad (4.88)$$

The control aim now is to stabilize the origin of system.

4.8.3 Control Design

Let us design the sliding surface as

$$\sigma \;=\; z_2 + K z_1 \qquad (4.89)$$

such that when the motion is restricted to the manifold, the reduced-order dynamics will have the desired performance

$$\dot{z}_1 \;=\; -b^2 K z_1$$

and the desired value will be tracked exponentially

$$z_1 \;=\; C_1 e^{-K b^2 t}$$

Secondly we want to change the state variables (z_1, z_2) to (z_1, σ). Taking into account (4.89) we can write z_2 as

$$z_2 \;=\; \sigma - K z_1 \qquad (4.90)$$

and $\dot{\sigma}$ as

$$\dot{\sigma} \;=\; \dot{z}_2 + K \dot{z}_1 \qquad (4.91)$$

Then using Eqs. (4.90) and (4.91) we can easily construct the input

$$u_1 = -K(-b^2 K z_1 + b^2 \sigma) + \frac{\kappa_1}{b} z_1 + v \qquad (4.92)$$

that will transform the system into

$$\dot{z}_1 = -b^2 K z_1 + b^2 \sigma \qquad (4.93)$$

$$\dot{\sigma} = v + w \qquad (4.94)$$

where v is the virtual control established in Eq. (4.65). To select the bounds for the disturbance, we can write (4.83) in terms of (z_1, σ) as

$$w(t, z_1) = \kappa_2 x_3(t) - \kappa_2 b z_1 \qquad (4.95)$$

where the term $x_3(t)$ is considered as exogenous, and only x_1 belongs to system (4.82). From Eq. (4.95) we can obtain the terms g_1 and g_2 as

$$g_1(z_1, \sigma, t) = 0 \tag{4.96}$$

$$g_2(z_1, t) = \kappa_2 x_3(t) - \kappa_2 b z_1 \tag{4.97}$$

$$\frac{dg_2(z_1, t)}{dt} = \kappa_2 x_4(t) - \kappa_2 b \dot{z}_1 \tag{4.98}$$

$$\frac{dg_2(z_1, t)}{dt} = \kappa_2 x_4(t) - \kappa_2 b(b^2 \sigma - b^2 K z_1) \tag{4.99}$$

Next ϱ_1 and ϱ_2 are selected to accomplish the restriction (4.66). Since $|\varphi_2(\sigma)| > \frac{1}{2}$ everywhere except on $\sigma = 0$ we can select ϱ_1 and ϱ_2 as follows $\varrho_1 = 0$ and

$$\varrho_2 = 2[\kappa_2 x_4(t) - \kappa_2 b(b^2 \sigma - b^2 K z_1)] \tag{4.100}$$

Finally we use the inverse transform

$$\begin{bmatrix} 0 \\ b \end{bmatrix} u_1 = T^{-1} \begin{bmatrix} 0 \\ 1 \end{bmatrix} u_1$$

4.8.4   Experimental Results

The total time of the experiment was $10[s]$ and the desired position was $x_d = 1[cm]$, This position is demanded when $t = 0.5[s]$. The parameters δ, β, and ϵ of the variable gains k_1 and k_2 and the fixed gain k_3 are selected as $\delta = 0.001$, $\beta = 4.1$, $\epsilon = 0.11$, and $k_3 = 8$, and the parameter K from the sliding surface (4.89) is selected as $K = 3$. The reference x_d is reached despite the disturbance as can be seen in Figs. 4.18 and 4.19. Chattering is completely eliminated (see Fig. 4.18). This result is achieved with a sampling time of $T_s = 1[ms]$. The behavior of σ is shown in Fig. 4.20 and the VGSTA output is shown in Fig. 4.21.

4.9   Notes and References

The twisting controller [75, 132] was historically the first 2-sliding mode controller to be proposed. The suboptimal controller appears first in [18, 20]. The controller with prescribed convergence law was proposed in [75, 132]. The quasi-continuous control algorithm is proposed in [127, 128]. In the particular case when

$$u = -\alpha \operatorname{sign}\left(\dot{\sigma} + \beta |\sigma|^{1/2} \operatorname{sign}(\sigma)\right), \quad \alpha, \beta > 0, \ \alpha K_m - C > \beta^2/2 \tag{4.101}$$

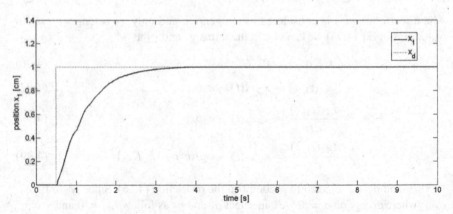

Fig. 4.18 Output of the system tracking $x_d = 1[cm]$

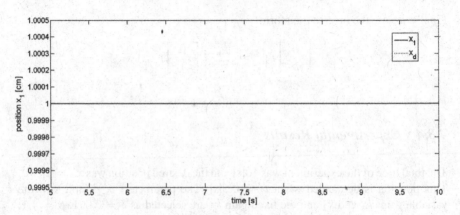

Fig. 4.19 Zoom of the system output tracking $x_d = 1[cm]$

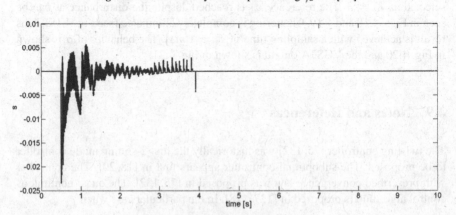

Fig. 4.20 The sliding surface

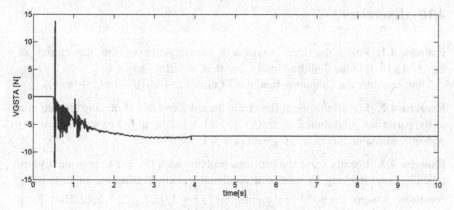

Fig. 4.21 The output of the VGSTA

the controller given in Eq. (4.101) is similar to so-called terminal sliding mode controllers [138]. An alterative detailed proof of Theorem 4.2 can be found in [17].

The long-standing concern associated with conventional sliding mode control is the attenuation of the so-called chattering effect. Many different approaches have been suggested: see, for example, [29,31,87–90,98,171]. However, 2-sliding mode control ideas provide effective tools for the reduction or even practical elimination of the chattering without compromising the benefits of conventional sliding modes: see, for example, [18,20,31,33–35,125,132]. Additional information about 2-sliding mode controllers and differentiators can be in [15, 16, 19, 122, 153].

Theorem 4.8 is based on the results presented in [123]. The accuracy estimations formulated in Theorem 4.8 remain valid in the presence of sufficiently small noise and/or sampling intervals. Note that although Theorem 4.9 is not formulated for arbitrary 2-sliding homogeneous controllers, it is valid for all standard 2-SM controllers [20, 132]. It can be shown that the chattering phenomenon is indeed mitigated by means of this procedure. Moreover, noise caused by unaccounted-for fast stable actuators and sensors does not produce chattering. Theorem 4.10 extends the Lyapunov-based design method from [142] for the standard twisting algorithm in order to include (i) linear (nonhomogeneous) terms and (ii) variable gains, in order to alleviate the drawbacks of the standard twisting algorithm. The use of the Lyapunov method is instrumental here, since neither geometric or homogeneity based proofs can be used to deal with these extensions [12, 126]. Section 4.9 presents the results of the paper [101]. It is a particular case of the Lyapunov-based approach to the second-order sliding mode control design presented by Moreno in [142,143]. Based on this approach fixed-time convergent controllers and differentiators [51, 152] are developed ensuring a uniform convergence time with respect to initial conditions. Such algorithms being applied to hybrid and switched systems with strictly positive dwell-time can ensure the convergence of observers and controllers before the system jumps of switches.

4.10 Exercises

Exercise 4.1. Prove the local asymptotic convergence of the trajectories of Eqs. (4.1), (4.3) to the 2-sliding mode $x = 0$, $u = 1/2$ for any $a > 2$.

Hint: consider the Lyapunov function $V(x) = |x| + \frac{1}{2a}(u - \sin(x + \pi/6))$.

Exercise 4.2. List all the controllers from Sect. 4.2 capable of making the output x of the perturbed pendulum $\ddot{x} = \sin(x) + f(t) + u$, $|f(t)| \leq 1$ exactly follow any real-time available function $x_c(t)$ with $|\ddot{x}_c| \leq 1$.

Exercise 4.3. Locally solve the previous problem with $|\dot{x}| < 2$ to provide asymptotically exact tracking by means of continuous control, using the super-twisting controller. Assume the additional conditions $\left|\dot{f}\right| \leq 1$, $|\ddot{x}_c| \leq 2$; $\dot{x}$ hold. Hint: Keep the constraint $\dot{x} + x = 0$.

Exercise 4.4. Solve the previous problem of asymptotically exact tracking, by means of continuous control, using the controllers from Sect. 4.2. Both $\dot{x}$ and $\ddot{x}$ are assumed to be available.

Exercise 4.5. Choose the parameters of the differentiator (4.38), (4.39) to facilitate the exact differentiation of the signal $f = \cos(2(3t - 5)) + 4t$.

Exercise 4.6. Check the solution of the previous problem by means of computer simulation. Introduce arbitrary noise of the magnitude 0.01 and check the robustness of the differentiator in the presence of nosie of different frequencies (e.g., one can take the "noise" $0.01 sin(\omega t)$ or any other periodic function, even discontinuous).

Exercise 4.7. Solve problems 4.2, 4.3 by means of output-feedback controllers.

Exercise 4.8. Solve problem 4.4 using a differentiator assuming that $\dot{x}$ is available.

Exercise 4.9. Verify the solutions of problems 4.7, 4.8 by computer simulation. In addition introduce small measurement noise.

Chapter 5
Analysis of Sliding Mode Controllers in the Frequency Domain

Conventional sliding mode control, studied in Chap. 2, and second-order sliding mode control (Chap. 4) are the most obvious choices in controlling systems with bounded matched disturbances/uncertainties. Sliding mode control laws allow us achieve to insensitivity of system's compensated dynamics to these perturbations. The ultimate price for this insensitivity is a high-frequency (that is equal to infinity in an ideal sliding mode) switching control function that after being filtered by the plant yields self-sustained oscillations of almost zero amplitude. The main advantage of higher (second-)order sliding mode control is its ability to guarantee higher accuracy of the sliding variable stabilization at zero than conventional sliding mode control. Both conventional and second-order sliding mode control laws are derived (see Chaps. 2 and 4) assuming that the relative degree is equal to one in the case of conventional sliding mode control and equal to two for the case of second-order sliding modes. However, any real plant and any real switching element will contain parasitic dynamics that are not taken into account in the system's mathematical model. These parasitic dynamics increase the system's relative degree and could yield control chattering, i.e., self-sustaining oscillations of lower frequency and nonzero amplitude, causing degradation in the system's performance. In this chapter the robustness of the system in a sliding mode with respect to parasitic dynamics is studied using the *describing function technique*. This approximate technique is useful from a practical point of view for the analysis of self-sustained oscillation (limit cycles) as soon as the transient response is over.

5.1 Introduction

The phenomenon of chattering is caused by the inevitable existence of so-called *parasitic* or *unmodeled* dynamics that exist along with the *principal dynamics* of the plant. The *principal dynamics* are the dynamics of the plant model that are used for the SMC design. However, to implement the designed control algorithms, devices such as actuators and sensors are needed. These devices bring into the system certain parasitic dynamics that are not accounted for during the SMC design and which are treated as *unmodeled* dynamics at this step. Since the *parasitic* dynamics are usually connected in series with the *principal* dynamics, the combined relative degree of the actuator–plant–sensor becomes equal to the sum of the plant relative degree, the actuator relative degree, and the sensor relative degree.

It is known that conventional SMC in systems with parasitic dynamics of relative degree two or higher exhibit chattering.[1] For the same reason, it is logical to expect a similar behavior from systems with 2-SM controllers, as the above-mentioned 2-SM algorithms contain the sign function or infinite gains. In this chapter systems controlled by conventional SMC and all principal 2-SM controllers are analyzed in the frequency domain, while the parasitic/unmodeled dynamics are taken into account. The goal of this study is to analyze the robustness of the conventional SMC and the major 2-SM controllers, including the continuous super-twisting algorithm (see Chaps. 2 and 4), to parasitic/unmodeled dynamics by detecting possible self-sustained oscillations (limit cycles) and estimating their parameters (such as amplitude and frequency) via *describing function (DF) techniques* (see Appendix B). It is worth noting that the DF method provides only an approximate solution.[2]

5.2   Conventional SMC Algorithm: DF Analysis

Conventional SMC techniques rest on the concept of switching the system's structure in order to force the trajectory of the system to a switching surface and to maintain the system's motion on this surface thereafter. The system's motion on the sliding surface appeared to be insensitive to bounded model uncertainties and external disturbances (see Chap. 2). However, the robustness of conventional SMC to parasitic/unmodeled dynamics requires special consideration. It is expected that these dynamics may cause control chattering, which yields limit cycles and degradation in the system's performance. In many systems chattering is highly undesirable especially after the transient response is over. An analysis of self-sustained oscillations (limit cycles) in linear time-invariant systems with SMC and parasitic dynamics, via DF technique, is presented in this section. The linear system's principal dynamics are assumed to be given in Brunovsky's canonical form:

---

[1] This agrees with the classical research work on relay feedback systems theory [9, 181], where it was proven that for the plant of order 3 and higher the equilibrium point in the origin cannot be stable.

[2] There exists a method named the locus of a perturbed relay system Locus of Perturbed Relay System (LPRS) [29] that gives an exact analysis of limit cycles in perturbed relay systems. However, the LPRS method is computationally much more intensive than DF method. Conversely, the DF method is simple and efficient and provides accuracy sufficient for the analysis of practical systems.

$$\dot{x}(t) = A_p x(t) + B_p u(t) \tag{5.1}$$

where

$$A_p = \begin{bmatrix} 0 & 1 & 0 & \ldots & 0 \\ 0 & 0 & 1 & \ldots & 0 \\ .. & .. & .. & \ldots & .. \\ -a_0 & -a_1 & -a_2 & \ldots & -a_{n-1} \end{bmatrix} \qquad B_p = \begin{bmatrix} 0 \\ 0 \\ \vdots \\ b \end{bmatrix}$$

It is assumed that the characteristic polynomial

$$P(\lambda) = \lambda^n + a_{n-1}\lambda^{n-1} + \ldots + a_1\lambda + a_0$$

of system in (5.1) is Hurwitz.

The sliding variable is designed to have the form

$$\sigma = x_n + c_{n-1}x_{n-1} + \cdots + c_1 x_1 \tag{5.2}$$

The relative degree of the sliding variable in (5.2) is equal to one. Therefore, any conventional SMC that drives the sliding variable σ to zero in finite time can be designed by satisfying the sliding mode existence condition:

$$\sigma\dot{\sigma} \leq -\rho|\sigma| \tag{5.3}$$

The sliding mode control law is chosen as

$$u(t) = -U_m \mathrm{sign}(\sigma) \tag{5.4}$$

where the modulation amplitude $U_m > 0$ is assumed to be chosen large enough. The system's (5.1) dynamics in the sliding mode $\sigma = 0$ are derived:

$$\begin{cases} \dot{x}_1 = x_2 \\ \dot{x}_2 = x_3 \\ \quad \cdots \\ \dot{x}_{n-1} = -c_1 x_1 - c_2 x_2 - \cdots - c_{n-1}x_{n-1} \\ x_n = -c_1 x_1 - c_2 x_2 - \cdots - c_{n-1}x_{n-1} \end{cases} \tag{5.5}$$

In an ideal sliding mode, the frequency of switching of the control function (5.2) and (5.4) is infinite, while the sliding variable (5.2) oscillates with zero amplitude, and the state variables are continuous. However, parasitic dynamics may constrain the system's motion only to some vicinity of the switching surface $|\sigma| \leq \varepsilon > 0$ (a real sliding mode). In a real sliding mode, systems with SMC and parasitic dynamics can exhibit lower frequency self-sustained oscillation (chattering).

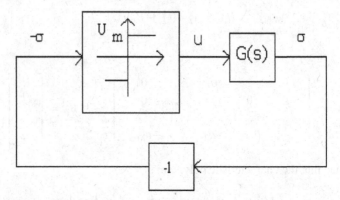

Fig. 5.1 Block diagram of a linear system with relay control and ideal sliding

Ideal Sliding Mode Analysis via DF Technique

The block diagram of the system (5.1), (5.2) and (5.4) is shown below in Fig. 5.1.
The transfer function of the plant is identified as

$$G(s) = \frac{\sigma(s)}{u(s)} = \frac{b\left(s^{n-1} + c_{n-1}s^{n-2} + \ldots + c_2 s + c_1\right)}{s^n + a_{n-1}s^{n-1} + \ldots + a_1 s + a_0} \tag{5.6}$$

Assume that there exists a periodic motion (self-sustained oscillations or a limit
cycle) with amplitude A_c and oscillation frequency ω_c in some vicinity of the
switching surface (5.2) $|\sigma| \leq \varepsilon$ where $\varepsilon > 0$ in the system. In the frequency domain
analysis we replace the Laplace variable s by $j\omega$, where ω is the frequency. Thus
we assume that the following equation is valid:

$$-\sigma = A_c \sin(\omega_c) \tag{5.7}$$

The amplitude A_c and the oscillation frequency ω_c have to satisfy the harmonic
balance equation (see Appendix B)

$$G(j\omega) = -\frac{1}{N(A, \omega)} \tag{5.8}$$

where the describing function of the relay nonlinearity is identified as follows:

$$N(A, \omega) = \frac{4U_m}{\pi A} \tag{5.9}$$

Definition 5.1. A strictly stable linear system[3] with a transfer function (5.6) will be
strictly passive if and only if

---

[3]For details see [100].

$$|\arg G(j\omega)| < \frac{\pi}{2} \quad \forall \omega \in [0, \infty) \tag{5.10}$$

It is obvious that the phase characteristic of the transfer function (5.6) satisfies the following equality:

$$\lim [\arg G(j\omega)] = -\frac{\pi}{2} \quad as \quad \omega \to \infty \tag{5.11}$$

$$\lim [\arg G(j\omega)] = 0 \quad as \quad \omega \to 0$$

Assume that the coefficients $c_1, c_2, \ldots, c_{n-1}$ have been selected so that the passivity condition (5.10) is met. Then there exists a unique solution: the magnitude $A_c = 0$ and the frequency $\omega_c \to \infty$ of Eqs. (5.8) and (5.9). This fact is illustrated in the following example.

Example 5.1. Let the plant be given by the following equations:

$$\dot{x}_1 = x_2$$

$$\dot{x}_2 = -x_1 - x_2 + u \tag{5.12}$$

$$\sigma = x_1 + x_2$$

with control

$$u = -\text{sign}(\sigma) \tag{5.13}$$

In accordance with Fig. 5.1

$$G(s) = \frac{s + 1}{s^2 + s + 1} \tag{5.14}$$

and the harmonic balance condition is given by two equations that equate the real and imaginary parts of Eq. (5.8):

$$\text{Re}\,[G(j\omega] = -\frac{\pi A}{4U_m} \tag{5.15}$$

i.e.,

$$\frac{1 - \omega + \omega^2}{(1 - \omega^2)^2 + \omega^2} = -\frac{\pi A}{4U_m}$$

and

$$\frac{\omega^2}{(1 - \omega^2)^2 + \omega^2} = 0 \tag{5.16}$$

The obvious solution of Eq. (5.15) and (5.16) is

$$A_c = 0, \; \omega_c \to \infty \tag{5.17}$$

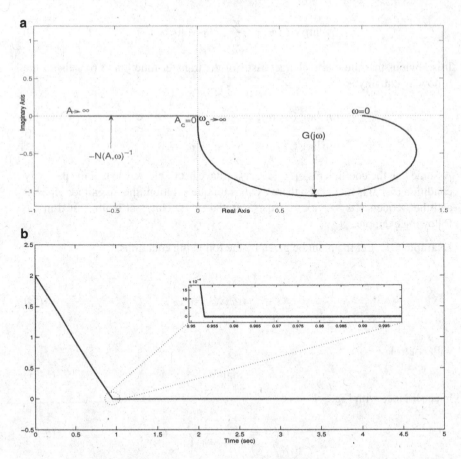

Fig. 5.2 Graphical solution of the harmonic balance equation for system $G(s)$. (**a**) Graphical solution. (**b**) Surface

This solution can be interpreted as a limit cycle with infinity frequency and zero amplitude that is expected in systems with ideal sliding mode control. The DF method does not give the opportunity to identify finite time or asymptotic convergence to the predicted limit cycle. A simple criterion for finite time convergence is achieved if the angle between the high-frequency asymptote of the Nyquist plot of the plant and the low-amplitude asymptote of the negative reciprocal of the describing function of the nonlinear element, the so-called *phase deficit*, is greater than zero. The graphical solution of the harmonic balance Eq. (5.8) that is presented in Fig. 5.2 confirms the solution given by (5.17) and also finite time convergence.

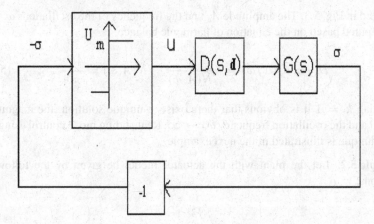

Fig. 5.3 Block diagram of linear system with relay control and parasitic dynamics

Real Sliding Mode Analysis via Describing Function Technique

Now system (5.1), (5.2) and (5.4) with parasitic dynamics is considered. The block diagram of this system is shown in Fig. 5.3 and the transfer function of the parasitic dynamics is

$$D(s, \mathbf{d}) = \frac{d_0}{s^k + d_{k-1}s^{k-1} + \ldots + d_1 s + d_0} \tag{5.18}$$

where the characteristic polynomial $P_d(\lambda) = \lambda^k + d_{k-1}\lambda^{k-1} + \ldots + d_1\lambda + d_0$ is Hurwitz.

In the system with parasitic dynamics given in (5.18) the relative degree of the sliding variable (5.2) is based on the combined principal (5.1) and parasitic dynamics (5.18) and is equal to $k + 1$. Consequently, the sliding mode existence condition (5.3) cannot be met, which is why only a real sliding mode can exist in such a system.

Suppose the coefficients in the equation of the switching surface (5.2) are chosen such that the passivity condition (5.10) is met. It is obvious that the phase frequency characteristic of the transfer function $D(s, \mathbf{d})G(s)$ satisfies the following:

$$\lim \left[\arg D(j\omega, \mathbf{d})G(j\omega) \right] = -\frac{\pi}{2}(k + 1) \quad as \quad \omega \to \infty \tag{5.19}$$

Therefore, the transfer function $D(s, \mathbf{d})G(s)$ is not passive for all $k \geq 1$. Assume that in the real sliding mode there exist self-sustained oscillations (limit cycle) with amplitude A_c and oscillation frequency ω_c in the system given by the block diagram presented in Fig. 5.3. The amplitude A_c and the frequency of the oscillations ω_c can be computed based on the equation of harmonic balance:

$$D(j\omega, \mathbf{d})G(j\omega) = -\frac{1}{N(A, \omega)}, \quad N(A, \omega) = \frac{4U_m}{\pi A} \qquad (5.20)$$

Assuming $k = 1$ it is obvious that there exists a unique solution: the magnitude $A_c = 0$ and the oscillation frequency $\omega_c \to \infty$. Real sliding mode control using the DF technique is illustrated in the next example.

Example 5.2. Let the plant with the actuator model be given by the following equations:

$$\dot{x}_1 = x_2$$

$$\dot{x}_2 = -x_1 - x_2 + u \qquad (5.21)$$

$$0.01\dot{u}_a = -u_a + u$$

$$\sigma = x_1 + x_2$$

and the controller given by

$$u = -\text{sign}(\sigma) \qquad (5.22)$$

In accordance with Fig. 5.3 we have

$$D(s, \mathbf{d})G(s) = \frac{s + 1}{(0.01s + 1)(s^2 + s + 1)} \qquad (5.23)$$

The solution for the harmonic balance equation for the system associated with $D(s, \mathbf{d})G(s)$ is presented in Fig. 5.4.

Based on Fig. 5.4 we can conclude that the existence of an asymptotic limit cycle with $A_c = 0$ and $\omega_c \to \infty$ is predicted.

Assuming $k = 2$, it is obvious that there exists a unique solution with magnitude $A_c > 0$ and frequency $0 < \omega_c = C < \infty$ to Eq. (5.20). This fact is graphically illustrated in the following example.

Example 5.3. Let the plant with the actuator be given by the following equations:

$$\dot{x}_1 = x_2$$

$$\dot{x}_2 = -x_1 - x_2 + u \qquad (5.24)$$

$$0.0001\ddot{u}_a = -0.01\dot{u}_a - u_a + u$$

$$\sigma = x_1 + x_2$$

The solution for the harmonic balance equation for system $D(s, \mathbf{d})G(s)$ is presented in Fig. 5.5a. Figure 5.5b shows a zoom of the intersection between $G(j\omega)$ and

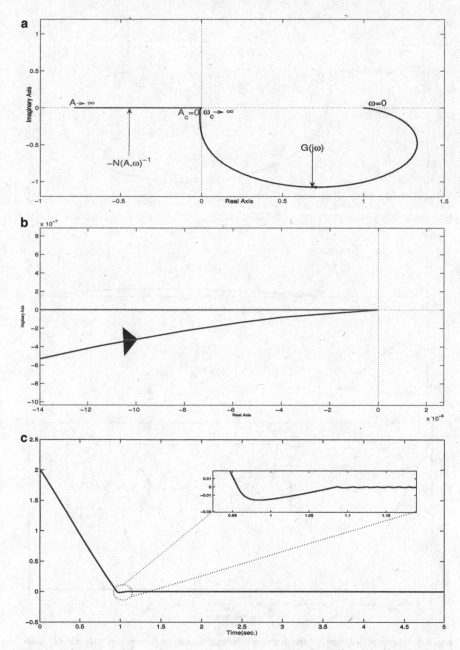

Fig. 5.4 Graphical solution of the harmonic balance equation for the system $D(s, d)G(s)$ with a first-order actuator. (**a**) Graphical solution. (**b**) Graphical solution zoom. (**c**) Surface

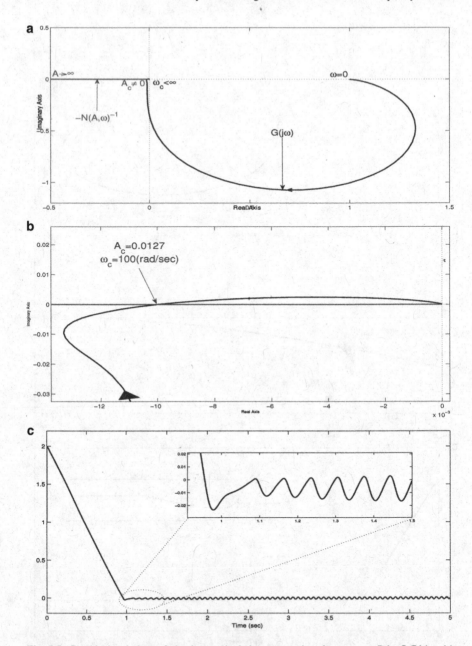

Fig. 5.5 Graphical solution of the harmonic balance equation for system $D(s, \boldsymbol{d})G(s)$ with second-order actuator. (**a**) Graphical solution. (**b**) Graphical solution zoom. (**c**) Surface

$N(A, \omega)^{-1}$ where we can see the value of the oscillation parameters. Based on Fig. 5.5b, c we can conclude the existence of a limit cycle with $A_c = 0.0127$ and $\omega_c = 100(rad/sec)$.

The results of the presented analysis are summarized in the following proposition:

Proposition 5.1. *If the switching surface (5.2) is designed to make the transfer function (5.6) strictly passive and the stable parasitic dynamics are given by Eq. (5.18), then in the system given by Eqs. (5.1), (5.2), (5.4) and (5.18) there exist:*

(a) An ideal sliding mode for $k = 0$
(b) An asymptotic sliding mode for $k = 1$
(c) A real sliding mode with chattering for $k \geq 2$

5.3 Twisting Algorithm: DF Analysis

The twisting algorithm (see Chap. 4) is one of the simplest and most popular algorithms among the second-order sliding mode algorithms. There are two ways of using the twisting algorithm: to apply it to the principal dynamics of a system of relative degree two or to apply it to the principal dynamics of a system of relative degree one and introduce an integrator in series with the plant (twisting as a filter). For the principal dynamics of relative degree two it can be formulated as follows. Let the plant (or the plant plus actuator) be given by the following differential equations:

$$\dot{x}(t) = Ax(t) + Bu(t), \quad \sigma = Cx \tag{5.25}$$

where A and B are matrices of appropriate dimensions, $x \in \mathbb{R}^n$, and $\sigma \in \mathbb{R}$ can be treated as either the sliding variable or the output of the plant. We assume that the plant is asymptotically stable, apart from some possible integrating terms, and is a low-pass filter. We shall also use the plant description in the form of a transfer function $W(s)$, which can be obtained from the formulas (5.25) as follows:

$$W(s) = C(Is - A)^{-1}B$$

Also, let the control u of the twisting algorithm be given as

$$u(t) = -\alpha_1 \text{sign}(\sigma) - \alpha_2 \text{sign}(\dot{\sigma}) \tag{5.26}$$

where α_1 and α_2 are positive values, $\alpha_1 > \alpha_2 > 0$.

Assume that a periodic motion occurs in the system with the twisting algorithm. The objective is to find the parameters of this periodic motion. The system will be analyzed with the use of the DF method. As normal in a DF analysis, we assume that the harmonic response of the plant is that of a low-pass filter, so that the output of the plant is a harmonic oscillation. To find the DF of the twisting algorithm as the first harmonic of the periodic control signal divided by the amplitude of $\sigma(t)$,

$$N = \frac{\omega}{\pi A} \int_0^{2\pi/\omega} u(t) \sin(\omega t) dt + j \frac{\omega}{\pi A} \int_0^{2\pi/\omega} u(t) \cos(\omega t) dt \qquad (5.27)$$

where A is the amplitude of the input to the nonlinearity (of $\sigma(t)$ in our case) and ω is the frequency of $\sigma(t)$. However, the twisting algorithm can be analyzed as the parallel connection of two ideal relays where the input to the first relay is the sliding variable and the input to the second relay is the derivative of the sliding variable. The DF for those nonlinearities are well known. For the first relay the DF is $N_1 = \frac{4\alpha_1}{\pi A}$ and for the second relay $N_2 = \frac{4\alpha_2}{\pi a_\sigma}$, where a_σ is the amplitude of $d\sigma/dt$. Also, we need to take into account the relationship between σ and $d\sigma/dt$ in the Laplace domain, which gives the relationship between the amplitudes A and a_σ as $a_\sigma = A\omega$, where ω is the frequency of the oscillation. Using the notation of the twisting algorithm we can write this as follows:

$$N = N_1 + sN_2 = \frac{4\alpha_1}{\pi A} + j\omega \frac{4\alpha_2}{\pi a_\sigma} = \frac{4}{\pi A}(\alpha_1 + j\alpha_2) \qquad (5.28)$$

where $s = j\omega$. Note that the DF of the twisting algorithm depends on the amplitude value only. This suggests finding the parameters of the limit cycle via the solution of the harmonic balance equation:

$$W(j\omega)N(A) = -1 \qquad (5.29)$$

where A is the generic amplitude of the oscillation at the input of the nonlinearity and $W(j\omega)$ is the complex frequency response characteristic (Nyquist plot) of the plant. Using the notation of the twisting algorithms this equation can be rewritten as follows:

$$W(j\omega) = -\frac{1}{N(A)} \qquad (5.30)$$

where the function at the right-hand side is given by

$$-\frac{1}{N} = \pi A \frac{-\alpha_1 + j\alpha_2}{4(\alpha_1^2 + \alpha_2^2)}$$

Equation (5.29) is equivalent to the condition of the complex frequency response characteristic of the open-loop system intersecting the real axis at the point $(-1, j0)$. A graphical illustration of the technique of solving Eq. (5.29) is given in Fig. 5.6. The function $-\frac{1}{N}$ is a straight line whose slope depends on the ratio α_2/α_1. This line is located in the second quadrant of the complex plane. The point of intersection of this function and the Nyquist plot $W(j\omega)$ provides the solution of the periodic problem. This point gives the frequency of the oscillation ω_c and the amplitude A_c. Therefore, if the transfer function of the plant (or plant plus actuator) has relative

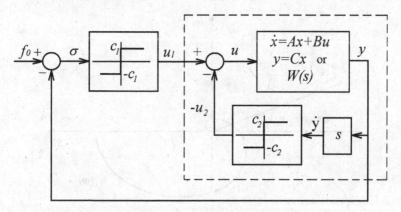

Fig. 5.6 Transformed system with twisting control

degree higher than *two* a periodic motion may occur in such a system. For that reason, if an actuator of first or higher order is added to the plant of relative degree *two* driven by the twisting controller, a periodic motion may occur in the system.

The conditions for the existence of a periodic solution in a system with the twisting controller can be derived from the analysis of Fig. 5.6. Obviously, every system with a plant of relative degree *three* and higher would have a point of intersection with the negative reciprocal of the DF of the twisting algorithm and, therefore, a periodic solution could exist.

Another modification of the twisting algorithm is its application to a plant with relative degree one with the introduction of an integrator (see Chap. 4). This is usually referred to as the "twisting as a filter" algorithm. The above reasoning is applicable in this case too. The introduction of the integrator in series with the plant makes the relative degree of this part of the system equal to two. As a result, any actuator introduced in the loop increases the overall relative degree to at least three. In this case, there always exists a point of intersection between the Nyquist plot of the serial connection of the actuator, the plant and the integrator, and of the negative reciprocal of the DF of the twisting algorithm (Fig. 5.7). Thus, if an actuator of first or higher order is added to the plant with relative degree one, a periodic motion may occur in the system with the "twisting as a filter algorithm.".

Remark 5.1. The asymptotic 2-SM relay controller

$$\ddot{x} = -a\dot{x} - bx - k\operatorname{sign}(x),\ a > 0, k > 0$$

can also be studied. It can be shown that this system is exponentially stable. With respect to our analysis, from Fig. 5.7, it also follows that the frequency of the periodic solution for the twisting algorithm is always higher than the frequency of the asymptotic second-order sliding mode relay controller, because the latter is determined by the point of intersection of the Nyquist plot and the real axis.

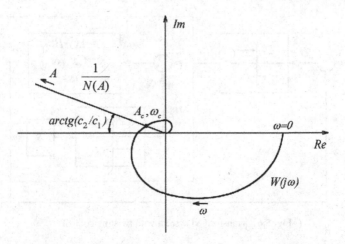

Fig. 5.7   Finding a periodic solution in system with twisting control

5.4   Super-Twisting Algorithm: DF Analysis

5.4.1   DF of Super-Twisting Algorithm

The super-twisting algorithm is probably one of the most popular second-order sliding mode algorithms. It is used for systems with principal dynamics of relative degree *one*. The control $u(t)$ for the super-twisting algorithm is given as a sum of two components:

$$u(t) = u_1(t) + u_2(t) \tag{5.31}$$

$$\dot{u}_1 = -\gamma \, \text{sign}(\sigma)$$

$$u_2 = \begin{cases} -\lambda |s_0|^\rho \, \text{sign}(\sigma), & \text{if} \quad |\sigma| > s_0 \\ -\lambda |\sigma|^\rho \, \text{sign}(\sigma), & \text{if} \quad |\sigma| \le s_0 \end{cases}$$

where α, ρ, and s_0 are design parameters, with $0.5 \le \rho < 1$.

The system under analysis can be represented in the form of the block diagram as in Fig. 5.8.

For an arbitrary value of the power ρ in (5.31), the formula for the DF of such a nonlinear function can be given as follows:

$$N_2 = \frac{2\lambda A^{\rho-1}}{\pi} \int_0^\pi (\sin(\psi))^{\rho+1} d\psi = \frac{2\lambda A^{\rho-1}}{\sqrt{\pi}} \frac{\Gamma(\frac{\rho}{2}+1)}{\Gamma(\frac{\rho}{2}+1.5)} \tag{5.32}$$

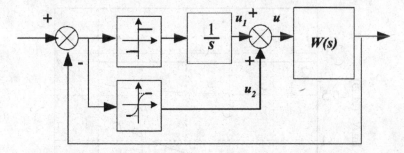

Fig. 5.8 Block diagram of a linear system with super-twisting control

where $0 < \rho < 1$, A is the amplitude of the variable y, $A \leq s_0$ (this is considered as the most important range of the amplitude values for the analysis of the steady state), and Γ is the gamma function.[4]

With the square root nonlinearity ($\rho = 0.5$) the DF formula can be derived as

$$N_2 = \frac{2\lambda}{\pi A} \int_0^{\pi} \sqrt{A \sin(\psi)} \sin(\psi) d\psi = \frac{2\lambda}{\sqrt{\pi A}} \frac{\Gamma(1.25)}{\Gamma(1.75)} \approx \frac{1.1128\lambda}{\sqrt{A}} \qquad (5.33)$$

The DF of the first component of the super-twisting algorithm can be written:

$$N_1 = \frac{4\gamma}{\pi A} \frac{1}{j\omega}$$

which is a result of the cascade connection of the ideal relay having a DF equal to $\frac{4\gamma}{\pi A}$ and an integrator with transfer function $1/s$ (for the harmonic signal, the Laplace variable s can be replaced with $j\omega$). Taking into account both control components, we can rewrite the DF of the super-twisting algorithms as

$$N = N_1 + N_2 = \frac{4\gamma}{\pi A} \frac{1}{j\omega} + \frac{1.1128\lambda}{\sqrt{A}} \qquad (5.34)$$

Note that the DF of the super-twisting algorithm depends on both the amplitude and the frequency values. The parameters of the limit cycle can be found via the solution of the harmonic balance Eq. (5.29), where the DF N is given by (5.34). The negative reciprocal of the DF can be represented by the following formula:

$$-\frac{1}{N} = \frac{1}{1.1128\frac{\lambda}{\sqrt{A}} + \frac{4\gamma}{\pi A}\frac{1}{j\omega}} = -\frac{0.8986\frac{\sqrt{A}}{\lambda} + j1.1329\frac{\gamma}{\lambda^2}\frac{1}{\omega}}{1 + 1.3092\frac{\gamma^2}{\lambda^2}\frac{1}{A\omega^2}}$$

---

[4]For details, see, for example,[39].

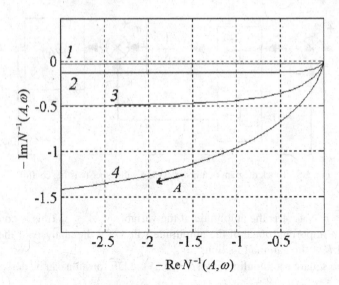

Fig. 5.9 Plots of the function $-1/N$ for super-twisting control

Now $-1/N$ is a function of two variables: the amplitude and the frequency. It can be depicted as a number of plots representing the amplitude dependence, with each of those plots corresponding to a certain frequency. The frequency range of interest lies below the frequency corresponding to the intersection of the Nyquist plot and the real axis. The plots of the function $-1/N$ are depicted in Fig. 5.9. Plots 1–4 correspond to four different frequencies, with the following relationship: $\omega_1 > \omega_2 > \omega_3 > \omega_4$. Each of these plots represents the dependence of the DF on the amplitude value.

The function $-N^{-1}(A)$ (where ω=const) has an asymptote as $A \to \infty$, which is the horizontal line $-j1.1329\frac{\gamma}{\lambda^2}\frac{1}{\omega}$. Also, it is easy to show that

$$\lim_{A \to 0} \arg(-N^{-1}(A), \omega) = -\pi/2$$

5.4.2 Existence of the Periodic Solutions

The solution of the harmonic balance Eq. (5.29) can be computed numerically by the application of various iterative techniques. However, the complex Eq. (5.29) with two unknown variables, A and ω, can be reduced to one real equation having only one unknown variable ω as follows. Write Eq. (5.29) in the form of $N(A) = -W^{-1}(j\omega)$, where $N(A)$ is given by (5.34):

$$\frac{4\gamma}{\pi A}\frac{1}{j\omega} + 1.1128\frac{\lambda}{\sqrt{A}} = -W^{-1}(j\omega) \qquad (5.35)$$

Considering the real part of both sides we can obtain

$$1.1128\frac{\lambda}{\sqrt{A}} = -\operatorname{Re} W^{-1}(j\omega) \tag{5.36}$$

Eliminating A from the Eqs. (5.35) and (5.36) an equation with one unknown variable ω can be obtained as follows:

$$\Psi(\omega) = \frac{4\gamma}{\pi\omega}\frac{1}{\operatorname{Im} W^{-1}(j\omega)} - \left(\frac{1.1128\lambda}{\operatorname{Re} W^{-1}(j\omega)}\right)^2 = 0 \tag{5.37}$$

Once Eq. (5.37) has been solved, the amplitude of oscillation A_c can be computed as follows:

$$A_c = \frac{4\gamma}{\pi\omega_c}\frac{1}{\operatorname{Im} W^{-1}(j\omega_c)} \tag{5.38}$$

where ω_c is the frequency of the oscillations.

Therefore, if a periodic motion occurs, its parameters can be estimated from (5.37) and (5.38).

Proposition 5.2. *If the relative degree of the plant is two or higher and the plant does not have double zero poles, then at least one periodic solution of the system with the super-twisting algorithm can be always predicted and estimated using the DF technique.*

Proof. We first prove Proposition 5.2 for plants of relative degree three and higher. It follows from the formula of the DF of the algorithm (5.34) that a periodic solution should always be looked for within the frequency range that corresponds to $-\pi/2$ and $-\pi$ of the phase characteristic of the plant (see also Fig. 5.9). Denote by ω_1 the frequency where the phase characteristic of the plant is $-\pi/2$ (assuming that there is only one such frequency). Similarly denote the frequency ω_2 as the frequency at which the phase characteristic of the plant is $\arg W(j\omega_2) = -\pi$. Both frequencies are finite. Find the following two limits of the function from (5.37): $\Psi(\omega_1+) = -\infty$ and $\Psi(\omega_2-) = \infty$. Now observe that the signs are different and that the function $\Psi(\omega)$ is continuous within the range $\omega \in (\omega_1, \omega_2)$ [This follows from (5.37).] Therefore, within the specified range, there exists at least one solution of Eqs. (5.25) and (5.31). Assume now that the relative degree of the plant is two. In this case, we can define the frequency ω_1 in the same manner as before, but the frequency ω_2 becomes infinite. Write the asymptotic representation of the plant transfer function for high-frequency inputs in a polynomial form:

$$W(s) \approx \frac{1}{a_2s^2 + a_1s + a_0}$$

Then, substituting $j\omega$ for s, we obtain $W^{-1}(j\omega) \approx -a_2\omega^2 + a_0 + ja_1\omega$, and it is easy to see that for sufficiently large ω,

$$\Psi(\omega) = \frac{4\gamma}{\pi}\frac{1}{a_1\omega^2} - \frac{1.1128^2\lambda^2}{(a_0 - a_2\omega^2)^2} > 0$$

Therefore, between ω_1 and ω_2, there always exists a certain frequency ω, which provides a solution to Eq. (5.37). This completes the proof. $\qquad\square$

Remark 5.2. It is important that the point of intersection is located in the third quadrant of the complex plane. Therefore, if the transfer function of the plant (or the plant plus actuator) has relative degree higher than one, a periodic motion may occur in such a system. For that reason, if *parasitic dynamics* of first or higher order are added to the *principal dynamics of* a relative degree one system driven by the super-twisting controller, a periodic motion may occur in the system. From Fig. 5.9, it also follows that the frequency of the periodic solution for the super-twisting algorithm is always lower than the frequency of the periodic motion in the system with the conventional first-order SM relay controller, because the latter is determined by the point of the intersection of the Nyquist plot and the real axis.

5.4.3   Stability of Periodic Solution

Proposition 5.3. *If the following inequality holds then the periodic solution given by Eq. (5.37) is locally stable:*

$$\text{Re}\,\frac{h_1(A,\omega)}{h_2(A,\omega) + N(A,\omega)\frac{\partial \ln W(s)}{\partial s}\big|_{s=j\omega}} < 0 \qquad (5.39)$$

where $h_1(A,\omega) = \frac{1.1128\lambda}{2A^{\frac{3}{2}}} - j\frac{4\gamma}{\pi\omega A^2}, h_2(A,\omega) = \frac{4\gamma}{\pi\omega^2 A}.$

Proof. To investigate the local stability of the solution of (5.37), we consider the system transients due to small perturbations of this solution when A is quasi-statically varied to $(A + \Delta A)$. Here we assume that the harmonic balance equation still holds for small perturbations, so that a damped oscillation of the complex frequency $j\omega + (\Delta\sigma + j\Delta\omega)$ corresponds to the modified amplitude $(A + \Delta A)$:

$$N(A + \Delta A, j\omega + (\Delta\sigma + j\Delta\omega))W(j\omega + (\Delta\sigma + j\Delta\omega)) = -1 \qquad (5.40)$$

where the DF $N(A,\omega)$ is given by Eq. (5.34). The nominal solution is determined by zero perturbations: $\Delta\sigma = \Delta\omega = \Delta A = 0$. By considering variations around the nominal solution defined by ω and A, let us find the conditions when $\Lambda = \Delta\sigma/\Delta A$ is negative. First write an equation for the amplitude perturbation ΔA. For that purpose take the derivative of (5.40) with respect to ΔA as follows:

$$\left\{ \frac{dN(\Delta A, \Delta \sigma, \Delta \omega)}{d\Delta A}\Big|_{\Delta A=0} W(j\omega) + \frac{dW(\Delta \sigma, \Delta \omega)}{d\Delta A}\Big|_{\Delta A=0} N(A, \omega) \right\} \Delta A = 0$$

$$(5.41)$$

Taking derivatives of N and W, and considering their composite functions, we obtain

$$\frac{dN(\Delta A, \Delta \sigma, \Delta \omega)}{d\Delta A}\Big|_{\Delta A=0} = -j\frac{4\gamma \omega}{\pi A^2} - \frac{1.1128\lambda}{2A^{\frac{3}{2}}} + \frac{4\gamma A}{\pi \omega^2}\left(\frac{d\Delta \sigma}{d\Delta A} + j\frac{d\Delta \omega}{d\Delta A}\right) \quad (5.42)$$

$$\frac{dW}{d\Delta A}\Big|_{\Delta A=0} = \frac{dW}{ds}\Big|_{s=j\omega}\left(\frac{d\Delta \sigma}{d\Delta A} + j\frac{d\Delta \omega}{d\Delta A}\right) \quad (5.43)$$

Solving (5.41) for $\left(\frac{d\Delta \sigma}{d\Delta A} + j\frac{d\Delta \omega}{d\Delta A}\right)$ and taking account of (5.42) and (5.44), we can obtain an analytical formula. Considering only the real part of this formula we obtain (5.39). This completes the proof. □

5.5 Prescribed Convergence Control Law: DF Analysis

The prescribed convergence control law is given as follows:

$$u = -\lambda \text{sign}(\dot{\sigma} + \beta |\sigma|^\rho \text{sign}(\sigma))$$

with $0.5 < \rho < 1$. The system under analysis can be represented in the form of the block diagram as in Fig. 5.10. In an autonomous mode (i.e., with no input signal), the error signal is equal to the negative output. Assuming that a periodic motion occurs, the objective is to find the parameters of this periodic motion. The controller has one input and, for that reason, a describing function of the algorithm can be obtained in accordance with the definition given in (5.27) above. As previously, assume that $\sigma(t) = A\sin(\omega t)$. Applying Eq. (5.27) to the nonlinear

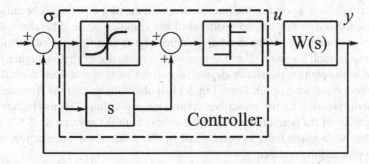

Fig. 5.10 Block diagram of a linear system with prescribed convergence control law

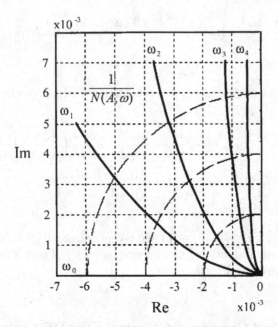

Fig. 5.11 Plots of the function $-1/N$ for the prescribed control law

function denoted as a dashed rectangle in Fig. 5.10, the describing function of the algorithm can be easily obtained. For the prescribed control law, the DF is a function of not only the amplitude but also of the frequency. The DF can be found only algorithmically. The values of the negative reciprocal of the DF are given in Fig. 5.11 as functions of the amplitude A for a few discrete values of the frequency $\omega (0 < \omega_1 < \omega_2 < \omega_3 < \omega_4, \omega_0 = 0)$. The function $-N_1^{-1}(A, \omega)$ (where ω =const) is completely located in the second quadrant. Two limiting cases for the frequencies $\omega \to 0$ and $\omega \to \infty$ correspond to the real half axis and to the imaginary half axis, respectively. Note that the magnitude of the describing function for the prescribed control law does not depend on frequency and is given by the following formula: $|N| = 4\lambda/(\pi A)$. In Fig. 5.11, the contours representing equal values of A are depicted as dash lines. The Nyquist plot of any plant of relative degree *three* or higher will have a point of the intersection with the plot $N_1^{-1}(A, \omega)$ because the former is located in the second quadrant of the complex plane. Therefore, if the transfer function of the plant has relative degree higher than *two*, a periodic motion may occur in such a system. For that reason, if an actuator of first or higher order is added to the plant with relative degree two driven by the prescribed control law, a periodic motion may occur. From Fig. 5.11, it also follows that the frequency of the periodic solution for the prescribed control law algorithm is always higher than the frequency of the periodic motion in the system with the asymptotic 2-SM relay controller, because the latter is determined by the point of the intersection of the Nyquist plot and the real axis.

5.6 Suboptimal Algorithm: DF Analysis

Now consider the generalized suboptimal algorithm given as follows:

$$u(t) = -\alpha(t)U_M \text{sign}(\sigma(t) - \beta\sigma_M(t)) \qquad (5.44)$$

$$\alpha(t) = \begin{cases} 1 & \text{if } \sigma_M(t)(y(t) - \beta\sigma_M(t)) \geq 0 \\ \alpha^*, & \text{if } \sigma_M(t)(y(t) - \beta\sigma_M(t)) < 0 \end{cases}$$

where $\sigma_M(t)$ is a piecewise constant function representing the value of the last
singular point of $\sigma(t)$ (i.e., the most recent value of $\sigma(t)$ satisfying the condition
$\dot{\sigma}(t) = 0$), U_M is the control amplitude, $\beta \in [0, 1)$ is the anticipation parameter, and
$\alpha^* \geq 1$ is the modulation parameter. Let us assume that the steady-state behavior
of the system (5.25) and (5.44) is a periodic, unimodal symmetric motion with zero
mean and shows that the motion under this assumption can exist. The sequence
of singular points of the variable $\sigma(t)$ is then an alternating sequence of positive
and negative values of the same magnitude: $\sigma_M^P, -\sigma_M^P, \beta\sigma_M^P, -\beta\sigma_M^P$ (where "P"
stands for periodic). The switching of the control occurs at the instants when the
plant output $\sigma(t)$ becomes equal either to σ_M^P or to $\beta\sigma_M^P$. This would correspond to
the following nonlinear characteristic of the controller (Fig. 5.12).

With this representation, the DF method can be conveniently used for analysis of
the system with the suboptimal algorithm. The usual assumption for applicability of
the DF method is the linear part (the combined *principal* and *parasitic* dynamics)

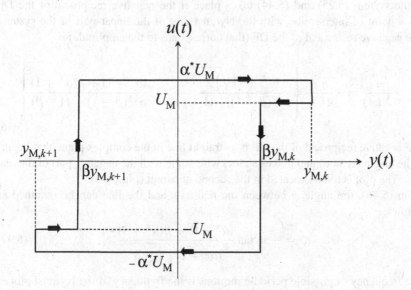

Fig. 5.12 Nonlinear characteristic of sub-optimal control

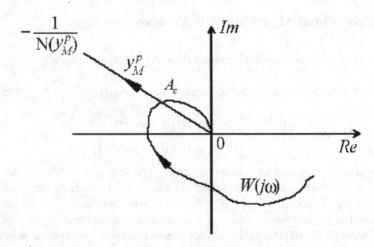

Fig. 5.13 Finding a periodic solution in a system with suboptimal control

satisfying the filtering hypothesis. As a result, the DF of the suboptimal algorithm can be given by the following equation:

$$N(A) = \frac{2U_M}{\pi A} \left\{ (\alpha^* + 1)\sqrt{1 - \beta^2} + j[(\alpha^* - 1) + \beta(\alpha^* + 1)] \right\} \qquad (5.45)$$

where $A = \sigma_M^P$ is the amplitude of the oscillations of the output. A periodic solution of the system (5.25) and (5.44) takes place if the negative reciprocal of the DF has a point of intersection with the Nyquist plot of the linear part of the system. The negative reciprocal of the DF (that corresponds to the amplitude) is

$$-\frac{1}{N(A)} = -\frac{\pi A}{4U_M} \left\{ \frac{(\alpha^* + 1)\sqrt{1 - \beta^2}}{\alpha^{*2}(\beta + 1) + (1 - \beta)} + j\frac{(\alpha^* - 1) + \beta(\alpha^* + 1)}{\alpha^{*2}(\beta + 1) + (1 - \beta)} \right\}$$

$$(5.46)$$

The negative reciprocal of the DF is a straight line in the complex plane that begins at the origin ($A = 0$) and makes a clockwise angle with the negative part of the real axis. The plot is totally located in the second quadrant (Fig. 5.13).
From (5.46), the angle ϕ between the real axis and the line can be obtained as follows:

$$\phi = -\arctan \frac{(\alpha^* - 1) + \beta(\alpha^* + 1)}{(\alpha^* + 1)\sqrt{1 - \beta^2}} \qquad (5.47)$$

The frequency of possible periodic motions is the frequency of the Nyquist plot of the plant at the point of intersection with the plot $-\frac{1}{N(A)}$. This can be expressed as the solution of the following nonlinear equation:

$$\arg W(j\omega) = -\pi - \phi \tag{5.48}$$

where ϕ is given by Eq. (5.47), ω is the frequency of the periodic solution, and $W(s)$ is the transfer function of the combined *parasitic* and *principal* dynamics.

The amplitude of this periodic motion depends on the magnitude of the frequency response $W(j\omega)$ of the plant at the frequency ω. Usually a larger value of ϕ provides a higher frequency and a smaller amplitude of the periodic motion. The amplitude A_c of the periodic motion can be evaluated as follows:

$$A_c = \frac{2\sqrt{2}U_m \sqrt{\alpha^{*2}(1+\beta)+1-\beta}}{\pi} |\arg W(j\omega_c)| \tag{5.49}$$

where ω_c is the frequency of the oscillations.

It follows from (5.47) that the range of ϕ values is $[0; \pi/2]$. Therefore, a periodic motion may occur if the phase characteristic of the combined *parasitic* and *principal* dynamics includes the range from $-\pi$ to -1.5π. This is only possible if the combined relative degree of the *parasitic* and *principal* dynamics is three or higher (however, there may be some cases when the phase characteristic a plant of relative degree lower than three can go below $-\pi$).

5.7 Comparisons of 2-Sliding Mode Control Algorithms

The examples which have been studied confirm the properties of the algorithms and also demonstrate high accuracy—the DF analysis matches very accurately the results of the simulations (see Tables 5.1–5.4). In the simulations the following values of control algorithm parameters have been used: for the twisting algorithm

Table 5.1 Results of computing and simulations for the system with the plant of relative degree two $W_p(s) = \frac{1}{s^2+s+1}$ and an actuator of relative degree one $W_a(s) = \frac{1}{0.01s+1}$ driven by the 2-SM control algorithms

	Twisting	Prescribed	Suboptimal
ω_c (DF analysis)	77.05	80.85	59.81
ω_c (simulations)	77.68	81.72	55.16
Amplitudes, A_c	1.67e-4	1.51e-4	3.6e-4

Table 5.2 Results of computing and simulations for the system with the plant of relative degree two $W_p(s) = \frac{1}{s^2+s+1}$ and an actuator of relative degree two $W_a(s) = \frac{1}{0.0001s^2+0.01s+1}$ driven by the 2-SM control algorithms

	Twisting	Prescribed	Suboptimal
ω_c (DF analysis)	54.64	63.10	50.83
ω_c (simulations)	54.53	61.60	50.67
Amplitudes, A_c	4.83e-4	3.67e-4	4.63e-4

Table 5.3 Results of computing and simulations for the system with the plant of relative degree one $W_p(s) = \frac{s+1}{s^2+s+1}$ and an actuator of relative degree one $W_a(s) = \frac{1}{0.01s+1}$ driven by the 2-SM control algorithms

	Twisting as a filter	Super-twisting	Conventional SMC
ω (DF analysis)	75.00	66.16	∞
ω (simulations)	75.51	64.96	Converge to ∞
Amplitudes, A_c	2.53e-6	2.33e-4	0

Table 5.4 Results of computing and simulations for the system with the plant of relative degree one $W_p(s) = \frac{s+1}{s^2+s+1}$ and an actuator of relative degree two $W_a(s) = \frac{1}{0.0001s^2+0.01s+1}$ driven by the 2-SM control algorithms

	Twisting as a filter	Super-twisting	Conventional SMC
ω (DF analysis)	53.52	55.18	100.00
ω (simulations)	53.41	54.26	99.26
Amplitudes, A_c	9.48e-6	4.81e-4	1.27e-4

$c_1 = 0.8, c_2 = .0.6$; for the prescribed control law $\lambda = \beta = 1, \rho = 0.5$; for the sub optimal algorithm $\alpha^* = 1, \beta = 0.5$; and for the super-twisting algorithm $\lambda = 0.6, \gamma = 0.8, \rho = 0.5$.

The presented analysis demonstrates that the 2-SM control *algorithms cannot be chattering-free although the algorithms are continuous*. A frequency domain analysis of the known 2-SM control algorithms proves that very convincingly. Furthermore the cause of chattering is the inevitable existence of *parasitic* (not accounted for in the 2-SM controller design) dynamics associated with the actuators, sensors, etc.

However, new algorithms may be developed in the future that may be claimed to be chattering-free. Of course, an individual frequency-domain analysis of each available algorithm can provide some clues, but a more detailed analysis of the principles the 2-SM is based on would be very useful and might prevent wishful but wrong conclusions in the future.

The frequency-domain analysis which has been presented leads to the formulation of *two* ideas or *two* frequency-domain interpretations of the 2-SM controller designs. Those ideas are (1) an *advance switching* and (2) *control smoothing* due to introduction of an integrator. The first principle is applied to plants of relative degree two, and the second principle is applied to the plants of relative degree one. Let us consider those two principles in more detail.

The *advance switching* idea is used in the twisting algorithm, prescribed control law, and suboptimal algorithm. The advance switching is different in comparison with the switching in the conventional relay feedback control and is achieved by introducing phase-lead elements (differentiators) into the system loop or from the use of the negative hysteresis of the relay nonlinearity. In the frequency domain it is manifested as the location of the negative reciprocal of the DF of the respective algorithm in the second quadrant of the complex plane (see Figs. 5.6, 5.11, and 5.13).

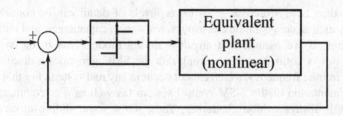

Fig. 5.14 Equivalent relay system

The *control smoothing* idea is realized via introducing an integrator in series with one of the 2-SM control algorithms which have advance switching or a relay nonlinearity (the super-twisting algorithm). This idea essentially exploits the stability margins of a relay control system with a plant of relative degree *one*, since the addition of one more relative degree does not cause finite frequency periodic motions in such systems. However, the introduction of the integrator in series with the *principal dynamics* of relative degree *one* makes the relative degree of this part of the system equal to *two*. As a result, any *parasitic dynamics* increases the overall relative degree to at least *three*. In this case, there always exists a point of intersection between the Nyquist plot of the serial connection of the actuator, the plant and the integrator, and the negative reciprocal of the DF of the SMC algorithm. Thus, if *parasitic dynamics* of relative degree *one* or higher are added to the *principal dynamics* of relative degree *one*, a periodic motion may occur in the system. Therefore, the existence of parasitic dynamics makes the existence of chattering unavoidable even in the case of the continuous control algorithms.

From the above analysis, the following observation can be formulated: regardless of the algorithm, if the system relative degree (that includes the actuator, plant, and the algorithm itself if the latter contains integrators) is three or higher, a periodic motion may occur in the system. Let us analyze this fact on the basis of the DF method. First note that any 2-SM control algorithm contains the relay (sign) function. After that let us transform the original system into the conventional relay feedback system notation. In Fig. 5.14, the equivalent plant, which now includes the original plant, actuator, and all parts of the 2-SM control algorithm except for the relay element, is nonlinear. However, the nonlinearities of this equivalent plant do not change the order and relative degree of the equivalent plant. Consequently they can be replaced with some gains (which is what was done in the DF analysis above). Therefore, a property which is known for the relay systems should also apply to the equivalent relay system in Fig. 5.14. It can be formulated as the possibility of the existence of a periodic motion if the relative degree is *three or higher* and the existence of an ideal sliding mode otherwise.

The nature of the parasitic dynamics is the dynamics of energy transformation from one form to another. For example, an actuator is supposed to transform an electrical signal into a force or mechanical motion, and a sensor has to transform mechanical motion into an equivalent electrical signal. It follows from first principles that the relative degree of the actuator–plant–sensor dynamics will always be higher than two. Moreover, even lower levels of detail can be considered, if necessary, such as delays in the electronics, resistance, capacitance, and inductance of the wires. If we assume that an ideal sliding mode exists in the system in Fig. 5.14, this is equivalent to assuming that the SMC generates a discontinuous control of infinite frequency, which cannot occur in any real system. For that reason, any implementation of the 2-SM control system (as well as a conventional SMC system) will always exhibit chattering. Thus, if for some sliding mode design, we obtained results corresponding to an ideal sliding mode that only means that the parasitic dynamics were either not considered or not considered in sufficient detail. Chattering is an inherent property of the sliding mode principal. The problem is therefore in controlling the parameters of the chattering and adjusting them to acceptable values. The approach presented here can be of significant help in achieving those goals.

5.8   Notes and References

DF analysis can give (predict) only the approximate limit cycle solution. The early results on the analysis of self-sustained oscillations in systems with conventional SMC can be found in [161]. A precise limit cycle analysis can be undertaken in 2-sliding systems using the LPRS technique presented in [29].

It is worth noting that the nonlinearity in Fig. 5.12 can be represented as a sum of two conventional hysteresis relays, for which the DFs are known [11].

The criterion for finite time convergence in terms of the angle between the high-frequency asymptote of the Nyquist plot of the plant and the low-amplitude asymptote of the negative reciprocal is given in [30].

A detailed analysis of self-sustained oscillations (chattering) in 2-SM systems is given in [28–30, 32, 34, 35, 88–90, 117, 150]. The analysis of the propagation of periodic signals for the systems governed by 2-SM control is studied in [36]. The self sustained oscillations in mechanical systems are studied in [4, 5].

5.9   Exercises

Exercise 5.1. Study a real sliding mode in the system

$$\dot{x}_1 = x_2$$
$$\dot{x}_2 = -2x_1 - 3x_2 + u_a$$
$$0.0001\ddot{u}_a = -0.01\dot{u}_a - u_a + u$$
$$\sigma = x_1 + x_2$$

with the controller

$$u = -\rho \text{sign}(\sigma)$$

(a) Predict the parameters of the limit cycle by solving the harmonic balance equation.
(b) Confirm the results via computer simulation.

Exercise 5.2. Repeat Exercise 5.1 assuming that the system is controlled by the twisting controller

$$u = -0.8 \text{sign}(\sigma) - 0.6 \text{sign}(\dot{\sigma})$$

Exercise 5.3. Repeat Exercise 5.1 assuming that the system is controlled by the super-twisting controller

$$u(t) = u_1(t) + u_2(t)$$

$$\dot{u}_1 = -\gamma \, \text{sign}(\sigma)$$

$$u_2 = \begin{cases} -\lambda |s_0|^\rho \, \text{sign}(\sigma), & \text{if } |\sigma| > s_0 \\ -\lambda |\sigma|^\rho \, \text{sign}(\sigma), & \text{if } |\sigma| \le s_0 \end{cases}$$

with $\gamma = 0.8$, $\lambda = 0.6$, and $\rho = 0.5$.

Exercise 5.4. Repeat Exercise 5.1 assuming that the system is controlled by the prescribed convergence control law

$$u = -\lambda \text{sign}(\dot{\sigma} + \beta |\sigma|^\rho \, \text{sign}(\sigma))$$

where $\lambda = 1$, $\rho = 0.5$ and $\beta = 1$.

Exercise 5.5. Repeat Exercise 5.1 assuming that the system is controlled by the sub-optimal controller

$$u(t) = -U_M \text{sign}(\sigma(t) - \beta \sigma_M(t)) \tag{5.50}$$

with $U_M = 0.1$ and $\beta = 0.2$.

Exercise 5.6. Consider the DC–DC converter presented in Fig. 5.15. The corresponding dynamic equations are given by

$$\begin{aligned} L\frac{di}{dt} &= -v + u_a V_{in} \\ C\frac{dv}{dt} &= i - \frac{v}{R} \end{aligned}$$

where i is the current through the inductor L, v is the voltage across the capacitor C, V_{in} is the input voltage, and $u \in \{0, 1\}$ is the switching control signal. The goal is to stabilize the output voltage v at the desired level v_d.

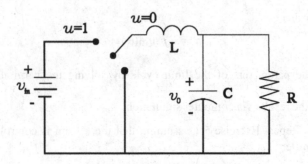

Fig. 5.15 DC–DC Buck converter

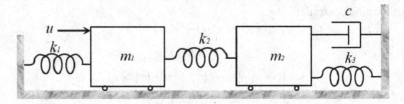

Fig. 5.16 Mass–spring–damper system

Assume that the switching element has the following dynamics:

$$\tau \dot{u}_a = -u_a + u$$

where $u = \frac{1}{2}(1 - \text{sign}(\sigma))$ and $\sigma = v - v_d$.

(a) Predict the parameters of the limit cycle by solving the harmonic balance equation.
(b) Confirm the results via computer simulations.

Exercise 5.7. Consider the mass–spring–damper system presented in Fig. 5.16. The dynamic model of the system is

$$m_1 \ddot{x}_1 + (k_1 + k_2)x_1 - k_2 x_2 = u$$
$$m_2 \ddot{x}_2 + c\dot{x}_2 + (k_3 + k_2)x_2 - k_2 x_1 = 0$$

where $m_1 = 1.28$, $m_2 = 1.08$, $k_1 = 190$, $k_2 = 450$, $k_3 = 190$, $c = 15$, and x_1 and x_2 are the positions of the masses m_1 and m_2, respectively. The goal of control is to stabilize the position of m_2 at the origin. Using the suboptimal controller

$$u = -0.1\text{sign}(\sigma - 0.5\sigma_M)$$

where $y = x_2$, compute the parameters of the limit cycle by solving the harmonic balance equation and confirm the results via computer simulation.

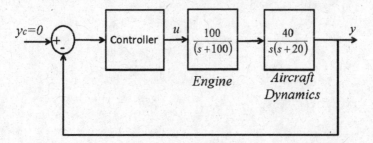

Fig. 5.17 Control system of the heading angle y of the bi-wing aircraft

Exercise 5.8. Consider the disk drive suspension system represented by the transfer functions

$$G_a(s) = \frac{5}{0.001s + 1}; \qquad G_p(s) = \frac{48.78}{s(s + 20)}$$

where $G_a(s)$ is the transfer function of the actuator, named the motor coil, and $G_p(s)$ is the transfer function of the plant, named the arm. Assume the input of the actuator is produced by the twisting algorithm

$$u = -0.8\text{sign}(\sigma) - 0.6\text{sign}(\dot{\sigma})$$

Then compute the parameters of the limit cycle by solving the harmonic balance equation, and confirm the results via a computer simulation.

Exercise 5.9. The heading angle y of a bi-wing aircraft is controlled by the prescribed convergence controller

$$u = -10\text{sign}(\dot{\sigma} + 2|\sigma|^{0.5}\,\text{sign}(\sigma))$$

The block diagram of the control system is given in Fig. 5.17, where the transfer function of the engine is considered as unmodeled dynamics. Compute the parameters of the limit cycle by solving the harmonic balance equation and confirm the results via computer simulation.

Chapter 6
Higher-Order Sliding Mode Controllers and Differentiators

In this chapter we further develop the ideas of Chap. 4 where second-order sliding modes were formulated. As we have seen, second-order sliding modes make the sliding variables vanish in finite time, when the relative degree of the variable equals two, and are able to solve the same problem by means of continuous control, if the relative degree is one. This helps to remove dangerous high-energy vibrations (the dangerous types of chattering). So-called higher-order sliding modes (HOSMs) solve these problems for arbitrary relative degrees. The realization of the scheme requires more information: usually one needs to calculate or measure a number of successive time derivatives of the sliding variables. However that problem is also solved within a similar framework. As a result, arbitrary-order exact robust differentiators are developed, having their own significance in terms of general observation theory. In particular, tracking problems are solved in finite time and with ideal accuracy, by means of smooth control, if the output relative degree is known. The accuracy remains high in the presence of small noises, switching inaccuracies and delays, etc.

While the theory presented in this chapter can be considered complicated, its application is very simple and straightforward. Suppose that the studied problem is to make a smooth scalar output σ vanish and to keep it at zero. Assume that the system is smooth and affine in the (scalar) control and also that when the output is successively differentiated with respect to time, the control appears for the first time in the rth total time derivative $\sigma^{(r)}$ and its functional coefficient is strictly positive. The number r is called the relative degree. The derivatives $\dot{\sigma}, \ldots, \sigma^{(r-1)}$ which are required to implement the controller can be calculated in real time by means of the HOSM differentiator. The differentiator is robust with respect to input noises and exact in their absence. If the coefficient of control in $\sigma^{(r)}$ is negative, one simply needs to change the sign of the listed controllers.

At that point one can just start to use the results, as applied in Sect. 6.10, and the reader is indeed invited to look at Sect. 6.10 before reading the preceding theoretical developments. However, the theory is required for deep understanding of the proposed method, its accuracy, and limitations. In particular, the HOSM problem

Y. Shtessel et al., *Sliding Mode Control and Observation*, Control Engineering,   213
DOI 10.1007/978-0-8176-4893-0_6, © Springer Science+Business Media New York 2014

is shown to be reduced to finite-time stabilization of a differential inclusion. The task is performed in the framework of homogeneity theory that not only simplifies the proofs and the control design but also readily calculates the accuracy of the controllers and differentiators.

6.1   Introduction

The main idea of sliding mode control is to keep a properly chosen constraint at zero in sliding mode. It is natural to classify sliding modes by the smoothness degree of the constraint function calculated along the system trajectories. More precisely, let the constraint be of the form $\sigma = 0$, where σ is some smooth function of the state and time, and let it be kept identically on Filippov trajectories of a discontinuous dynamic system. There is no sense to consider the function σ on the sliding mode trajectories, for it is just identical zero. On the other hand the sliding mode $\sigma \equiv 0$ may be classified by the first total derivative $\sigma^{(r)}$ which contains a discontinuity in a small vicinity of the sliding mode trajectories. That number r is called the sliding order. The formal definition is following.

Definition 6.1. Consider a discontinuous differential equation $\dot{x} = f(x)$ (Filippov differential inclusion $\dot{x} \in F(x)$) with a smooth output function $\sigma = \sigma(x)$, and let it be understood in the Filippov sense.[1] Then, if

1. The total time derivatives $\sigma, \dot{\sigma}, \ldots, \sigma^{(r-1)}$ are continuous functions of x
2. The set

$$\sigma = \dot{\sigma} = \ldots = \sigma^{(r-1)} = 0 \tag{6.1}$$

   is a nonempty integral set (i.e., consists of Filippov trajectories)
3. The Filippov set of admissible velocities at the r-sliding points Eq. (6.1) contains more than one vector

the motion on the set Eq. (6.1) is said to exist in an r-sliding (rth-order sliding) mode. The set Eq. (6.1) is called the r-sliding set. The nonautonomous case is reduced to the one considered above by introducing the fictitious equation $\dot{t} = 1$.

The words "rth-order sliding" are often abridged to "r-sliding," and the term "r-sliding controller" replaces the expression "finite-time-convergent r-sliding mode controller." In the case when σ is a vector function, the same definition is applied to each scalar component of σ, and results in a vector sliding order.

As we will immediately see, the sliding order is closely connected to the notion of relative degree. In order to introduce the concept we need the notion of a Lie derivative, which is actually the derivative of a function along the trajectories of a vector field.

Let $\sigma(x)$ be a differentiable function, where $x \in \mathbb{R}^n$ and $f(x)$ is a vector field (i.e., an n-dimensional vector function). Then *the Lie derivative* of σ with respect to f at the point x_0 is defined as

[1]For details, see Sect. 2.1.1.

$$L_f \sigma(x_0) = \nabla \sigma(x_0) \cdot f(x_0) = \sum_{i=1}^{n} \frac{\partial}{\partial x_i} \sigma(x_0) f_i(x_0)$$

Consider a smooth dynamic system

$$\dot{x} = a(x) + b(x)u, \ x \in \mathbb{R}^n, \ u \in \mathbb{R} \qquad (6.2)$$

with a scalar input (control) $u \in \mathbb{R}$ and a smooth scalar output $\sigma(x)$. The total derivative of the output σ is defined as the time derivative of σ along the trajectory of the system

$$\dot{\sigma} = \frac{d}{dt} \sigma(x(t)) = \nabla \sigma(x)(a(x) + b(x)u) = L_a \sigma(x) + L_b \sigma(x)u \qquad (6.3)$$

The relative degree of the output σ with respect to the input u at the point x is defined as the order of the total derivative of σ, at which the control appears with a nonzero coefficient for the first time, provided the control coefficients are identically zero in all the lower-order derivatives. In particular, the relative degree equals 1 at the given point x_0, if $L_b \sigma(x_0) \neq 0$. Suppose that $L_b \sigma(x) \equiv 0$ in a vicinity of x_0 then differentiating Eq. (6.3) yields

$$\ddot{\sigma} = L_a^2 \sigma(x) + L_b L_a \sigma(x)u$$

Continuing the calculation in this way we obtain the following formal definition by Isidori.

Definition 6.2. The number r is called the relative degree of the output σ of the system (6.2) with respect to the input u at the point x_0, if the conditions

$$L_b \sigma(x) = L_b L_a \sigma(x) = \ldots = L_b L_a^{r-2} \sigma(x) = 0, \ L_b L_a^{r-1} \sigma(x) \neq 0$$

hold in some vicinity of the point x_0.

The case when the system depends on time so that

$$\dot{x} = a(t, x) + b(t, x)u, \ x \in \mathbb{R}^n, \ u \in \mathbb{R}, \ \sigma = \sigma(t, x) \in \mathbb{R} \qquad (6.4)$$

can be handled by introducing the fictitious equation $\dot{t} = 1$ so that one obtains the system

$$\begin{pmatrix} \dot{x} \\ \dot{t} \end{pmatrix} = \begin{pmatrix} a(t, x) \\ 1 \end{pmatrix} + \begin{pmatrix} b(t, x) \\ 0 \end{pmatrix} u = \tilde{a}(t, x) + \tilde{b}(t, x)u$$

and the definition above then can be applied.

Definition 6.3. The number r is called the relative degree of the output σ of the system (6.4) with respect to the input u at the point (t_0, x_0), if the conditions

$$L_{\tilde{b}}\sigma = L_{\tilde{b}}L_{\tilde{a}}\sigma = \ldots = L_{\tilde{b}}L_{\tilde{a}}^{r-2}\sigma = 0, \ L_{\tilde{b}}L_{\tilde{a}}^{r-1}\sigma \neq 0$$

hold in some vicinity of the point (t_0, x_0).

Thus, if the relative degree equals r, the output of Eq. (6.4) satisfies the equations

$$\sigma^{(i)} = L_{\tilde{a}}^i\sigma, \ i = 1, 2, \ldots r - 1; \ \ \sigma^{(r)} = L_{\tilde{a}}^r\sigma + \left(L_{\tilde{b}}L_{\tilde{a}}^{r-1}\sigma\right)u$$

Therefore, $\sigma^{(i)}$, $i = 1, 2, \ldots r - 1$, are continuous functions of t, x, and keeping $\sigma \equiv 0$ by means of a permanently switching control is only possible in the r-sliding mode.

6.2   Single-Input Single-Output Regulation Problem

Consider a dynamic system of the form (6.4), where $x \in \mathbb{R}^n$, a, b, and $\sigma : \mathbb{R}^{n+1} \to \mathbb{R}$ are unknown smooth functions and even the dimension n might be uncertain. Suppose only measurements of σ are available in real time. The task is to force in finite time $\sigma \equiv 0$ by means of an appropriate control $u \in \mathbb{R}$. Note that the stated problem can be considered as a general tracking problem for a black-box uncertain process.

The relative degree of the system is assumed to be constant and known and given by r so that

$$\sigma^{(r)} = h(t, x) + g(t, x)u \tag{6.5}$$

where $h(t, x) = L_{\tilde{a}}^r\sigma$ and $g(t, x) = L_{\tilde{b}}L_{\tilde{a}}^{r-1}\sigma \neq 0$. It is assumed that $g(t, x)$ is positive and for some $K_m, K_M, C > 0$ the inequalities

$$0 < K_m \leq g(t, x) \leq K_M, \quad |h(t, x)| \leq C \tag{6.6}$$

hold—which is always true at least locally. If $g(t, x)$ is negative one just needs to change the sign of the controller which will be developed. The trajectories of Eq. (6.4) are assumed infinitely extendible in time for any Lebesgue-measurable bounded control $u(t, x)$.

Thus, the task is actually to finite-time stabilize the states $\sigma \dot{\sigma}, \ldots, \sigma^{(r-1)}$ of the uncertain system (6.5) depending on the variable parameter $x(t)$ under the conditions in Eq. (6.6). Note that while the relative degree r is indeed needed to be known, the exact knowledge of the parameters K_m, K_M, C is not really necessary in practice, for they only influence the magnitude of the control to be designed. Indeed, in any case, probably the best way to determine the control parameters is by extensive simulation of system (6.4).

Unfortunately no static continuous output-feedback control is capable of stabilizing system (6.5) under the uncertainty conditions considered. Indeed, any continuous control

$$u = \varphi(\sigma, \dot{\sigma}, \ldots, \sigma^{(r-1)}) \tag{6.7}$$

forcing $\sigma \equiv 0$, would need to satisfy the equality $\varphi(0, 0, .., 0) = -h(t, x)/g(t, x)$, whenever Eq. (6.5) holds. Obviously, $\varphi(0, 0, .., 0) \in [-C/K_m, C/K_m]$. Consider two trivial systems of the form (6.5) satisfying conditions (6.6):

$$\sigma^{(r)} = C + K_m u, \ \sigma \in \mathbb{R}, \ u \in \mathbb{R}, \ x = (\sigma, \dot{\sigma}, \ldots, \sigma^{(r-1)})^T \in \mathbb{R}^r \tag{6.8}$$

$$\sigma^{(r)} = -C + K_m u, \ \sigma \in \mathbb{R}, \ u \in \mathbb{R}, \ x = (\sigma, \dot{\sigma}, \ldots, \sigma^{(r-1)})^T \in \mathbb{R}^r \tag{6.9}$$

It is impossible to find a continuous control law to simultaneously satisfy Eqs. (6.8) and (6.9) while $\sigma^{(r)} = 0$. Thus, *the control* defined in Eq. (6.7) *has to be discontinuous at least on the set* Eq. (6.1). In other words, an r-sliding mode $\sigma = 0$ is to be established.

From Eqs. (6.5) and (6.6), the solutions of Eq. (6.5) satisfy the inclusion

$$\sigma^{(r)} \in [-C, C] + [K_m, K_M] u \tag{6.10}$$

In contrast to the uncertain system (6.5), the inclusion in Eq. (6.10) is perfectly known. However it does not "remember" about system (6.4) except the constants r, C, K_m, K_M. Thus, provided Eq. (6.6) holds, the finite-time stabilization of Eq. (6.10) at the origin simultaneously solves the stated problem for all possible systems (6.4).

The closed differential inclusion Eqs. (6.7), (6.10) is understood in the Filippov sense, which means that the control Eq. (6.7) should be locally bounded and the right-hand-side vector set is enlarged at the discontinuity points of Eq. (6.7) in order to satisfy the convexity and semi-continuity conditions from Definition 2.2. For this end the Filippov procedure from Definition 2.3 is applied to the function (6.7), and the resulting scalar set is substituted for u in Eq. (6.10).

Note that implementing the HOSM controller given by Eq. (6.7) requires observation of the derivatives $\dot{\sigma}, \ldots, \sigma^{(r-1)}$ or real-time differentiation of the output. The problem remains complicated, even if the derivatives $\dot{\sigma}, \ldots, \sigma^{(r-1)}$ are available. One of the main approaches to this problem is based on homogeneity theory. The controllers, which we will design, are *r-sliding homogeneous*. Let us now introduce the necessary notions.

6.3 Homogeneity, Finite-Time Stability, and Accuracy

Assign a weight (*the homogeneity degree*) m_i to each coordinate $x_i \in \mathbb{R}$, where $m_i > 0$. We will write

$$\deg(x_i) = m_i$$

The corresponding simple linear transformation

$$d_\kappa : (x_1, x_2, \ldots, x_n) \mapsto (\kappa^{m_1} x_1, \kappa^{m_2} x_2, \ldots, \kappa^{m_n} x_n) \qquad (6.11)$$

is called the *homogeneity dilation*, and $\kappa > 0$ is called its parameter.

Definition 6.4. A function $f : \mathbb{R}^n \to \mathbb{R}$ is called homogeneous of the degree (weight) $q \in \mathbb{R}$ with the dilation Eq. (6.11) and written as $\deg(f) = q$, if for any $\kappa > 0$, the identity $f(d_\kappa x) = \kappa^q f(x)$ holds.

For instance, in accordance with Definition 6.4, quadratic and cubic forms are homogeneous functions of weights 2 and 3, respectively, corresponding to the trivial homogeneity weights $\deg x_i = 1$. Next, let us try $\deg x_1 = 2$, $\deg x_2 = 3$; then

$$\deg \left(x_1^3 - x_2^2 \text{sign}(x_1) + 2 \,|x_1|^{3/2} \, x_2 \right) = 6$$

Indeed

$$\left(\kappa^2 x_1\right)^3 - \left(\kappa^3 x_2\right)^2 \text{sign}(\kappa^2 x_1) + 2 \left|\kappa^2 x_1\right|^{3/2} \kappa^3 x_2 = \kappa^6 \left(x_1^3 - x_2^2 \text{sign}(x_1) + 2 \,|x_1|^{3/2} \, x_2 \right)$$

The following are the simple rules of homogeneous arithmetic. The homogeneity degree of zero is not defined. Let A and B be some homogeneous functions of $x \in \mathbb{R}^n$ different from identical zero, and let λ be a real number; then

1. The sum of A and B is a homogeneous function only if $\deg A = \deg B$
2. $\forall \lambda \neq 0 \deg \lambda = 0$
3. $\deg AB = \deg A + \deg B$
4. $\deg (A/B) = \deg A - \deg B$
5. $\deg (\lambda A) = \deg A$
6. $\deg \frac{\partial}{\partial x_i} A = \deg A - \deg x_i$, if $\frac{\partial}{\partial x_i} A$ is not identical zero

To verify the last equality it can be seen that

$$\frac{\partial}{\partial \kappa^{m_i} x_i} A(d_\kappa x) = \kappa^{-m_i} \frac{\partial}{\partial x_i} \kappa^{\deg A} A(x) = \kappa^{\deg A - m_i} \frac{\partial}{\partial x_i} A(x)$$

Definition 6.5. A vector field $f : \mathbb{R}^n \to \mathbb{R}^n$, $f = (f_1, f_2, \ldots, f_n)^T$, is called *homogeneous of degree* $q \in \mathbb{R}$ with the dilation Eq. (6.11) and written as $\deg f = q$, if all its components f_i are homogeneous and the identities

$$\deg f_i = \deg x_i + \deg f = \deg x_i + q, \quad i = 1, 2, \ldots, n \qquad (6.12)$$

hold.

In order to better understand the above definition, consider the differential equation

$$\dot{x} = f(x) \qquad (6.13)$$

and introduce a homogeneity weight of the time variable. Let $\deg t = -q$, $q \in \mathbb{R}$; then Eq. (6.12) reads as

$$\deg \dot{x}_i = \deg x_i - \deg t = \deg f_i, \ i = 1, 2, \ldots, n$$

and means that Eq. (6.13) is invariant with respect to the linear time-coordinate transformation

$$G_\kappa : (t, x) \mapsto (\kappa^{-q} t . d_\kappa x), \ \kappa > 0 \tag{6.14}$$

Therefore, we will also call differential equation (6.13) homogeneous with homogeneity degree q. The equalities (6.12) can be rewritten as one vector equality:

$$f(x) = \kappa^{-q} d_\kappa^{-1} f(d_\kappa x)$$

Note that in the scalar case $x \in \mathbb{R}$ any function $f : \mathbb{R} \to \mathbb{R}$ can be considered as a vector field as well. The homogeneity degrees of the function f and of the vector field f are different in that case. Indeed, let for example $\deg x = 1$, then the homogeneity degree of the function x^2 equals to 2, while the homogeneity degree of the corresponding vector field is 1. The ambiguity disappears if we speak about the homogeneity of the differential equation $\dot{x} = x^2$. Indeed, $\deg x^2 = 2$, but on the other hand, due to homogeneity, $\deg x^2 = \deg x - \deg t$, which means that $\deg t = -1$ and the homogeneity degree of the differential equation is 1.

Any linear time-invariant homogeneous differential equation is indeed homogeneous with homogeneity degree 0, and $\deg x_i = 1, i = 1, 2, \ldots n$. Let the weights of x_1 and x_2 be 3 and 2, respectively. Then the system of differential equations

$$\begin{cases} \dot{x}_1 = x_2 \\ \dot{x}_2 = -x_1^{1/3} - |x_2|^{1/2} \text{sign}(x_2) \end{cases} \tag{6.15}$$

is homogeneous of the degree -1, $\deg t = 1$.

The nonzero homogeneity degree q of a vector field can always be scaled to ± 1 by an appropriate proportional change to the weights of the coordinates and time. Indeed, substituting $\kappa = \lambda^s$, where $\lambda, s > 0$, one obtains $\kappa^{m_i} x_i = \lambda^{s m_i} x_i$, $\kappa^{-q} t = \lambda^{-sq} t$.

Definition 6.6. A vector-set field $F(x) \subset \mathbb{R}^n$, $x \in \mathbb{R}^n$, and the differential inclusion

$$\dot{x} \in F(x) \tag{6.16}$$

are called homogeneous of the degree $q \in \mathbb{R}$ with the dilation Eq. (6.11), which is written as $\deg F = q$, if Eq. (6.16) is invariant with respect to the time-coordinate transformation Eq. (6.14).

The invariance is equivalent to the identity

$$F(x) = \kappa^{-q} d_\kappa^{-1} F(d_\kappa x), \ \kappa > 0 \tag{6.17}$$

Indeed, the necessary equivalence of $\dot{x} \in F(x)$ and $\frac{d(d_\kappa x)}{d(\kappa^{-q}t)} \in F(d_\kappa x)$ implies Eq. (6.17). Definition 6.5 is a particular case of Definition 6.6. Also here the nonzero homogeneity degree can always be scaled to ± 1.

Let once more the weights of x_1 and x_2 be 3 and 2, respectively. Then the differential inequality

$$|\dot{x}_1| + \dot{x}_2^{4/3} \leq x_1^{4/3} + x_2^2$$

corresponds to the homogeneous differential inclusion

$$(\dot{x}_1, \dot{x}_2) \in \{(z_1, z_2) \,|\, |z_1| + z_2^{4/3} \leqslant x_1^{4/3} + x_2^2, \; z_1, z_2 \in \mathbb{R}^2\}$$

and has degree $+1$. Indeed, let $\deg t = -1$; then $\deg |\dot{x}_1| = \deg \dot{x}_2^{4/3} = \deg x_1^{4/3} = \deg x_2^2 = 4$.

Definition 6.7. A differential inclusion $\dot{x} \in F(x)$ (equation $\dot{x} = f(x)$) is further called *globally asymptotically stable* at 0, if it is Lyapunov stable, and all solutions are infinitely extended in time and converge to 0 as $t \to \infty$.

Definition 6.8. A differential inclusion $\dot{x} \in F(x)$ (equation $\dot{x} = f(x)$) is called *globally uniformly asymptotically stable* at 0, if it is Lyapunov stable, and for any $R > 0, \varepsilon > 0$ there exists $T > 0$ such that any trajectory starting within the disk $||x|| < R$ enters the disk $||x|| < \varepsilon$ in time T and stays there forever.

It can be shown that in the case of a homogeneous continuous differential equation the last two definitions are equivalent.

Definition 6.9. A differential inclusion $\dot{x} \in F(x)$ (equation $\dot{x} = f(x)$) is further called *globally uniformly finite-time stable* at 0, if it is Lyapunov stable, and for any $R > 0$ there exists $T > 0$ such that any trajectory starting within the disk $||x|| < R$ stabilizes at zero in the time T.

The importance of the notions introduced is justified by the following remarkable result: asymptotic stability of a (continuous) homogeneous differential equation (6.13) with negative homogeneous degree is equivalent to its global uniform finite-time stability.

For example, system (6.15) is finite-time stable. Indeed, consider the Lyapunov function

$$V = \tfrac{3}{4}x_1^{4/3} + \tfrac{1}{2}x_2^2, \dot{V} = -|x_2|^{3/2}$$

Then, according to LaSalle's theorem, the trajectories asymptotically converge to the largest invariant subset of the set $\dot{V} = 0$, i.e., to the origin. Thus, the system is asymptotically stable. Since its homogeneity degree is -1, we also get finite-time stability.

Definition 6.10. A set D is called *dilation retractable* if $d_\kappa D \subset D$ for any $\kappa \in [0, 1]$. In other words for each point x it contains the whole line $d_\kappa x, \kappa \in [0, 1]$.

Definition 6.11. A homogeneous differential inclusion $\dot{x} \in F(x)$ (equation $\dot{x} = f(x)$) is called *contractive* if there are two compact sets D_1 and D_2 and a number $T > 0$, such that D_2 lies in the interior of D_1 and contains the origin; D_1 is dilation retractable; and all trajectories starting at the time 0 within D_1 remain in D_2 at the time moment T.

The properties of homogenous Filippov inclusions are studied in Theorems 6.1, 6.2, and 6.3 presented below.

Theorem 6.1. *Let $\dot{x} \in F(x)$ be a homogeneous Filippov inclusion with a negative homogeneous degree $-p$, i.e., $\deg t = p$; then Definitions 6.8, 6.9, and 6.11 are equivalent and the maximal settling time is a locally bounded homogeneous function of the initial conditions of the degree p.*

Sketch of the proof. Obviously, both Definitions 6.8 and 6.9 imply Definition 6.11 and Definition 6.9 implies 6.8. Thus, it is enough to prove that 6.11 implies 6.9. All trajectories starting in the set D_1 concentrate in a smaller set D_2 in time T. Applying the homogeneity transformation it can be verified that the same is true with respect to the sets $d_\kappa D_1$, $d_\kappa D_2$ and the time κT for any $\kappa > 0$. An infinite collapsing chain of embedded regions is now constructed, such that any point belongs to one of the regions and the resulting convergence time is majored by a geometric series.

Due to the continuous dependence of solutions of the Filippov inclusion $\dot{x} \in F(x)$ on its graph $\Gamma = \{(x, y) \mid y \in F(x)\}$, the contraction feature is obviously robust with respect to perturbations causing small changes of the inclusion graph in some fixed vicinity of the origin.

Corollary 6.1 *The global uniform finite-time stability of homogeneous differential equations (Filippov inclusions) with negative homogeneous degree is robust with respect to locally small homogeneous perturbations.*

Let $\dot{x} \in F(x)$ be a homogeneous Filippov differential inclusion with $\deg x_i = m_i > 0$ and $\deg t = p > 0$. Consider the case of "noisy measurements" for x_i with the magnitude $\beta_i \tau^{m_i}$, $\beta_i, \tau > 0$ that is

$$\dot{x} \in F(x_1 + \beta_1 [-1, 1] \tau^{m_1}, \ldots, x_n + \beta_n [-1, 1] \tau^{m_n})$$

Successively applying the global closure of the right-hand-side graph and the convex closure at each point x, a new Filippov differential inclusion $\dot{x} \in F_\tau(x)$ can be obtained.

Theorem 6.2. *Let $\dot{x} \in F(x)$ be a globally uniformly finite-time stable homogeneous Filippov inclusion with homogeneity weights $m_1, \ldots, m_n$ and degree $-p < 0$, and let $\tau > 0$. Suppose that a continuous function $x(t)$ is defined for any $t \geq -\tau p$ satisfying some initial conditions $x(t) = \xi(t)$, $t \in [-\tau^p, 0]$. Then, if $x(t)$ is any solution of the disturbed inclusion*

$$\dot{x} \in F_\tau(x(t - \tau^p, 0)), \; 0 < t < \infty$$

the inequalities $|x_i| < \gamma_i \tau^{m_i}$ are established in finite time with some positive constants γ_i independent of τ and ξ.

Note that Theorem 6.2 covers the cases of retarded or discrete noisy measurements of all or some of the coordinates and any mixed cases. In particular, infinitely extendible solutions certainly exist in the case of noisy discrete measurements of some variables or in the constant time-delay case. For example, with small delays of the order of τ introduced in the right-hand side of Eq. (6.15), an accuracy of $x_1 = O(\tau^3)$, $\dot{x}_1 = x_2 = O(\tau^2)$, and $\dot{x}_2 = O(\tau)$ is obtained. From Corollary 6.1, for a sufficiently small ε, the addition of the term $\varepsilon x_1^{2/3}$ in the first equation of Eq. (6.15) disturbs neither the finite-time stability nor the asymptotic accuracy.

6.4   Homogeneous Sliding Modes

Suppose that the feedback in Eq. (6.7) imparts homogeneity properties to the closed-loop inclusion Eqs. (6.7), (6.10). Due to the term $[-C, C]$, the right-hand side of Eq. (6.10) can only have homogeneity degree 0 with $C \neq 0$. Indeed, with a positive degree, the right-hand side of Eqs. (6.7), (6.10) approaches zero near the origin, which is not possible with $C \neq 0$. With a negative degree it is not bounded near the origin, which contradicts the local boundedness of φ. Thus, $\deg \sigma^{(r)} = \deg \sigma^{(r-1)} - \deg t = 0$, and the homogeneity degree of $\sigma^{(r-1)}$ has to be opposite to the degree of the whole differential inclusion.

Scaling the inclusion homogeneity degree to -1, get that the homogeneity weights of $t, \sigma, \dot{\sigma}, \ldots, \sigma^{(r-1)}$ are $1, r, r-1, \ldots, 1$, respectively. This homogeneity is called *r-sliding homogeneity*.

Definition 6.12. The inclusion Eqs. (6.7), (6.10) is called *r-sliding homogeneous* if for any $\kappa > 0$ the combined time-coordinate transformation

$$G_\kappa : \left(t, \sigma, \dot{\sigma}, \ldots, \sigma^{(r-1)} \right) \mapsto \left(\kappa t, \kappa^r \sigma, \kappa^{r-1} \dot{\sigma}, \ldots, \kappa \sigma^{(r-1)} \right) \qquad (6.18)$$

preserves the closed-loop inclusion Eqs. (6.7), (6.10).

Note that the Filippov differential inclusion corresponding to the closed-loop inclusion Eqs. (6.7), (6.10) is also *r*-sliding homogeneous.

Transformation Eq. (6.18) transfers Eqs. (6.7), (6.10) into

$$\frac{d^r (\kappa^r \sigma)}{(d \kappa t)^r} \in [-C, C] + [K_m, K_M] \cdot \varphi \left(\kappa^r \sigma, \kappa^{r-1} \dot{\sigma}, \ldots, \kappa \sigma^{(r-1)} \right)$$

and hence, Eqs. (6.7), (6.10) is *r*-sliding homogeneous if

$$\varphi \left(\kappa^r \sigma, \kappa^{r-1} \dot{\sigma}, \ldots, \kappa \sigma^{(r-1)} \right) = \varphi \left(\sigma, \dot{\sigma}, \ldots, \sigma^{(r-1)} \right) \qquad (6.19)$$

Definition 6.13. Controller Eq. (6.7) is called r-sliding homogeneous (rth-order sliding homogeneous) if Eq. (6.19) holds for any $(\sigma, \dot{\sigma}, \ldots, \sigma^{(r-1)})$ and $\kappa > 0$. The corresponding sliding mode is also called homogeneous (if it exists).

Such an r-sliding homogeneous controller is inevitably discontinuous at the origin $(0, \ldots, 0)$, unless φ is a constant function. It is also uniformly bounded, since it is locally bounded and takes on all its values in any vicinity of the origin. Recall that the values of φ on any zero-measure set do not affect the corresponding Filippov inclusion.

A controller is called r-sliding homogeneous *in the broader sense* if Eq. (6.18) preserves the resulting trajectories of Eq. (6.10). Thus, the suboptimal 2-sliding controller (see Sect. 4.2.3)

$$u = -r_1\text{sign}(\sigma - \sigma^*/2) + r_2\text{sign}(\sigma^*), \ r_1 > r_2 > 0 \tag{6.20}$$

where

$$r_1 - r_2 > \frac{C}{K_m}; \quad r_1 + r_2 > \frac{4C + K_M(r_1 - r_2)}{3K_m} \tag{6.21}$$

is homogeneous, although it does not have the feedback form (6.7).

6.5 Accuracy of Homogeneous 2-Sliding Modes

With $r = 2$ we get $\deg \sigma = 2$, $\deg \dot{\sigma} = 1$. As follows from Theorem 6.2 the discrete-sampling versions of 2-sliding homogeneous controllers with the sampling interval τ provide the accuracy level $\sigma = O(\tau^2)$, $\dot{\sigma} = O(\tau)$. Similarly, noisy measurements lead to the accuracy $\sigma = O(\varepsilon)$, $\dot{\sigma} = O(\varepsilon^{1/2})$, if the maximal errors arising from sampling σ and $\dot{\sigma}$ are of the order of ε and $\varepsilon^{1/2}$, respectively.

Almost all known 2-sliding controllers are 2-sliding homogeneous. Let us check the homogeneity of the 2-sliding controllers studied in Chap. 4. In other words, condition (6.19) is to be checked:

- The twisting controller (Sect. 4.2.2) satisfies Eq. (6.19); indeed

$$u = -(r_1\text{sign}(\sigma) + r_2\text{sign}(\dot{\sigma})) = -(r_1\text{sign}(\kappa^2\sigma) + r_2\text{sign}(\kappa\dot{\sigma}))$$

has the convergence conditions

$$(r_1 + r_2)K_m - C > (r_1 - r_2)K_M + C, \ (r_1 - r_2)K_m > C$$

Its typical trajectory in the plane $\sigma, \dot{\sigma}$ is shown in Fig. 6.1.
- The controller with the prescribed convergence law Eq. (4.21) satisfies

$$u = -\alpha\,\text{sign}\left(\dot{\sigma} + \beta\,|\sigma|^{1/2}\,\text{sign}(\sigma)\right) = -\alpha\,\text{sign}\left(\kappa\dot{\sigma} + \beta\,|\kappa^2\sigma|^{1/2}\,\text{sign}(\kappa^2\sigma)\right)$$

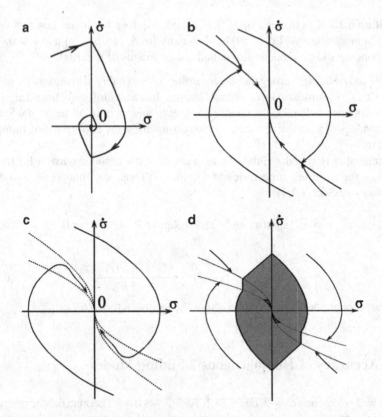

Fig. 6.1 Convergence of various 2-sliding homogeneous controllers (**a**) Twisting controller (**b**) Prescribed convergence law (**c**) Quasi-continuous controller (**d**) Contractivity

- The quasi-continuous controller Eq. (4.21) satisfies

$$u = -\alpha \frac{\dot{\sigma} + \beta|\sigma|^{1/2}\text{sign}(\sigma)}{|\dot{\sigma}| + \beta|\sigma|^{1/2}} = -\alpha \frac{\kappa\dot{\sigma} + \beta|\kappa^2\sigma|^{1/2}\text{sign}(\kappa^2\sigma)}{|\kappa\dot{\sigma}| + \beta|\kappa^2\sigma|^{1/2}}$$

Next consider the super-twisting controller Eq. (4.36) applied to a system of relative degree 1, satisfying Eqs. (4.34) and (4.35) (see the end of the proof of Theorem 4.5). Starting from the moment when $|u_1| < U_M$ is established, the closed-loop system solutions satisfy the inclusion

$$\dot{\sigma} \in -[K_m, K_M]|\sigma|^{1/2}\text{sign}(\sigma) + \xi$$

$$\dot{\xi} \in [-C, C] - \alpha[K_m, K_M]\text{sign}(\sigma)$$

where $\xi = h(t,x) + g(t,x)u_1$. The inclusion is homogeneous with homogeneity degree -1, $\deg \sigma = 2$, $\deg \xi = 1$. Since it is finite-time stable, we get the accuracies $\sigma = O(\tau^2)$, $\dot{\sigma} = \xi = O(\tau)$, and $\sigma = O(\varepsilon)$, $\dot{\sigma} = \xi = O(\varepsilon^{1/2})$, which have been already announced, but not proved in Sect. 4.2.5. The same accuracy is also obtained for the first-order differentiator (Sect. 4.3.2).

6.6 Arbitrary-Order Sliding Mode Controllers

The following are the two most well-known r-sliding controller families. The controllers of the form

$$u = -\alpha \Psi_{r-1,r}\left(\sigma, \dot{\sigma}, \ldots, \sigma^{(r-1)}\right) \qquad (6.22)$$

are defined by recursive procedures, have magnitude $\alpha > 0$, and solve the general output regulation problem for system (6.10). The parameters of the controllers can be chosen in advance for each relative degree r. Only the magnitude α is to be adjusted for any fixed C, K_m, K_M, most conveniently by computer simulation, thus avoiding complicated and redundantly large estimations. Obviously, α has to be negative with $\frac{\partial}{\partial u}(\sigma^{(r)}) < 0$. In the following, $\beta_1, \ldots, \beta_r > 0$ are the controller parameters, which define the convergence rate.

6.6.1 Nested Sliding Controllers

The following procedure defines the "nested" r-sliding controllers, based on a pseudo-nested structure of 1-sliding modes. Let $q \in \mathbb{N}$ and $q > 1$. The controllers are built by the following recursive procedure:

$$N_{i,r} = \left(|\sigma|^{q/r} + |\dot{\sigma}|^{q/(r-1)} + \ldots + |\sigma^{(i-1)}|^{q/(r-i+1)}\right)^{1/q}$$
$$\Psi_{0,r} = \text{sign}(\sigma), \quad \Psi_{i,r} = \text{sign}\left(\sigma^{(i)} + \beta_i N_{i,r} \Psi_{i-1,r}\right)$$

What follows are the nested sliding mode controllers (of the first family) for $r \leq 4$ with q being the least multiple of $1, \ldots, r$:

1. $u = -\alpha \, \text{sign}(\sigma)$
2. $u = -\alpha \, \text{sign}\left(\dot{\sigma} + |\sigma|^{1/2} \, \text{sign}(\sigma)\right)$
3. $u = -\alpha \, \text{sign}\left(\ddot{\sigma} + 2\left(|\dot{\sigma}|^3 + |\sigma|^2\right)^{1/6} \times \text{sign}\left(\dot{\sigma} + |\sigma|^{2/3} \, \text{sign}(\sigma)\right)\right)$
4. $u = -\alpha \, \text{sign}\left(\dddot{\sigma} + 3\left(\ddot{\sigma}^6 + \dot{\sigma}^4 + |\sigma|^3\right)^{1/12} \times \right.$

$$\left. \text{sign}\left(\ddot{\sigma} + \left(\dot{\sigma}^4 + |\sigma|^3\right)^{1/6} \, \text{sign}\left(\dot{\sigma} + 0.5\,|\sigma|^{3/4} \, \text{sign}(\sigma)\right)\right)\right)$$

These controllers can be given an intuitive inexact explanation based on recursively nested standard sliding modes. For example, consider the 3-sliding controller

$$u = -\alpha \, \text{sign} \left(\ddot{\sigma} + \beta_2 \left(|\dot{\sigma}|^3 + |\sigma|^2 \right)^{1/6} \times \text{sign} \left(\dot{\sigma} + \beta_1 |\sigma|^{2/3} \, \text{sign}(\sigma) \right) \right) \quad (6.23)$$

which is applied to the corresponding inclusion

$$\dddot{\sigma} \in [-C, C] + [K_m, K_M] u \qquad (6.24)$$

First of all, we check its 3-sliding homogeneity. The homogeneity weights are deg $\ddot{\sigma} = 3$, deg $\dot{\sigma} = 2$, deg $\sigma = 1$, and

$$\text{sign} \left(\kappa \ddot{\sigma} + \beta_2 \left(|\kappa^2 \dot{\sigma}|^3 + |\kappa^3 \sigma|^2 \right)^{1/6} \times \text{sign} \left(\kappa^2 \dot{\sigma} + \beta_1 |\kappa^3 \sigma|^{2/3} \, \text{sign}(\kappa^3 \sigma) \right) \right)$$

$$= \text{sign} \left(\ddot{\sigma} + \beta_2 \left(|\dot{\sigma}|^3 + |\sigma|^2 \right)^{1/6} \times \text{sign} \left(\dot{\sigma} + \beta_1 |\sigma|^{2/3} \, \text{sign}(\sigma) \right) \right)$$

The idea is that for a sufficiently large α, a 1-sliding mode is established on the surface

$$\ddot{\sigma} + \beta_2 \left(|\dot{\sigma}|^3 + |\sigma|^2 \right)^{1/6} \times \text{sign} \left(\dot{\sigma} + \beta_1 |\sigma|^{2/3} \, \text{sign}(\sigma) \right) = 0 \qquad (6.25)$$

In turn Eq. (6.25) leads to the establishment of the sliding mode

$$\dot{\sigma} + \beta_1 |\sigma|^{2/3} \, \text{sign}(\sigma) = 0 \qquad (6.26)$$

and solutions of Eq. (6.26) converge to identical zero.

Unfortunately, this explanation is not correct. Indeed, no sliding modes are possible on the discontinuous surface Eq. (6.25). In reality, the sliding mode is established on the continuous components of the surface Eq. (6.25) and is lost when the sliding trajectory reaches the discontinuity set Eq. (6.26). It leaves the surface Eq. (6.25) towards its other continuity component. Once it hits the other continuous component of Eq. (6.25), a 1-sliding mode is once more established on Eq. (6.25). The 1-sliding trajectory once more reaches the set Eq. (6.26) and so on. As a result, a motion arises around the set Eq. (6.26) in the 3-dimensional space σ, $\dot{\sigma}$, $\ddot{\sigma}$. The larger β_2, the faster the trajectories arrive at another continuity component of Eq. (6.25), and the smaller the resulting 3-dimensional vicinity of Eq. (6.26). In the projection to the coordinates σ, $\dot{\sigma}$ we get a disturbed Eq. (6.26). Obviously its solutions converge to some vicinity of $\sigma = 0$, $\dot{\sigma} = 0$, which corresponds to some vicinity of $\sigma = 0$, $\dot{\sigma} = 0$, $\ddot{\sigma} = 0$ in the 3-dimensional space.

The above reasoning means that starting from the unite ball centered at the origin $\sigma = 0$, $\dot{\sigma} = 0$, $\ddot{\sigma} = 0$, with a sufficiently large β_2 and a correspondingly large α, the trajectories concentrate in a small vicinity of the origin. Thus we have the contractivity feature of the 3-sliding homogeneous inclusion Eqs. (6.23), (6.24). Now from Theorem 6.2 finite-time stability is ensured.

6.6.2 Quasi-continuous Sliding Controllers

The obvious disadvantage of the nested sliding mode controllers is the high-frequency leaping of transient trajectories when they approach the HOSM. Although the leaping magnitude tends to zero during the transient, it may cause problems in practice.

As we have seen in Sect. 6.2 the control law in Eq. (6.22) should be discontinuous at least at the HOSM:

$$\sigma = \dot{\sigma} = \ldots = \sigma^{(r-1)} = 0 \tag{6.27}$$

An r-sliding controller Eq. (6.7) is called *quasi-continuous*, if it is continuous everywhere except at Eq. (6.27). In practice with $r > 1$, such a control law remains continuous, since due to noise, delays, and other imperfections, the r equalities in Eq. (6.27) never hold simultaneously.

The following procedure defines a family of quasi-continuous controllers:

$$\varphi_{0,r} = \sigma, \ N_{0,r} = |\sigma|, \ \Psi_{0,r} = \frac{\varphi_{0,r}}{N_{0,r}} = \text{sign}(\sigma)$$

$$\varphi_{i,r} = \sigma^{(i)} + \beta_i \, N_{i-1,r}^{(r-i)/(r-i+1)} \, \Psi_{i-1,r}$$

$$N_{i,r} = \left|\sigma^{(i)}\right| + \beta_i \, N_{i-1,r}^{(r-i)/(r-i+1)}, \ \Psi_{i,r} = \frac{\varphi_{i,r}}{N_{i,r}}$$

The following are quasi-continuous controllers with $r \leq 4$, which have simulation-tested β_i:

1. $u = -\alpha \, \text{sign}(\sigma)$
2. $u = -\alpha \left(\dot{\sigma} + |\sigma|^{1/2} \text{sign}(\sigma) \right) / \left(|\dot{\sigma}| + |\sigma|^{1/2} \right)$
3. $u = -\alpha \dfrac{\ddot{\sigma} + 2\left(|\dot{\sigma}| + |\sigma|^{2/3}\right)^{-1/2} \left(\dot{\sigma} + |\sigma|^{2/3} \text{sign}(\sigma) \right)}{|\ddot{\sigma}| + 2\left(|\dot{\sigma}| + |\sigma|^{2/3}\right)^{1/2}}$
4. $u = -\alpha \, \varphi_{3,4} / N_{3,4}$ where

$$\varphi_{3,4} = \dddot{\sigma} + 3 \left[\ddot{\sigma} + \left(|\dot{\sigma}| + 0.5 |\sigma|^{3/4} \right)^{-1/3} \left(\dot{\sigma} + 0.5 |\sigma|^{3/4} \text{sign}(\sigma) \right) \right] \times$$

$$\left[|\ddot{\sigma}| + \left(|\dot{\sigma}| + 0.5 |\sigma|^{3/4} \right)^{2/3} \right]^{1/2}$$

and

$$N_{3,4} = |\ddot{\sigma}| + 3\left[|\ddot{\sigma}| + \left(|\dot{\sigma}| + 0.5\,|\sigma|^{3/4}\right)^{2/3}\right]^{1/2}$$

It is easy to see that the sets of parameters β_i are the same for both the quasi-continuous and nested controllers with $r \leq 4$. Note that while enlarging α increases the class Eq. (6.5), Eq. (6.6) of systems, to which the controller is applicable, the parameters β_i are tuned to provide the required convergence rate. The authors consider the latter family as the best one. In addition to the reduced chattering, another advantage of these controllers is the simplicity in terms of adjustment of their coefficients.

The idea of the quasi-continuous controllers is a generalization of the idea of the 2-sliding controller Eq. (4.31) in Theorem 4.4.

Theorem 6.3. *Each representative of order r in the two families of arbitrary-order sliding mode controllers above is r-sliding homogeneous and finite-time stabilizing.*

Sketch of the proof. The proof of the theorem is based on Theorem 6.1, i.e., on the proof of the contractivity property. Asymptotic accuracies of these controllers are readily obtained from Theorem 6.2. In particular $\sigma^{(i)} = O(\tau^{r-i})$, $i = 0, 1, \ldots, r-1$, if the sampling interval equals τ.

6.7 Arbitrary-Order Robust Exact Differentiation

Any r-sliding homogeneous controller can be complemented by an $(r-1)$th order differentiator producing an output-feedback controller. In order to preserve the demonstrated exactness, finite-time stability, and the corresponding asymptotic properties, the natural way to calculate $\dot{\sigma}, \ldots, \sigma^{(r-1)}$ in real time is by means of a robust finite-time-convergent exact *homogeneous* differentiator. Its application is possible due to the boundedness of $\sigma^{(r)}$ provided by the boundedness of the feedback control Eq. (6.7).

Let the input signal $f(t)$ be a function defined on $[0, \infty)$ consisting of a bounded Lebesgue-measurable noise with unknown features and of an unknown base signal $f_0(t)$, whose kth derivative has a known Lipschitz constant $L > 0$. The problem of finding real-time robust estimations of $\dot{f}_0(t), \ddot{f}_0(t), \ldots, f_0^{(k)}(t)$ which are exact in the absence of measurement noises is solved by the differentiator

$$
\begin{aligned}
\dot{z}_0 &= v_0, \; v_0 = -\lambda_k L^{1/(k+1)} |z_0 - f(t)|^{k/(k+1)} \operatorname{sign}(z_0 - f(t)) + z_1 \\
\dot{z}_1 &= v_1, \; v_1 = -\lambda_{k-1} L^{1/k} |z_1 - v_0|^{(k-1)/k} \operatorname{sign}(z_1 - v_0) + z_2 \\
&\;\vdots \\
\dot{z}_{k-1} &= v_{k-1}, \; v_{k-1} = -\lambda_1 L^{1/2} |z_{k-1} - v_{k-2}|^{1/2} \operatorname{sign}(z_{k-1} - v_{k-2}) + z_k \\
\dot{z}_k &= -\lambda_0 L \operatorname{sign}(z_k - v_{k-1})
\end{aligned}
$$

$$(6.28)$$

If the parameters $\lambda_0, \lambda_1, \ldots, \lambda_k > 0$ are properly chosen, the following equalities are true in the absence of input noise, after a finite-time transient process:

$$z_0 = f_0(t), \ldots, z_i = v_{i-1} = f_0^{(i-1)}(t), \ i = 1, 2, \ldots, i$$

Note that the differentiator has a recursive structure. Once the $\lambda_0, \lambda_1, \ldots, \lambda_{k-1}$ are properly chosen for the $(k-1)$th-order differentiator with Lipschitz constant L, only one parameter λ_k needs to be tuned for the kth-order differentiator with the same Lipschitz constant and the parameter λ_k just has to be taken sufficiently large. Any $\lambda_0 > 1$ can be used to start this process.

Theorem 6.4. *For any given $\lambda_0 > 1$ there exists an infinite positive sequence $\{\lambda_n\}$, such that for each natural k the parameters $\lambda_0, \lambda_1, \ldots, \lambda_k$ provide finite-time convergence of the kth-order differentiator Eq. (6.28).*

Sketch of the proof. Denote $\sigma_i = \left(z_i - f^{(i)}(t)\right)/L$. Then dividing all the equations by L and subtracting $f^{(i+1)}(t)/L$ from both sides of the equation with $\dot{z}_i$ on the left, $i = 0, \ldots, k$, one obtains

$$\dot{\sigma}_0 = -\lambda_k |\sigma_0|^{k/(k+1)} \operatorname{sign}(\sigma_0) + \sigma_1,$$

$$\dot{\sigma}_1 = -\lambda_{k-1} |\sigma_1 - \dot{\sigma}_0|^{(k-1)/k} \operatorname{sign}(\sigma_1 - \dot{\sigma}_0) + \sigma_2,$$

$$\vdots$$

$$\dot{\sigma}_{k-1} = -\lambda_1 |\sigma_{k-1} - \dot{\sigma}_{k-2}|^{1/2} \operatorname{sign}(\sigma_{k-1} - \dot{\sigma}_{k-2}) + \sigma_k,$$

$$\dot{\sigma}_k \in -\lambda_0 L \operatorname{sign}(\sigma_k - \dot{\sigma}_{k-1}) + [-1, 1]$$

where the inclusion $f^{(i+1)}(t)/L \in [-1, 1]$ is used in the last line. This differential inclusion is homogeneous with homogeneity degree -1 and weights $k+1, k, \ldots, 1$ of $\sigma_0, \sigma_1, \ldots, \sigma_k$, respectively. The finite-time convergence of the differentiator follows from the contractivity property of this inclusion and Theorem 6.1. $\qquad\square$

Remark 6.1. A possible choice of the differentiator parameters for $k \leq 5$ is $\lambda_0 = 1.1, \lambda_1 = 1.5, \lambda_2 = 3, \lambda_3 = 5, \lambda_4 = 8, \lambda_5 = 12$. Another possible choice is $\lambda_0 = 1.1, \lambda_1 = 1.5, \lambda_2 = 2, \lambda_3 = 3, \lambda_4 = 5, \lambda_5 = 8$.

Theorem 6.2 gives the asymptotic accuracy of the differentiator. Let the measurement noise be any Lebesgue-measurable function with magnitude not exceeding ε. Then an accuracy of $\left|z_i(t) - f_0^{(i)}(t)\right| = O(\varepsilon^{(k+1-i)/(k+1)})$ is obtained. That accuracy is shown to be asymptotically the best possible.

Note that the differentiator in Eq. (6.28) is not presented in the standard form of dynamic system, i.e., $\dot{z} = f(t, z)$. When implementing it in a computer, one should check that all auxiliary variables v_i are evaluated using the same current values of z_i, $i = 0, 1, \ldots, k$. One can also substitute the value of v_0 from the first line of Eq. (6.28) into the second line, then substitute the value of v_1 from the second line into the third line, etc., to obtain the non-recursive form of the differentiator

$$
\begin{aligned}
\dot{z}_0 &= -\tilde{\lambda}_k L^{1/(k+1)} |z_0 - f(t)|^{k/(k+1)} \operatorname{sign}(z_0 - f(t)) + z_1 \\
\dot{z}_1 &= -\tilde{\lambda}_{k-1} L^{2/(k+1)} |z_0 - f(t)|^{(k-1)/(k+1)} \operatorname{sign}(z_0 - f(t)) + z_2 \\
&\ \ \vdots \\
\dot{z}_{k-1} &= -\tilde{\lambda}_1 L^{k/(k+1)} |z_0 - f(t)|^{1/(k+1)} \operatorname{sign}(z_0 - f(t)) + z_k \\
\dot{z}_k &= -\tilde{\lambda}_0 L \operatorname{sign}(z_0 - f(t))
\end{aligned}
\tag{6.29}
$$

It can be verified that $\tilde{\lambda}_k = \lambda_k$, $\tilde{\lambda}_i = \lambda_i \tilde{\lambda}_{i+1}^{i/(i+1)}$, $i = k-1, k-2, \ldots, 0$. In particular, with the choice $\lambda_0 = 1.1, \lambda_1 = 1.5, \lambda_2 = 2, \lambda_3 = 3, \lambda_4 = 5, \lambda_5 = 8$ for the coefficients λ_i, for $k = 1$ one obtains $\tilde{\lambda}_0 = 1.1, \tilde{\lambda}_1 = 1.5$; for $k = 2$ one obtains $\tilde{\lambda}_0 = 1.1, \tilde{\lambda}_1 = 2.12, \tilde{\lambda}_2 = 2$; for $k = 3$ one obtains $\tilde{\lambda}_0 = 1.1, \tilde{\lambda}_1 = 3.06, \tilde{\lambda}_2 = 4.16, \tilde{\lambda}_3 = 3$; for $k = 4$ one obtains $\tilde{\lambda}_0 = 1.1, \tilde{\lambda}_1 = 4.57, \tilde{\lambda}_2 = 9.3, \tilde{\lambda}_3 = 10.03, \tilde{\lambda}_4 = 5$; for $k = 5$ one obtains $\tilde{\lambda}_0 = 1.1, \tilde{\lambda}_1 = 6.93, \tilde{\lambda}_2 = 21.4, \tilde{\lambda}_3 = 34.9, \tilde{\lambda}_4 = 26.4, \tilde{\lambda}_5 = 8$. Note that these parameter values can be rounded to 2 meaningful digits without any loss of convergence.

Example 6.1. The following is the fifth-order differentiator:

$$
\begin{aligned}
\dot{z}_0 &= v_0, \ v_0 = -8L^{1/6} |z_0 - f(t)|^{5/6} \operatorname{sign}(z_0 - f(t)) + z_1 \\
\dot{z}_1 &= v_1, \ v_1 = -5L^{1/5} |z_1 - v_0|^{4/5} \operatorname{sign}(z_1 - v_0) + z_2 \\
\dot{z}_2 &= v_2, \ v_2 = -3L^{1/4} |z_2 - v_1|^{3/4} \operatorname{sign}(z_2 - v_1) + z_3 \\
\dot{z}_3 &= v_3, \ v_3 = -2L^{1/3} |z_3 - v_2|^{2/3} \operatorname{sign}(z_3 - v_2) + z_4 \\
\dot{z}_4 &= v_4, \ v_4 = -1.5L^{1/2} |z_4 - v_3|^{1/2} \operatorname{sign}(z_4 - v_3) + z_4 \\
\dot{z}_5 &= -1.1L \operatorname{sign}(z_5 - v_4), \quad \left| f^{(6)}(t) \right| \le L
\end{aligned}
$$

It is applied, with $L = 1$, for differentiating the function

$$
f(t) = \sin(0.5t) + \cos(0.5t) \quad \left| f^{(6)}(t) \right| \le 1
$$

The initial values of the differentiator variables are taken zero. In practice it is reasonable to take the initial value of z_0 equal to the current sampled value of $f(t)$, significantly shortening the transient. Convergence of the differentiator is demonstrated in Fig. 6.2. The fifth derivative is not exact due to the software restrictions (the number of decimal digits of the mantissa). In fact, higher-order differentiation requires special software development.

6.8   Output-Feedback Control

Introducing the differentiator of order $r - 1$ given above in the feedback loop, one obtains an output-feedback r-sliding controller

$$
u = -\alpha \, \varphi(z_0, z_1, \ldots, z_{r-1})
\tag{6.30}
$$

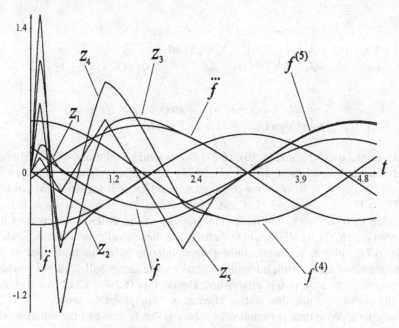

Fig. 6.2 Fifth-order differentiation

where

$$\dot{z}_0 = v_0, \; v_0 = -\lambda_{r-1} L^{1/r} |z_0 - \sigma|^{(r-1)/r} \, \text{sign} \, (z_0 - \sigma) + z_1$$
$$\dot{z}_1 = v_1, \; v_1 = -\lambda_{r-2} L^{1/(r-1)} |z_1 - v_0|^{(r-2)/(r-1)} \, \text{sign} \, (z_1 - v_0) + z_2$$
$$\vdots$$
$$\dot{z}_{r-2} = v_{r-2}, \; v_{r-2} = -\lambda_1 L^{1/2} |z_{r-2} - v_{r-3}|^{1/2} \, \text{sign} \, (z_{r-2} - v_{r-3}) + z_{r-1}$$
$$\dot{z}_{r-1} = -\lambda_0 L \, \text{sign} \, (z_{r-1} - v_{r-2})$$

$$(6.31)$$

with $L \geq C + \sup |\varphi| \; K_M$ and the parameters λ_i of differentiator Eq. (6.31) are chosen in advance (Sect. 6.7).

Theorem 6.5. *Let controller Eq. (6.30) be r-sliding homogeneous and finite-time stable, and the parameters of the differentiator in Eq. (6.31) be properly chosen with respect to the upper bound of $|\varphi|$. Then the output-feedback controller Eqs. (6.30), (6.31) provides the finite-time convergence of each trajectory to the r-sliding mode $\sigma \equiv 0$.*

Proof. Denote $s_i = z_i - \sigma^{(i)}$. Then using $\sigma^{(r)} \in [-L, L]$, the controller Eqs. (6.30), (6.31) can be rewritten as

$$u = -\alpha \, \varphi \, (s_0 + \sigma, s_1 + \dot{\sigma}, \dots, s_{r-1} + \sigma^{(r-1)}) \qquad (6.32)$$

where

$$
\begin{aligned}
\dot{s}_0 &= -\lambda_{r-1}L^{1/r}|s_0|^{(r-1)/r}\,\text{sign}\,(s_0)+s_1 \\
\dot{s}_1 &= -\lambda_{r-2}L^{1/(r-1)}|s_1-\dot{s}_0|^{(r-2)/(r-1)}\,\text{sign}\,(s_1-\dot{s}_0)+s_2 \\
&\vdots \\
\dot{s}_{r-2} &= -\lambda_1 L^{1/2}|s_{r-2}-\dot{s}_{r-3}|^{1/2}\,\text{sign}\,(s_{r-2}-\dot{s}_{r-3})+s_{r-1} \\
\dot{s}_{r-1} &\in -\lambda_0 L\text{sign}\,(s_{r-1}-\dot{s}_{r-2})+[-L,L]
\end{aligned}
\tag{6.33}
$$

Solutions of Eqs. (6.5), and (6.30), (6.31) correspond to solutions of the Filippov differential inclusion Eqs. (6.10), (6.32), (6.33). Assign the weights $r-i$ to $s_i,\sigma^{(i)}$, $i=0,1,\ldots,r-1$, to obtain a homogeneous differential inclusion Eqs. (6.10), (6.32), (6.33) of degree -1. Let the initial conditions belong to some ball in the space $s_i,\sigma^{(i)}$. Due to the finite-time stability of the differentiator part of the inclusion Eq. (6.33), it collapses in finite time, and the controller becomes equivalent to Eq. (6.7), which is uniformly finite-time stabilizing by assumption. Due to the boundedness of the control, no solution leaves some larger ball until the moment when $s_0\equiv\cdots\equiv s_{r-1}\equiv 0$ is established. Hence, Eqs. (6.10), (6.32), (6.33) is also globally uniformly finite-time stable. Theorem 6.3 completes the proof.   □

The convergence time is bounded by a continuous function of the initial conditions in the space $\sigma,\dot{\sigma},\ldots,\sigma^{(r-1)},s_0,s_1,\ldots,s_{r-1}$. This function is homogeneous of weight 1 and vanishes at the origin (Theorem 6.1).

Let the σ measurements be obtained with a sampling interval τ, or let them be corrupted by noise given by an unknown bounded Lebesgue-measurable function of time of magnitude ε; then the solutions of Eqs. (6.10), (6.32), (6.33) are infinitely extendible in time under the assumptions of Sect. 6.2, and the following theorem is a simple consequence of Theorem 6.2.

Theorem 6.6. *In the absence of measurement noise the discrete-measurement version of the controller Eqs. (6.32), (6.33) with a sampling interval τ ensures*

$$
|\sigma|\le\gamma_0\tau^r,\ |\dot{\sigma}|\le\gamma_1\tau^{r-1},\ldots,\left|\sigma^{(r-1)}\right|\le\gamma_{r-1}\tau
\tag{6.34}
$$

for some $\gamma_0,\gamma_1,\ldots,\gamma_{r-1}>0$. With continuous measurements in the presence of measurement noise of magnitude ε, the accuracies

$$
|\sigma|<\delta_0\varepsilon,\ |\dot{\sigma}|<\delta_1\varepsilon^{(r-1)/r},\ldots,\left|\sigma^{(r-1)}\right|<\delta_{r-1}\varepsilon^{1/r}
$$

are obtained for some $\delta_0,\delta_1,\ldots,\delta_{r-1}>0$. Here $\gamma_0,\gamma_1,\ldots,\gamma_{r-1}$, and $\delta_0,\delta_1,\ldots,\delta_{r-1}$ are positive constants depending only on the parameters from the problem statement Eq. (6.6) and the controller Eqs. (6.30), (6.31).

The asymptotic accuracy Eq. (6.34) provided by Theorem 6.2, is asymptotically the best possible with discontinuous $\sigma^{(r)}$ and discrete sampling. Note that with homogeneous sliding modes, the missing derivatives can be also estimated by means of divided finite differences. The results of this section are also valid for the suboptimal controller. Hence, the problem stated at the beginning of Sect. 6.2 is solved.

6.9 Tuning of the Controllers

Tuning of the controller parameters and possible adaptation of the control magnitude are needed to regulate the convergence rate and to overcome restriction Eq. (6.6).

6.9.1 Control Magnitude Tuning

For simplicity full information about the system state is assumed to be available. In particular, t, x, σ and its $r - 1$ successive derivatives are measured. Consider the controller

$$u = -\alpha \, \Phi(t, x)\Psi_{r-1,r}\left(\sigma, \dot{\sigma}, \ldots, \sigma^{(r-1)}\right) \qquad (6.35)$$

where $\alpha > 0$ and $\Psi_{r-1,r}$ is one of the two r-sliding homogeneous controllers introduced in Sects. 6.6.1 and 6.6.2. The function Φ is any Lebesgue-measurable locally bounded strictly positive function. *Any increase in the gain function $\Phi(t, x)$ does not interfere with the convergence.*

While the function Φ can be chosen large to control exploding systems, it is also reasonable to make the function Φ decrease significantly, when approaching the system operational point, thereby reducing chattering.

6.9.2 Parametric Tuning

The controller parameters presented in Sect. 6.6 provide a formal solution of the stated problem. Nevertheless, in practice one often needs to adjust the convergence rate, either to slow it down (relaxing the burden on the actuators) or to accelerate it in order to meet some system requirements. Note, in this context, redundantly enlarging the magnitude parameter α does not accelerate the convergence, but only increases the chattering, while its reduction may lead to loss of convergence.

A better approach is to take the controller

$$u = \lambda^r \alpha \Psi_{r-1,r}\left(\sigma, \dot{\sigma}/\lambda, \ldots, \sigma^{(r-1)}/\lambda^{r-1}\right), \ \lambda > 0$$

instead of

$$u = -\alpha \Psi_{r-1,r}\left(\sigma, \dot{\sigma}, \ldots, \sigma^{(r-1)}\right), \ \lambda > 0$$

which provide approximately a λ times reduction in the convergence time.

In the case of quasi-continuous controllers (Sect. 6.6.2) the form of the controller is preserved. The new parameters $\tilde{\beta}_1, \ldots, \tilde{\beta}_{r-1}, \tilde{\alpha}$ are calculated according to the formulas $\tilde{\beta}_1 = \lambda\beta_1, \tilde{\beta}_2 = \lambda^{r/(r-1)}\beta_2, \ldots, \tilde{\beta}_{r-1} = \lambda^{r/2}\beta_{r-1}, \tilde{\alpha} = \lambda_r\alpha$. The following are the resulting quasi-continuous controllers for $r \leq 4$, with simulation-tested β_i and a general gain function Φ:

1. $u = -\alpha \Phi \operatorname{sign}(\sigma)$
2. $u = -\alpha \Phi \left(\dot{\sigma} + \lambda |\sigma|^{1/2} \operatorname{sign}(\sigma) \right) / \left(|\dot{\sigma}| + \lambda |\sigma|^{1/2} \right)$
3. $u = -\alpha \Phi \dfrac{\ddot{\sigma} + 2\lambda^{3/2} \left(|\dot{\sigma}| + \lambda |\sigma|^{2/3} \right)^{-1/2} \left(\dot{\sigma} + \lambda |\sigma|^{2/3} \operatorname{sign}(\sigma) \right)}{|\ddot{\sigma}| + 2\lambda^{3/2} \left(|\dot{\sigma}| + \lambda |\sigma|^{2/3} \right)^{1/2}}$
4. $u = -\alpha \varphi_{3,4} / N_{3,4}$

where

$$\varphi_{3,4} = \ddot{\sigma} + 3\lambda^2 \left[\ddot{\sigma} + \lambda^{4/3} \left(|\dot{\sigma}| + 0.5\lambda |\sigma|^{3/4} \right)^{-1/3} \left(\dot{\sigma} + 0.5\lambda |\sigma|^{3/4} \operatorname{sign}(\sigma) \right) \right]$$
$$\times \left[|\ddot{\sigma}| + \lambda^{4/3} \left(|\dot{\sigma}| + 0.5\lambda |\sigma|^{3/4} \right)^{2/3} \right]^{1/2}$$

and

$$N_{3,4} = |\ddot{\sigma}| + 3\lambda^2 \left[|\ddot{\sigma}| + \lambda^{4/3} \left(|\dot{\sigma}| + 0.5\lambda |\sigma|^{3/4} \right)^{2/3} \right]^{1/2}$$

6.10   Case Study: Car Steering Control

Consider a simple kinematic model of car control

$$\dot{x} = v\cos(\varphi) \tag{6.36}$$
$$\dot{y} = v\sin(\varphi) \tag{6.37}$$
$$\dot{\varphi} = v/l \ \tan(\theta) \tag{6.38}$$
$$\dot{\theta} = u \tag{6.39}$$

where x and y are the Cartesian coordinates of the middle point of the rear axle, φ is the orientation angle, v is the longitudinal velocity, l is the length between the two axles, and θ is the steering angle (i.e., the control input) (Fig. 6.3). The task is to steer the car from a given initial position to the trajectory $y = g(x)$, where $g(x)$ and y are assumed to be available in real time.

Define $\sigma = y - g(x)$. Let $v = const = 10\,m/s$, $l = 5\,m$, $x = y = \varphi = \theta = 0$ at $t = 0$, $g(x) = 10\sin(0.05x) + 5$. Obviously the control appears for the first time explicitly in the third derivatives of x and y. Thus the relative degree of the system is 3 and both the nested (Sect. 6.6.1) and quasi-continuous (Sect. 6.6.2) 3-sliding controllers solve the problem. The controllers are of the form (6.30), (6.31). Choosing $\alpha = 20$ the listed nested controller takes the form

$$u = -20 \operatorname{sign}\left(z_2 + 2 \left(|z_1|^3 + |z_0|^2 \right)^{1/6} \cdot \operatorname{sign}\left(z_1 + |z_0|^{2/3} \operatorname{sign}(z_0) \right) \right)$$

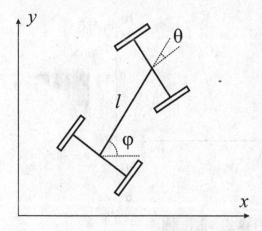

Fig. 6.3 Kinematic car model

where z_0, z_1, z_2 are the outputs of the differentiator Eq. (6.31) with $L = 400$. For the quasi-continuous controller given earlier $\alpha = 1$ has been chosen so that

$$u = -\frac{\left[z_2 + 2\left(|z_1| + |z_0|^{2/3}\right)^{-1/2} \left(z_1 + |z_0|^{2/3}\,\mathrm{sign}(z_0)\right)\right]}{\left[|z_2| + 2\left(|z_1| + |z_0|^{2/3}\right)^{1/2}\right]}$$

The differentiator Eq. (6.31) takes the form

$$\dot{z}_0 = v_0, \quad v_0 = -14.7361\,|z_0 - \sigma|^{2/3}\,\mathrm{sign}\,(z_0 - \sigma) + z_1$$
$$\dot{z}_1 = v_1, \quad v_1 = -30\,|z_1 - v_0|^{1/2}\,\mathrm{sign}\,(z_1 - v_0) + z_2$$
$$\dot{z}_2 = -400\,\mathrm{sign}\,(z_2 - v_1)$$

in the both cases.

The controller parameter α has conveniently been found by simulation. The differentiator parameter $L = 400$ is chosen deliberately large, in order to provide better performance in the presence of measurement errors ($L = 40$ is also sufficient but is worse with sampling noises). The control was applied only from time $t = 1$, in order to provide some time for the differentiator to converge.

The integration was carried out using the Euler method (the only reliable integration method with discontinuous dynamics), the sampling step being equal to the integration step $\tau = 10^{-4}$ s. It is seen from Fig. 6.4 that the nested controller indeed demonstrates additional chattering during the transient as predicted.

Using the quasi-continuous controller, in the absence of noise, the tracking accuracies $|\sigma| \le 5.4 \times 10^{-7}$, $|\dot{\sigma}| \le 2.4 \times 10^{-4}$, $|\ddot{\sigma}| \le 0.042$ were obtained. With $\tau = 10^{-5}$ s the accuracies $|\sigma| \le 5.6 \times 10^{-10}$, $|\dot{\sigma}| \le 1.4 \times 10^{-5}$, $|\ddot{\sigma}| \le 0.0042$

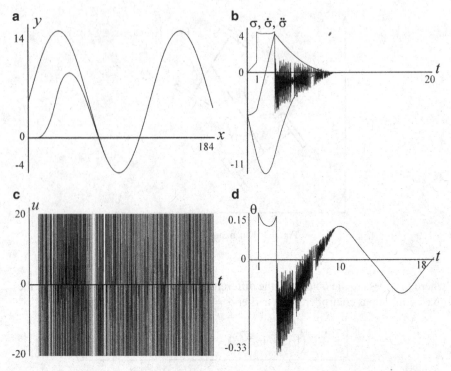

Fig. 6.4 Quasi-continuous 3-sliding car control (**a**) Car trajectory tracking (**b**) 3-sliding tracking deviations (**c**) Steering angle derivative (control) (**d**) Steering angle

were attained. The car trajectory, 3-sliding tracking errors, the steering angle θ, and its derivative u are shown in Fig. 6.5a–d respectively. It is seen from Fig. 6.5c that the control u remains continuous until the 3-sliding mode is obtained. The steering angle θ remains smooth and is quite practical.

In the presence of output noise of magnitude 0.01 m, tracking accuracy levels of $|\sigma| \leq 0.02, |\dot{\sigma}| \leq 0.14, |\ddot{\sigma}| \leq 1.3$ were obtained. With measurement noise of magnitude 0.1 m, the accuracy changes to $|\sigma| \leq 0.20, |\dot{\sigma}| \leq 0.62, |\ddot{\sigma}| \leq 2.8$. The performance of the controller with a measurement error magnitude of 0.1 m is shown in Fig. 6.6. It is seen from Fig. 6.6c that the control u is a continuous function of t. The steering angle vibrations have a magnitude of about 7 degrees and a frequency of 1 rad/s, which is also quite feasible. The performance does not significantly change, when the frequency of the noise varies in the range 100–100,000 rad/s.

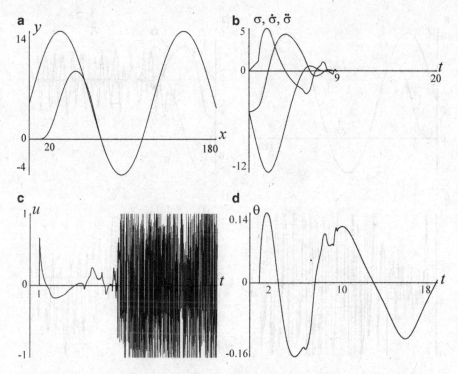

Fig. 6.5 Performance with the input noise magnitude 0.1 m (**a**) Car trajectory tracking (**b**) 3-sliding tracking deviations (**c**) Steering angle derivative (control) (**d**) Steering angle

6.11   Case Study: Blood Glucose Regulation

6.11.1   *Introduction to Diabetes*

The normal blood glucose concentration level in a human lies in a narrow range, 70–110 mg/dl. Different factors including food intake, rate of digestion, and exercise can affect this concentration. Two pancreatic endocrine hormones, glucagon and insulin, are responsible for regulating the blood glucose level. They are secreted from α- and β-cells of pancreas, respectively. These two hormones form two feedback loops in controlling the blood sugar level that function inversely. Insulin is secreted to lower the blood glucose concentration, e.g., after food intake, and glucagon responsible for increasing the glucose in blood, e.g., in fasting periods.

Diabetes is a disease characterized by an abnormal high glucose concentration caused by an impaired secretion (or action) of the insulin. It is diagnosed by a fasting blood glucose concentration higher than 126 mg/dl, and it can be treated by different schemes—one of them is the *insulin therapy* that consists of the injection of exogenous insulin.

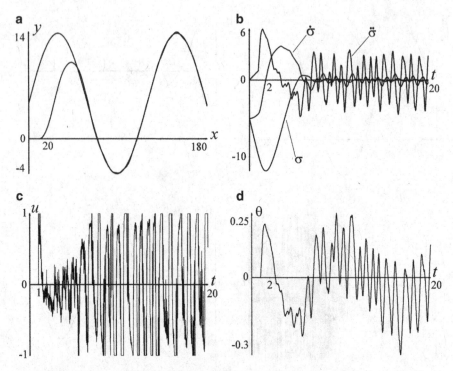

Fig. 6.6 3-sliding nested car control (**a**) Car trajectory tracking (**b**) 3-sliding tracking deviations. (**c**) Steering angle derivative (control) (**d**) Steering angle

Implementing tight glucose control in critically ill patients is the most important issue in diabetes management. The current medical guidelines for insulin therapy suggest three to four daily glucose measurements and an equivalent number of intramuscular insulin injections. Finding less invasive and less frequent methods has been the subject of interest for many researchers who are working in this area.

An alternative approach is to deliver insulin using a closed-loop device like a pump, which works like an artificial pancreas. This closed-loop device would include a glucose sensor imbedded under the skin and an insulin pump implanted in the abdomen. The sensor can measure blood glucose concentration and pass the information to a feedback control system that would calculate the necessary insulin delivery rate to keep the patient under metabolic control. A control signal will be sent to an insulin pump by the controller, to deliver the calculated amount of insulin. The pump infuses insulin through a catheter placed under the patient's skin. Since the latest generation of the implantable pumps allows different infusion rates of insulin, the feedback control system mimics the normal function of a pancreas more closely.

In general, using pumps is preferable to the frequent injection of insulin since it is more reliable in maintaining the correct level of sugar in the blood and also is

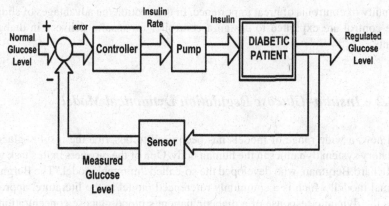

Fig. 6.7 Closed-loop control of diabetic patients using insulin pumps

closer to the normal action of the pancreas. However, creating a device, which would accurately replace three or more insulin injections per day for approximately 3-year lifespan, is not an easy task. It should be made from biocompatible materials and as small as possible. Ways to make the function of the implantable insulin pumps more ideal are currently under research. Figure 6.7 shows the block diagram of a closed-loop control system of a diabetic patient using an insulin pump.

In modeling drug delivery to the human body, certain requirements like finite reaching time and robustness to uncertainties should be satisfied. The human body is very sensitive to deviations of blood glucose concentration from the basal level. A small perturbation for a long period of time can cause in irreversible brain, kidney, and microcirculatory damage. So the time required to achieve the glucose regulation is of great importance. On the other hand, the primary drawback of any detailed physiological model is identifying a "nominal patient" to implement the model parameters. It is obvious that physical characteristics vary from person to person and so different patients have different responses to the same treatment, which in turn can cause parameter variations in the system. It must be taken into consideration that a small change in some of the parameters can dramatically affect the closed-loop performance and even result in the patient's death. Therefore, it is vital for the patients that controller used in designing the closed-loop system be robust to any kind of perturbations and disturbances.

In this case study, a relative degree approach together with a HOSM controller is used in designing the feedback algorithm for the glucose–insulin regulatory system. While in MPC, the controller is designed based on the previous information about the system; the HOSM control utilizes the current real-time information processed by HOSM observers/differentiators. Insensitivity to internal and external disturbances, ultimate accuracy and robustness as well as finite-time convergence that are main features of HOSM control, make it a suitable choice for control algorithms related to the human body, where extreme robustness to the patient parameters and external disturbances such as food intake and workout as well as continuity of control is of great importance. In this section the advantages of sliding mode control are exploited to design a controller for insulin delivery in diabetic patients.

6.11.2 Insulin–Glucose Regulation Dynamical Model

Until now, a wide range of models has been used to describe the insulin–glucose regulatory system dynamics in the human body. One of the pioneers in this task was Dr. Richard Bergman, who developed the so-called "minimal model."The Bergman minimal model, which is a commonly referenced model in the literature, approximates the dynamic response of a diabetic patient's blood glucose concentration to the insulin injection, using the following nonlinear differential equations:

$$\dot{G}(t) = -p_1[G(t) - G_b] - X(t)G(t) + D(t)$$
$$\dot{X}(t) = -p_2 X(t) + p_3[I(t) - I_b] \tag{6.40}$$
$$\dot{I}(t) = -n[I(t) - I_b] + \gamma[G(t) - h]^+ t + u(t)$$

where $t = 0$ indicates the time at which the glucose enters blood and the "+" sign shows the positive reflection to glucose intake. The variables in the equations above have the following physical meanings:

- $G(t)$, the glucose concentration in the blood plasma (mg/dl)
- $X(t)$, the insulin's effect on the net glucose disappearance, the insulin concentration in the remote compartments (1/min)
- $I(t)$, the insulin concentration in plasma at time t (μU/ml)
- G_b, the basal preinjection level of glucose (mg/dl)
- I_b, the basal preinjection level of insulin (μU/ml)
- p_1, the insulin-independent rate constant of glucose uptake in muscles and liver (1/min)
- p_2, the rate for decrease in tissue glucose uptake ability (1/min)
- p_3, the insulin-dependent increase in glucose uptake ability in tissue per unit of insulin concentration above the basal level [(μU/ml) min^{-2}]
- n, the first-order decay rate for insulin in blood (1/min)
- h, the threshold value of glucose above which the pancreatic β-cells; release insulin (mg/dl)
- γ, the rate at which the pancreatic β-cells' release insulin after the glucose injection when the glucose concentration is above the threshold [(μU/ml min^{-2}(mg/dl)$^{-1}$]

Bergman also introduces two other factors that play an important role in the regulation of glucose inside the body:

- S_I, insulin sensitivity, which is the capability of insulin to increase glucose disposal to muscles and liver and is the ratio of p_3/p_2
- S_G, glucose effectiveness, which is the ability of glucose to enhance its own disposal at basal insulin level and is equal to p_1

These two factors are part of the so-called metabolic portrait for every person and are important indicators of how glucose and insulin act inside that person's body. The term $\gamma[G(t) - h]^+$ in the third equation of the model acts as an internal regulatory function that formulates the insulin secretion in the body, which does not exist in diabetic patients that use insulin therapy. It has been also argued[2] that for diabetic subjects, the value of p_1 will be significantly reduced and can be approximated as zero. It is worth noting that all the values are calculated for a person of average weight and these are not constant numbers and vary from patient to patient, which makes the design of the controller a more challenging task.

To show the complete dynamics of the glucose–insulin regulatory system, two other terms are considered in Eq. (6.40). The term $D(t)$ shows the rate at which glucose is absorbed into the blood from the intestine, following food intake. Since in diabetic patients the normal insulin regulatory system does not exist, this glucose absorption is considered as a disturbance for the system dynamics. This disturbance can be modeled by a decaying exponential function of the form

$$D(t) = A\exp(-Bt), B > 0 \tag{6.41}$$

where t is in (min) and $D(t)$ is in (mg/dl/min). The signal $u(t)$, which is the controller, defines the insulin injection rate and replaces the normal insulin regulatory system of the body (which does not exist in diabetic patients). Therefore, the goal is to employ HOSM techniques to design the appropriate control function, $u(t)$ to compensate for the uncertainties and disturbances, and to stabilize the blood plasma glucose concentration of a diabetic patient at the basal level. It should be mentioned that the dynamics of the insulin pump is neglected in the model introduced in Eq. (6.40).

6.11.3 Higher-Order Sliding Mode Controller Design

The system introduced in Eq. (6.40) can be rewritten in state-space form as follows:

$$\begin{aligned}
\dot{x}_1 &= -p_1[x_1 - G_b] - x_1 x_2 + D(t) \\
\dot{x}_2 &= -p_2 x_2 + p_3[x_3 - I_b] \\
\dot{x}_3 &= -n[x_3 - I_b] + \gamma[x_1 - h]^+ t + u(t)
\end{aligned} \tag{6.42}$$

---

[2]For details see [82].

where x_1, x_2, and x_3 are blood plasma glucose concentration (mg/dl), the insulin's effect on the net glucose disappearance (1/min), and the insulin concentration in plasma (μU/ml), respectively.

Stabilizing the glucose concentration in the diabetic patient's blood at the basal/reference level G_b is an output tracking problem; thus, the tracking error is defined as the difference between the glucose concentration level and its basal value in the diabetic patient's blood as

$$e = G_b - G(t) = G_b - x_1 \tag{6.43}$$

Given the dynamical system introduced in Eq. (6.42), the controller $u(t)$ must be designed such that $e \to 0$ in the presence of uncertainties, parameter variations, and disturbances like oral food intake, $D(t)$.

First the relative degree of the system must be defined. Assuming $y = x_1$, the relative degree could be defined as the required number of successive differentiations of $y = x_1$ until the control appears in the equation. Using Eq. (6.42), the control function appears in the equations after the third differentiation, i.e.,

$$x_1^{(3)} = \phi(x, t) - p_3 x_1 u(t) \tag{6.44}$$

where

$$
\begin{aligned}
\phi(x, t) = {} & x_1[-p_1(p_1^2 + 3p_3 I_b) - p_3 I_b(p_2 + n) - p_3\gamma(x_1 - h)^+ t] \\
& + x_2[-p_1^2(1 + G_b) + p_1 p_2(2G_b - 1) + 2D(p_1 + p_2)] \\
& + x_3[-2p_3(p_1 + D)] + x_1 x_2[-(p_1 + p_2)^2 - 3p_3 I_b] \\
& + x_1 x_3[p_3(3p_1 + p_2 + n)] + x_1 x_2^2[-3(p_1 + p_2)] \\
& + x_2^2(p_1 G_b + D) + 3p_3 x_1 x_2 x_3 - x_1 x_2^3 \\
& + \ddot{D} + (p_1 G_b + D)(p_1^2 + 2p_3 I_b)
\end{aligned}
\tag{6.45}
$$

Since $p_3 \neq 0, x_1 \neq 0$, and $p_3 x_1 \in [1.2 \times 10^{-4}, 3 \times 10^{-2}]$, the system (6.44) has a well-defined relative degree, $r = 3$. This allows us to design the controller for the system in Eq. (6.42) that ensures $e \to 0$.

Nested HOSM Control Design

Arbitrary-order sliding mode controllers with finite-time convergence, that generate continuous controls, which were demonstrated and described earlier in this chapter, will be employed.

Introducing the sliding variable

$$\sigma = e = G_b - x_1 \tag{6.46}$$

the sliding variable dynamics are derived as

$$\begin{cases} \sigma^{(3)} = -\phi(x,t) + p_3 x_1 u(t) \\ \dot{u}(t) = v \end{cases} \tag{6.47}$$

where v is a virtual control. Then

$$\sigma^{(4)} = -\phi_1(x,t) + p_3 x_1 v \tag{6.48}$$

with

$$\phi_1(x,t) = -\dot{\phi}(x,t) + p_3 \dot{x}_1 u(t) = -\dot{\phi}(x,t) + p_3 u(t)[-p_1(x_1 - G_b) - x_1 x_2 + D(t)] \tag{6.49}$$

It is assumed in the design procedure that Eq. (6.49) is bounded by some positive value, i.e., $|\phi_1| \leq L$. A HOSM controller that stabilizes σ at zero in a finite time is taken as

$$v = -\alpha \, \text{sign} \left(\dddot{\sigma} + 3(\ddot{\sigma}^6 + \dot{\sigma}^4 + |\sigma|^3)^{1/12} \times \right.$$

$$\left. \text{sign} \left(\ddot{\sigma} + (\dot{\sigma}^4 + |\sigma|^3)^{1/6} \text{sign}(\dot{\sigma} + 0.5 |\sigma|^{3/4} \, \text{sign}(\sigma)) \right) \right) \tag{6.50}$$

It is obvious that introducing the virtual controller adds one further differentiation and increases the relative degree of the system. Now we have $r = 4$ instead of 3.

In order to compute $\sigma^{(k)}, k = 1, 2, 3$, the HOSM differentiator in Eq. (6.28) is employed.

Remark 6.2. Increasing the relative degree from 3 to 4 yields high-frequency switching in the virtual control, v, while the original control, u, is continuous since $u = \int v \, d\tau$. Also a fourth-order quasi-continuous controller can be used instead of the fourth-order nested controller given by Eq. (6.50). Chattering attenuation would be expected to be better.

HOSM (Quasi-Continuous) Control Design

A quasi-continuous HOSM that stabilizes the sliding variable Eq. (6.46), whose dynamics are given in Eq. (6.48), is

$$v = -\alpha \, \varphi_{3,4} / N_{3,4} \tag{6.51}$$

where

$$\varphi_{3,4} = \dddot{\sigma} + 3\lambda^2 \left[\ddot{\sigma} + \lambda^{4/3} \left(|\dot{\sigma}| + 0.5\lambda \, |\sigma|^{3/4} \right)^{-1/3} \left(\dot{\sigma} + 0.5\lambda \, |\sigma|^{3/4} \, \text{sign}(\sigma) \right) \right]$$

$$\times \left[|\ddot{\sigma}| + \lambda^{4/3} \left(|\dot{\sigma}| + 0.5\lambda \, |\sigma|^{3/4} \right)^{2/3} \right]^{1/2}$$

$$N_{3,4} = |\dddot{\sigma}| + 3 \left[|\ddot{\sigma}| + \left(|\dot{\sigma}| + 0.5 \, |\sigma|^{3/4} \right)^{2/3} \right]^{1/2}$$

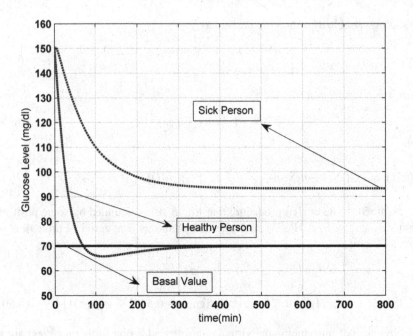

Fig. 6.8 Open-loop glucose regulatory system

The controller in Eq. (6.51) is continuous everywhere except on the r-sliding mode, i.e., when $\sigma \equiv 0$ or when $\sigma = \dot{\sigma} = \ddot{\sigma} = \sigma^{(3)} = 0$.

Remark 6.3. Although controller Eq. (6.51) is continuous everywhere except for $\sigma = \dot{\sigma} = \ddot{\sigma} = \sigma^{(3)} = 0$, its discontinuity and high-frequency switching in the 4-sliding mode makes Eq. (6.50) preferable for controlling the system (6.42), thus providing continuity for the control function u.

6.11.4 Simulation

The quasi-continuous control algorithm in Eq. (6.51) has been used for controlling the system (6.42) in the simulations. The third-order ($k = 3$) differentiator Eq. (6.28) is used to estimate $\dot{\sigma}$, $\ddot{\sigma}$, and $\dddot{\sigma}$ while computing the control function (6.51).

MATLAB is used to simulate the closed-loop system in order to show the validity of the proposed design. The first part of simulation has been done for the system of equations in Eq. (6.42), assuming there is no controller. Figure 6.8 shows the

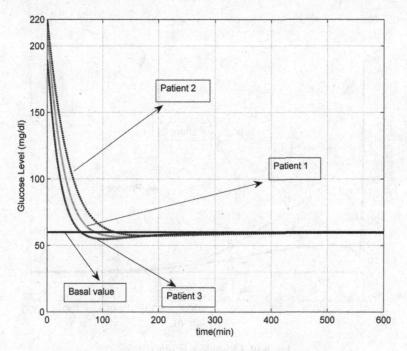

Fig. 6.9 Closed-loop glucose regulatory system using the controller

response of a healthy person and a sick person, to show the difference between their glucose regulatory systems. It is easy to see that the glucose value of the healthy person is finally stabilized back at the basal level in the presence of the meal disturbance, but the sick person's glucose level stays dangerously out of range.

To validate the proposed algorithm in Eq. (6.51), the control function is applied to the system (6.42) and the response of a sick person is examined. To check the robustness of the control algorithm to parameter variations, three sets of parameters for three different "patients" have been used. Figure 6.9 shows the results obtained from the simulation. It is obvious that the transient responses of the different patients to the same controller are different, but in all three cases, the glucose is stabilized at the basal level in a reasonable time interval. Figures 6.10 and 6.11 show the insulin profile and the control function for the three patients. The values that have been used in implementing the model and its parameters are given in Table 6.1. It should be noted that the dynamics of the pump is neglected in the simulations (Fig. 6.7).

Remark 6.4. A chattering analysis of the quasi-continuous control Eq. (6.51) in the presence of the actuator (insulin pump) unmodeled dynamics could be performed using the describing function techniques from Chap. 5.

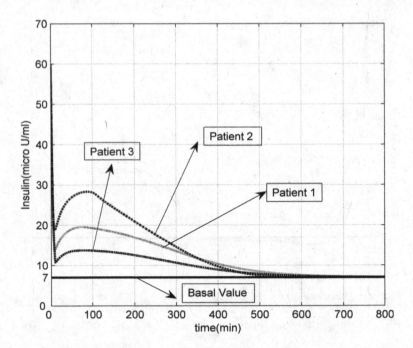

Fig. 6.10 Closed-loop insulin profile

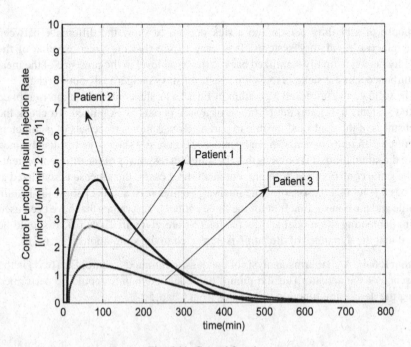

Fig. 6.11 Control function

Table 6.1 Minimal Bergman model parameters

	Normal	Patient 1	Patient 2	Patient 3
p_1	0.0317	0	0	0
p_2	0.0123	0.02	0.0072	0.0142
p_3	4.92×10^{-6}	5.3×10^{-6}	2.16×10^{-6}	9.94×10^{-5}
γ	0.0039	0	0	0
n	0.2659	0.3	0.2465	0.2814
h	79.0353	78	77.5783	82.9370
G_b	70	70	70	70
I_b	7	7	7	7
G_0	291.2	220	200	180
I_0	364.8	50	55	60

6.12 Notes and References

Chattering is highly undesirable especially if the energy associated with it cannot be neglected, i.e., if the energy does not vanish with the gradual vanishing of chattering-producing factors (noises, delays, small singular perturbation parameters, etc.). Corresponding formal notions of chattering were introduced in [130]. The standard chattering attenuation procedure is to consider the control derivative as a new control input, thus increasing the relative degree [18, 132]. It is proved in [130, 131] that the resulting systems are robust with respect to the presence of unaccounted-for fast stable actuators and sensors and no dangerous chattering is generated either by additional dynamics or by noise and delays. That also remains true when output feedback controllers are constructed, as in Sect. 6.8.

The formal definition of relative degree is given by Isidori in [112]. For details about homogeneity and in particular the definition of homogeneity degree, homogeneity dilation, etc., see [12]. Finite-time stabilization of smooth systems at an equilibrium point by means of continuous control is considered in [22]. The result that asymptotic stability of a continuous homogeneous differential equation with negative homogeneous degree is equivalent to global uniform finite-time stability is due to Bhat and Bernstein [22]. The homogeneous degree of a differential inclusion and global uniform finite-time stability of the Filippov inclusion appear in [126]. The properties of the homogenous Filippov inclusion that are studied in Theorems 6.1, 6.2, and 6.3 are from [126]. The qualitative proof of Theorem 6.2 follows the rigorous proof from [125]. Finite-time stability of homogeneous discontinuous differential equations is also considered in [129, 145].

The two families of r-sliding mode controllers, (nested and quasi-continuous) were originally proposed in [125, 127, 128]. The HOSM approach for systems with unmatched uncertainties is developed in [76, 77].

Details of the construction and proofs of the robust finite-time-convergent exact homogeneous differentiators are presented in [123, 125]. The accuracy of the HOSM differentiators coincides with the Kolmogorov estimations [120]. The list of gains for the arbitrary-order exact differentiator given in Remark 6.1 is from [125–127]. High-gain observers for nonlinear systems are studied in [10]. The formulas in Sect. 6.9.2 for the HOSM controllers with the tuned parameters appear in [133]. The proof of Theorem 6.4 is given in [125].

For further reading on the traditional treatment of diabetes see [55] and [144]. The Bergman insulin–glucose regulation model and the values of its parameters are introduced in [82]. Details on the delivery of insulin using a closed-loop device like an insulin pump are given in [114]. The relative degree approach and the HOSM control used to design the control algorithm for the glucose–insulin regulatory system is discussed in [115].

The study of adaptive HOSM control is an interesting direction of on-going research for HOSM control and improves its implementability [168, 170]. In systems of relative degree 1 with absolutely continuous perturbations the super-twisting algorithm can be used for finite-time compensation. If the perturbation is r times differentiable, an $(r+1)$th-order HOSM controller can be designed producing a control signal with an rth order of smoothness that theoretically exactly compensates the perturbation. Thus the decision about the sliding mode order depends on the perturbation smoothness. Therefore, the only theoretical restriction on the smoothness of HOSM-based controllers the smoothness of the perturbations to be rejected.

On the other hand the smoothness of the control law is an important requirement for controller realization due to the properties of some actuators. Moreover, the complexity of HOSM algorithms increases with their order. Therefore, the order of sliding is a compromise between (a) the smoothness of the perturbations; (b) the requirements associated with the actuators; (c) the complexity of the algorithms; (d) the level of noise; and finally (e) the discretization step. Suppose that we have made a decision about the order of the HOSM algorithm. Such a controller will experience chattering due to the discontinuity terms in the rth output derivative. If the estimation of the perturbation boundary is known, then the HOSM controllers can be easily designed. If the bound of the rth derivative of the perturbation is known, then, an r-sliding controller can be designed with overestimated control gains. However this leads to increased chattering and the waste of control energy. Hence, it is important to adapt a HOSM controller to the real perturbations that affect the system. The adaptive-gain 2-SMC, in particular, adaptive twisting and super-twisting controllers, for systems with matched bounded disturbances with unknown bounds are studied in [118, 119, 168, 170].

6.13   Exercises

Exercise 6.1. Prove the homogeneity of the nested sliding mode controllers presented in Sect. 6.6.1.

Exercise 6.2. Prove the homogeneity of the quasi-continuous sliding mode controllers presented in Sect. 6.6.1.

Exercise 6.3. Prove that the quasi-continuous controllers (Sect. 6.6.2) are during continuous everywhere except during the HOSM.

Exercise 6.4. What is the differentiator, which provides the 3 exact derivatives of the signal $\sin(3t)$?

Exercise 6.5. Check by computer simulation its performance in the presence of noises $0.000001\cos(1000t)$ and $0.001\cos(1000t)$, $0.01\cos(1000t)$.

Exercise 6.6. Consider the system

$$\dot{x}_1 = \cos(t) - e^{\cos(t)}x_2$$
$$\dot{x}_2 = a(x_1, x_2, x_3, t) + (2 + \cos(x_3))u$$
$$\dot{x}_3 = \sin(2x_1 + x_3 + 4t) - x_3 + u$$

where x_i, $u \in \mathbb{R}$. The output is $y = x_1$. The command signal $y_c(t)$ is smooth and is assumed to be unknown in advance, i.e., its exact formula should not be used in the controller. In addition the same controller should also work for both $a = \cos(x_1 + x_3)$ and $a = \cos(x_1 - x_3)$ as well. Design a controller to force $y = y_c(t)$ by means of a continuous bounded control and in finite time for zero initial conditions $x_i = 0, i = 1, 2, 3$:

a. Using full measurements, including the required derivatives of $y_c(t)$
b. Using only real-time measurements of $y(t) - y_c(t)$, while x_2, x_3 are not available
c. To check the robustness of the designed output-feedback controller with respect to small sampling noises of $y(t) - y_c(t)$

Exercise 6.7. Design a car control to provide tracking of the trajectory $y = 12\sin(0.05x) - 5\cos(0.02x)$. Check the robustness of the controller with respect to small measurement errors.

Exercise 6.8. Consider the insulin–glucose dynamics of patient 1, modeled by the Bergman minimal model in Eq. (6.42) with the parameters given in Table 6.1. Assume that the diabetic patient had a meal at 100 min after the start of the treatment: a corresponding disturbance term is given by Eq. (6.41) with $A = 50$ and $B = 3$. Design a nested HOSM controller given by Eq. (6.50). Tune the parameters of the HOSM controller. Study the robustness of the designed HOSM controller to the meal disturbance, via computer simulations.

Chapter 7
Observation and Identification via HOSM Observers

Control systems normally perform under uncertainties/disturbances and with measurement signals corrupted by noise. For systems with reliable models and noisy measurements, a filtration approach (Kalman filters, for example) is efficient. However, as shown in Chap. 3, sliding mode observers based on first-order sliding modes are effective in the presence of uncertainties/disturbances. Nevertheless, as discussed in that chapter, they are only applicable when the relative degree of the outputs with respect to the uncertainties/disturbances is one, and differentiation of noisy outputs signals is not needed.

Unfortunately, even for observation of mechanical systems with measured positions, the estimation of velocities—i.e., the derivatives of position—is necessary. The uncertainties/disturbances in mechanical systems are in the equations for accelerations and have relative degree two with respect to the measured positions. This means that differentiators which can provide the best possible accuracy in the presence of sampling steps and noise are needed for the general case of observation of control systems working under uncertainties/disturbances. HOSM differentiators are one class of such differentiators. In this chapter we will show how to design these HOSM observers for different types of systems.

In this chapter sliding mode based observers are presented as an alternative to the problem of observation of perturbed systems. In particular, high-order sliding mode (HOSM) based observers can be considered as a successful technique for the state observation of perturbed systems, due to their high precision and robust behavior with respect to parametric uncertainties. The existence of a direct relationship between differentiation and the observability problem makes sliding mode based differentiators a technique that can be applied directly for state reconstruction. Even when the differentiators appear as a natural solution to the observation problem, the use of the system knowledge for the design of an observation strategy results in a reduction in the magnitude of the gains for the sliding mode compensation terms. Moreover, complete or partial knowledge of the system model facilitates the application of the techniques to parametric reconstruction or disturbance reconstruction.

Y. Shtessel et al., *Sliding Mode Control and Observation*, Control Engineering,
DOI 10.1007/978-0-8176-4893-0_7, © Springer Science+Business Media New York 2014

7.1   Observation/Identification of Mechanical Systems

This section will begin by focusing on observation and identification of mechanical systems, which have been the focus of many studies throughout the years. Recent research on these systems has produced many important applications such as telesurgery with the aid of robotic manipulators, missile guidance and defense, and space shuttle control. The general model of second-order mechanical systems is derived from the Euler–Lagrange equations which are obtained from an energy analysis of such systems. They are commonly expressed in matrix form as

$$M(\mathbf{q})\ddot{\mathbf{q}} + C(\mathbf{q},\dot{\mathbf{q}})\dot{\mathbf{q}} + P(\dot{\mathbf{q}}) + G(\mathbf{q}) + \Delta(t,\mathbf{q},\dot{\mathbf{q}}) = \tau \qquad (7.1)$$

where $\mathbf{q} \in \mathbb{R}^n$ is a vector of generalized coordinates, $M(\mathbf{q})$ is the inertia matrix, $C(\mathbf{q},\dot{\mathbf{q}})$ is the matrix of Coriolis and centrifugal forces, $P(\dot{\mathbf{q}})$ is Coulomb friction, which possibly contains relay terms depending on $\dot{\mathbf{q}}$, $G(\mathbf{q})$ is the term associated with the gravitational forces, $\Delta(t,\mathbf{q},\dot{\mathbf{q}})$ is an uncertainty term; and τ is the generalized torque/force produced by the actuators. The control input τ is assumed to be given by some known feedback function. Note that $M(\mathbf{q})$ is invertible since $M(\mathbf{q}) = M^T(\mathbf{q})$ is strictly positive definite. All the other terms are supposed to be uncertain, but the corresponding nominal functions $M_n(\mathbf{q})$, $C_n(\mathbf{q},\dot{\mathbf{q}})$, and $P_n(\dot{\mathbf{q}})$, $G_n(\mathbf{q})$ are assumed known.

Introducing new variables $x_1 = \mathbf{q}$, $x_2 = \dot{\mathbf{q}}$, $u = \tau$, the model (7.1) can be rewritten in the state-space form

$$\begin{aligned}
\dot{x}_1 &= x_2, \\
\dot{x}_2 &= f(t,x_1,x_2,u) + \xi(t,x_1,x_2,u), \quad u = U(t,x_1,x_2), \qquad (7.2) \\
y &= x_1,
\end{aligned}$$

where the nominal part of the system dynamics is represented by the function

$$f(t,x_1,x_2,u) = -M_n^{-1}(x_1)[C_n(x_1,x_2)x_2 + P_n(x_2) + G_n(x_1) - u]$$

containing the known nominal functions M_n, C_n, G_n, P_n, while the uncertainties are lumped in the term $\xi(t,x_1,x_2,u)$. The solutions to system (7.3) are understood in a Filippov sense. It is assumed that the function $f(t,x_1,x_2,U(t,x_1,x_2))$ and the uncertainty $\xi(t,x_1,x_2,U(t,x_1,x_2))$ are Lebesgue measurable function of t and uniformly bounded in any compact region of the state space x_1, x_2.

In order to apply a state-feedback controller or to simply perform system monitoring, knowledge of the coordinate x_2 is required. Moreover, in the general case, for the design of the controller, it is necessary to know the parameters of the system. The tasks considered in this section are to design a finite-time convergent observer of the velocity $\dot{\mathbf{q}}$ for the original system (7.1), when only the position $\mathbf{q}$ and the nominal model are available, as well as the development of an identification algorithm to obtain the system parameters through knowledge of only the state x_1 (i.e., $\mathbf{q}$) and the input $u(t)$. Only the scalar case $x_1, x_2 \in R$ is considered for the sake of simplicity.

7.1.1 Super-Twisting Observer

One of the popular second-order sliding mode algorithms which offer a finite reaching time and which can be used for sliding mode based observation is the super-twisting algorithm considered in Chap. 4. The proposed super-twisting observer has the form

$$\dot{\hat{x}}_1 = \hat{x}_2 + z_1$$
$$\dot{\hat{x}}_2 = f(t, x_1, \hat{x}_2, u) + z_2 \qquad (7.3)$$

where $\hat{x}_1$ and $\hat{x}_2$ are the state estimates while the correction variables z_1 and z_2 are output error injections of the form

$$z_1 = \lambda |x_1 - \hat{x}_1|^{1/2} \operatorname{sign}(x_1 - \hat{x}_1)$$
$$z_2 = \alpha \operatorname{sign}(x_1 - \hat{x}_1) \qquad (7.4)$$

Taking $\tilde{x}_1 = x_1 - \hat{x}_1$ and $\tilde{x}_2 = x_2 - \hat{x}_2$ we obtain the error equations

$$\dot{\tilde{x}}_1 = \tilde{x}_2 - \lambda |\tilde{x}_1|^{1/2} \operatorname{sign}(\tilde{x}_1)$$
$$\dot{\tilde{x}}_2 = F(t, x_1, x_2, \hat{x}_2) - \alpha \operatorname{sign}(\tilde{x}_1) \qquad (7.5)$$

where

$$F(t, x_1, x_2, \hat{x}_2) = f(t, x_1, x_2, u) - f(t, x_1, \hat{x}_2, u) \qquad (7.6)$$

$$+ \xi(t, x_1, x_2, y) \qquad (7.7)$$

Suppose that the system states are bounded, then the existence of a constant f^+ is ensured, such that the inequality

$$|F(t, x_1, x_2, \hat{x}_2)| < f^+ \qquad (7.8)$$

holds for any possible t, x_1, x_2 and $|\hat{x}_2| \leq 2 \sup |x_2|$.

According to Sect. 4.3.2 the parameters of observer α and λ could be selected as $\alpha = a_1 f^+$ and $\lambda = a_2 (f^+)^{1/2}$, where $a_1 = 1.1$, $a_2 = 1.5$. Convergence of the observer states $(\hat{x}_1, \hat{x}_2)$ from Eqs. (7.3) and (7.4) to the system state variables (x_1, x_2) in Eq. (7.3) occurs in finite time, from the theorem in Sect. 4.3.2. All other theorems from Sect. 4.3.2 are also true for the observer Eqs. (7.3) and (7.4).

The standard 2-sliding-mode-based differentiator from Sect. 4.3.2 could be also implemented here to estimate the velocity. In this case, if the accelerations in the mechanical system are bounded, the constant f^+ can be found as the double maximal possible acceleration of the system. For the proposed observer, the design of the gains α and λ is based on an estimate of $F(t, x_1, x_2, \hat{x}_2, u)$. This means that the observer design in Eqs. (7.3) and (7.4) takes into account (partial) knowledge of systems dynamics and is more accurate.

A pendulum, a classical example of a mechanical system, is now used to illustrate the effectiveness of the proposed observer Eqs. (7.3) and (7.4).

Example 7.1. Consider a pendulum system with Coulomb friction and external perturbation given by the equation

$$\ddot{\theta} = \frac{1}{J}\tau - \frac{g}{L}\sin(\theta) - \frac{V_s}{J}\dot{\theta} - \frac{P_s}{J}\text{sign}(\dot{\theta}) + v \qquad (7.9)$$

where the values $m = 1.1, g = 9.815, L = 0.9, J = mL^2 = 0.891, V_S = 0.18$ and $P_s = 0.45$ are the system parameters for simulation purposes, and v is an uncertain external perturbation satisfying $|v| \leq 1$. The function $v = 0.5\sin(2t) + 0.5\cos(5t)$ was used in simulation. Now let Eq. (7.9) be driven by the twisting controller

$$\tau = -30\,\text{sign}(\theta - \theta_d) - 15\,\text{sign}(\dot{\theta} - \dot{\theta}_d) \qquad (7.10)$$

where $\theta_d = \sin(t)$ and $\dot{\theta}_d = \cos(t)$ are the reference signals. The system can be rewritten as

$$\dot{x}_1 = x_2,$$
$$\dot{x}_2 = \frac{1}{J}\tau - \frac{g}{L}\sin(x_1) - \frac{V_s}{J}x_2 - \frac{P_s}{J}\text{sign}(x_2) + v$$

Thus, the proposed velocity observer has the form

$$\dot{\hat{x}}_1 = \hat{x}_2 + 1.5(f^+)^{1/2}|\tilde{x}_1|^{1/2}\text{sign}(x_1 - \hat{x}_1)$$
$$\dot{\hat{x}}_2 = \frac{1}{J_n}\tau - \frac{g}{L_n}\sin(x_1) - \frac{V_{sn}}{J_n}\hat{x}_2 + 1.1f^+\,\text{sign}(x_1 - \hat{x}_1)$$

where $m_n = 1, L_n = 1, J_n = m_n L_n^2 = 1, V_{sn} = 0.2$, and $P_{sn} = 0.5$ are the "known" nominal values of the parameters and f^+ is to be assigned. Assume also that it is known that the real parameters differ from the assumed known values by not more than 10 %. The initial values $\theta = x_1 = \hat{x}_1 = 0$ and $\dot{\theta} = x_2 = 1, \hat{x}_2 = 0$ were taken at $t = 0$. Noting that $0 \leq \theta \leq 2\pi$, θ belongs to a compact set (a ring), and obviously the dynamic system in (7.9) is BIBS stable.

Easy calculations show that the given controller yields $|\tau| \leq 45$ and the inequality $|\dot{\theta}| \leq 70$ is guaranteed when the nominal values of the parameters and their maximal possible deviations are taken into account. Taking $|x_2| \leq 70$, $|\hat{x}_2| \leq 140$ it follows that $|F| = |\frac{1}{J}\tau - \frac{g}{L}\sin(x_1) - \frac{V_s}{J}x_2 - \frac{P_s}{J}\text{sign}(x_2) + v - \frac{1}{J_n}\tau + \frac{g}{L_n}\sin(x_1) + \frac{V_{sn}}{J_n}\hat{x}_2| < 60 = f^+$. Therefore, the observer parameters $\alpha = 66$ and $\lambda = 11.7$ were chosen. Simulations show that $f^+ = 6$ and the respective values $\alpha = 6.6$ and $\lambda = 4$ are sufficient. Note that the terms $\frac{mgL}{J}\sin(x_1)$ and $\frac{1}{J}\tau$ would have to be taken into account when selecting the differentiator parameters, when using the techniques from Chap. 4, causing much larger coefficients to be

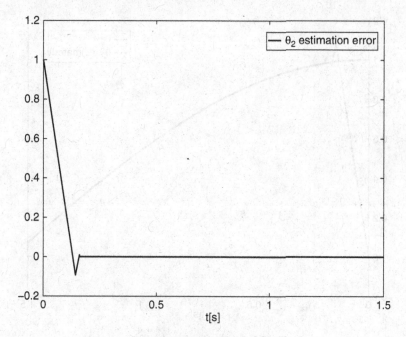

Fig. 7.1 Estimation error for x_2

used. The finite-time convergence of the velocity observation error to the origin is demonstrated in Fig. 7.1. Figure 7.2 illustrates the same convergence by comparing the estimated velocity to the real one. Finally, Fig. 7.3 shows the convergence of the observer dynamics in the $\tilde{x}_1$ versus $\tilde{x}_2$ plane.

7.1.2 Equivalent Output Injection Analysis

It is a well-known fact (see Chaps. 1–3) that the equivalent injection term contains information about the disturbances/uncertainties and unknown inputs in a system and can therefore be used for their reconstruction. This important concept is used in the analysis below to reconstruct unknown inputs in mechanical systems. Moreover, the problem of parameter estimation is also addressed in the latter part of this subsection.

Equivalent Output Injection

The finite-time convergence to the second-order sliding mode set ensures that there exists a time $t_0 > 0$ such that for all $t \geq t_0$ the following identity holds:

$$0 \equiv \dot{\tilde{x}}_2 \equiv \Delta F(t, x_1, x_2, \hat{x}_2, u) + \xi(t, x_1, x_2, u) - (\alpha_1 \text{sign}(\tilde{x}_1))_{eq}$$

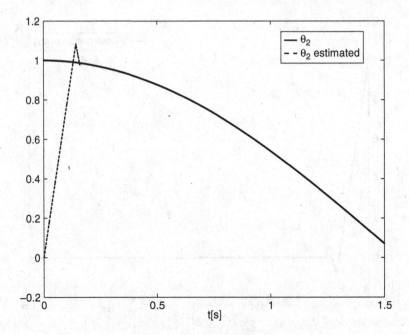

Fig. 7.2 Real and estimated velocity

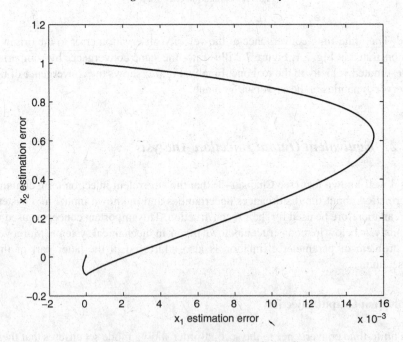

Fig. 7.3 Convergence to 0 of both $\tilde{x}_1$ and $\tilde{x}_2$

Notice that $\Delta F(t, x_1, x_2, \hat{x}_2, u) = f(t, x_1, x_2, u) - f(t, x_1, \hat{x}_2, u) = 0$ because $\hat{x}_2 = x_2$. Then the equivalent output injection z_{eq} is given by

$$z_{eq}(t) \equiv (\alpha_1 \text{sign}(\tilde{x}_1))_{eq} \equiv \xi(t, x_1, x_2, u) \tag{7.11}$$

Recall that the term $\xi(t, x_1, x_2, u)$ is composed of uncertainties and perturbations. This term may be written as

$$\xi(t, x_1, x_2, u) = \zeta(t) + \Delta F(t, x_1, x_2, u) \tag{7.12}$$

where $\zeta(t)$ is an external perturbation term and $\Delta F(t, x_1, x_2, u)$ embodies the parameter uncertainties.

Theoretically, the equivalent output injection is the result of an infinite frequency switching of the discontinuous term $\alpha_1 \text{sign}(\tilde{x}_1)$. Nevertheless, the realization of the observer produces high (finite) switching frequency making the application of a filter necessary. To eliminate the high-frequency component we will use the filter of the form:

$$\tau \dot{\bar{z}}_{eq}(t) = -\bar{z}_{eq}(t) + \alpha_1 \text{sign}(\tilde{x}_1) \tag{7.13}$$

where $\tau \in \mathbb{R}$ and $h \ll \tau \ll 1$ with h being the sampling step. It is possible to rewrite z_{eq} as the result of a filtering process in the following form

$$z_{eq}(t) = \bar{z}_{eq}(t) + \varepsilon(t) \tag{7.14}$$

where $\varepsilon(t) \in \mathbb{R}^n$ is the difference caused by the filtration process and $\bar{z}_{eq}(t)$ is the filtered version of $z_{eq}(t)$. It can be shown the $\lim_{\tau \to 0, h/\tau \to 0} \bar{z}_{eq}(\tau, h) = z_{eq}(t)$. In other words, the equivalent injection can be obtained by appropriate low-pass filtering of the discontinuous injection signals.

Perturbation Identification

Consider the case where the nominal model is totally known, i.e., for all $t > t_0$ the uncertain part $\Delta F(t, x_1, x_2, u) = 0$. The equivalent output injection takes the form

$$\bar{z}_{eq}(t) = (\alpha_1 \text{sign}(\tilde{x}_1))_{eq} = \zeta(t) \tag{7.15}$$

The result of the filtering process satisfies $\lim_{\tau \to 0, h/\tau \to 0} \bar{z}_{eq}(\tau) = \zeta(t)$. Then, any bounded perturbation can be identified, even in the case of discontinuous perturbations, by directly using the output of the filter. This is illustrated in the next example where a smooth continuous signal and a discontinuous perturbation are identified.

Example 7.2. Consider the mathematical model of the pendulum in Example 7.1 given by

$$\ddot{\theta} = \frac{1}{J} u - \frac{MgL}{2J} \sin(\theta) - \frac{V_s}{J} \dot{\theta} + v(t)$$

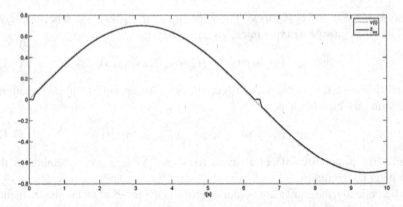

Fig. 7.4 Sinusoidal external perturbation identification

where $m = 1.1$ is the pendulum mass, $g = 9.815$ is the gravitational force, $L = 0.9$ is the pendulum length, $J = mL^2 = 0.891$ is the arm inertia, $V_S = 0.18$ is the pendulum viscous friction coefficient, and $v(t)$ is a bounded disturbance term. Assume that the angle θ is available for measurement. Introducing the variables $x_1 = \theta$, $x_2 = \dot{\theta}$ and the measured output $y = \theta$, the pendulum equation can be written in the state-space form as

$$\dot{x}_1 = x_2$$
$$\dot{x}_2 = \frac{1}{J}u - \frac{mgL}{2J}\sin(x_1) - \frac{V_s}{J}x_2 + v(t)$$
$$y = x_1$$

Suppose that all the system parameters ($m = 1.1$, $g = 9.815$, $L = 0.9$, $J = mL^2 = 0.891$, $V_S = 0.18$) are well known. The super-twisting observer for this system has the form

$$\dot{\hat{x}}_1 = \hat{x}_2 + \alpha_2 |\tilde{x}_1|^{1/2} \operatorname{sign}(\tilde{x}_1)$$
$$\dot{\hat{x}}_2 = \frac{1}{J}u - \frac{mgL}{2J}\sin(x_1) - \frac{V_s}{J}\hat{x}_2 + \alpha_1 \operatorname{sign}(\tilde{x}_1)$$
$$\tilde{x}_1 = y - \hat{x}_1$$

and the equivalent output injection in this case is given by

$$z_{eq} = (\alpha_1 \operatorname{sign}(\tilde{x}_1))_{eq} = v(t)$$

using a low-pass filter with $\bar{\tau} = 0.02[s]$. For a sinusoidal external perturbation, the identification is shown in Fig. 7.4. Using a filter $\bar{\tau} = 0.002[s]$ the perturbation identification for a discontinuous signal is shown in Fig. 7.5.

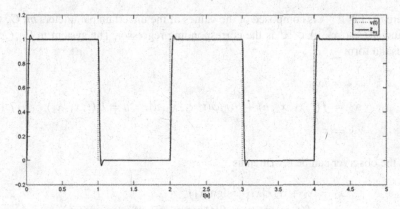

Fig. 7.5 Discontinuous perturbation identification

7.1.3 Parameter Identification

A problem that often arises in many control tasks is the uncertainty associated
with the values of certain parameters, or even, in some cases, a complete lack
of knowledge. In this situation, schemes to provide estimates of the unknown
parameters at each instant are required. Although many algorithms have been
developed to generate these estimates, the first step is usually to obtain a parametric
model in which the desired parameters are concentrated in what is called the
unknown parameter vector (denoted in most literature by θ). The interaction of these
parameters within the system can then be expressed in *regressor form* as a linear
combination of θ and a *regressor* which is a vector of known linear or nonlinear
functions.

Regressor Form

Consider the nominal case when the system is not affected by disturbances and the
only perturbations present are in the form of parametric uncertainties, i.e., $\zeta(t) =
0$ and $\xi(t, x_1, x_2, u) = \Delta F(t, x_1, x_2, u)$. The system acceleration (i.e., $\dot{x}_2$) can be
represented as the sum of a well-known part and an uncertain part:

$$\dot{x}_2 = f(t, x_1, x_2, u) + \Delta F(t, x_1, x_2, u)$$

where $f(t, x_1, x_2, u) \in \mathbb{R}^n$ is the known part of the system and $\Delta F(t, x_1, x_2, u)$ is
the uncertain part. Using the regressor notation[1] we can write the uncertain part as

$$\Delta F(t, x_1, x_2, u) = \theta(t)\varphi(t, x_1, x_2, u)$$

---

[1]For details see [173].

where $\theta(t) \in \mathbb{R}^{n \times l}$ is composed of the values of the uncertain parameters m, C, G, P and $\varphi(t, x_1, x_2, u) \in \mathbb{R}^l$ is the corresponding regressor. The system in Eq. (7.3) takes the form

$$
\begin{aligned}
\dot{x}_1 &= 7x_2 \\
\dot{x}_2 &= f(t, x_1, x_2, u) + \theta(t)\varphi(t, x_1, x_2, u), \quad u = U(t, x_1, \hat{x}_2) \quad (7.16) \\
y &= x_1
\end{aligned}
$$

and the observer can be rewritten as

$$
\begin{aligned}
\dot{\hat{x}}_1 &= \hat{x}_2 + \alpha_2 \lambda |\tilde{x}_1|^{1/2} \operatorname{sign}(\tilde{x}_1) \\
\dot{\hat{x}}_2 &= f(t, x_1, \hat{x}_2, u) + \bar{\theta}(t)\varphi(t, x_1, \hat{x}_2, u) + \alpha_1 \operatorname{sign}(\tilde{x}_1)
\end{aligned} \quad (7.17)
$$

where $\bar{\theta} \in \mathbb{R}^{n \times l}$ is a matrix of nominal values of the matrix $\theta(t)$. The error dynamics becomes

$$
\begin{aligned}
\dot{\tilde{x}}_1 &= \tilde{x}_2 - \alpha_2 \lambda(\tilde{x}_1)\operatorname{sign}(\tilde{x}_1) \\
\dot{\tilde{x}}_2 &= \theta(t) - \varphi(t, x_1, x_2, u) - \bar{\theta}(t))\varphi(t, x_1, \hat{x}_2, u) - \alpha_1 \operatorname{sign}(\tilde{x}_1)
\end{aligned} \quad (7.18)
$$

The task is to design an algorithm which provides parameter identification for the original system (7.1), when only the position x_1 is measurable and the nominal model $\bar{\theta}(t)\varphi(t, x_1, x_2, u)$ is known.

Time-Invariant Parameter Identification

Consider the case when the system parameters are time invariant, i.e., $\theta(t) = \theta$. During the sliding motion, the equivalent output injection can be represented in the form

$$
z_{eq}(t) = (\alpha_1 \operatorname{sign}(\tilde{x}_1))_{eq} = (\theta - \bar{\theta})\varphi(t, x_1, x_2, u) \quad (7.19)
$$

Notice that $\alpha_1 \operatorname{sign}(\tilde{x}_1)$ is a known term and finite-time convergence of the observer guarantees $\varphi(t, x_1, \hat{x}_2, u) = \varphi(t, x_1, x_2, u)$ for all $t \geq t_0$. Equation (7.19) represents a linear regression model where the vector parameters to be estimated are $(\theta - \bar{\theta})$. To obtain the real system parameters θ a linear regression algorithm could be proposed from Eq. (7.19).

The recursive least-squares algorithm applied for parameter identification of dynamical systems is usually designed using discretization of the regressor and derivative of the states in order to obtain the regressor form. Then the algorithm is applied in discrete form.

In mechanical system observation and identification, we deal with a data set of a continuous-time nature. That is why the implementation of any standard discretization scheme is related to unavoidable losses of existing information. This produces a systematic error—basically caused by the estimation of the derivatives of the process. As shown above, the proposed second-order sliding mode technique provides an estimation of the derivatives, converging in finite time, that avoids any additional errors arising from any standard discretization scheme implementation. Below we present a continuous-time version of the least-squares algorithm based on the proposed second-order sliding mode observation scheme. Notice that the proposed algorithm can be implemented in analog devices directly. Defining $\Delta_\theta := \theta - \hat{\theta}$, post-multiplying Eq. (7.19) by $\varphi^T(t, x_1, x_2, u)$ (written for notational simplicity as $\varphi^T(t)$). Now, using the auxiliary variable σ for integration in time, the average values of Eq. (7.19) take the form

$$\frac{1}{t}\int_0^t \bar{z}_{eq}(\sigma)\varphi^T(\sigma)d\sigma = \Delta_\theta \frac{1}{t}\int_0^t \varphi(\sigma)\varphi(\sigma)^T d\sigma \tag{7.20}$$

where $\bar{z}_{eq}$ is obtained from Eq. (7.13). Therefore, the system parameters can be estimated from Eq. (7.20) by

$$\hat{\Delta}_\theta = \left[\int_0^t \bar{z}_{eq}(\sigma)\varphi^T(\sigma)d\sigma\right]\left[\int_0^t \varphi(\sigma)\varphi^T(\sigma)d\sigma\right]^{-1} \tag{7.21}$$

where $\hat{\Delta}_\theta$ is the estimate of Δ_θ. For any square matrix the following equalities hold

$$\begin{aligned} \Gamma^{-1}(t)\Gamma(t) &= I, \\ \Gamma^{-1}(t)\dot{\Gamma}(t) + \dot{\Gamma}^{-1}(t)\Gamma(t) &= 0 \end{aligned} \tag{7.22}$$

Let us define $\Gamma(t) = \left[\int_0^t \varphi(\sigma)\varphi^T(\sigma)d\sigma\right]^{-1}$; then using Eq. (7.22) we can rewrite Eq. (7.21) in the form

$$\dot{\hat{\Delta}}_\theta = \left[\int_0^t \bar{z}_{eq}(\sigma)\varphi^T(\sigma)d\sigma\right]\dot{\Gamma}(t) + \bar{z}_{eq}(t)\varphi^T(t)\Gamma(t)$$

Now, using Eq. (7.20) we can write

$$\dot{\hat{\Delta}}_\theta = \hat{\Delta}_\theta\Gamma^{-1}(t)\dot{\Gamma}(t) + \bar{z}_{eq}(t)\varphi^T(t)\Gamma(t)$$

The equalities in Eq. (7.22) allow us to write a dynamic expression for estimating Δ_θ as

$$\dot{\hat{\Delta}}_\theta = \left[-\hat{\Delta}_\theta\varphi(t) + \bar{z}_{eq}(t)\right]\varphi^T(t)\Gamma(t) \tag{7.23}$$

In the same way, a dynamic form to find $\Gamma(t)$ is given by

$$\dot{\Gamma}(t) = -\Gamma(t)\varphi(t)\varphi^T(t)\Gamma(t) \tag{7.24}$$

The average values of the real $z_{eq}(t)$, without filtering, satisfy the equality

$$\int_0^t z_{eq}(\sigma)\varphi^T(\sigma)d\sigma = \Delta_\theta \int_0^t \varphi(\sigma)\varphi^T(\sigma)d\sigma$$

then

$$\Delta_\theta = \left[\int_0^t z_{eq}(\sigma)\varphi^T(\sigma)d\sigma\right]\Gamma(t)$$

Substituting from Eq. (7.14), the real values of the parameter vector Δ_θ satisfies

$$\Delta_\theta = \left[\int_0^t \bar{z}_{eq}(\sigma)\varphi^T(\sigma)d\sigma + \int_0^t \varepsilon(\sigma)\varphi^T(\sigma)d\sigma\right]\Gamma(t) \tag{7.25}$$

Let us assume $\bar{z}_{eq}(t) = \hat{\Delta}_\theta\varphi(t)$. In this case Eq. (7.25) becomes

$$\Delta_\theta = \left[\hat{\Delta}_\theta\int_0^t \varphi(\sigma)\varphi^T(\sigma)d\sigma + \int_0^t \varepsilon(\sigma)\varphi^T(\sigma)d\sigma\right]\Gamma(t)$$

which can be written as

$$\Delta_\theta = \hat{\Delta}_\theta + \left[\int_0^t \varepsilon(\sigma)\varphi^T(\sigma)d\sigma\right]\Gamma(t) \tag{7.26}$$

From Eqs. (7.21) and (7.26) it is possible to define the convergence conditions:

$$\sup\|t\,\Gamma(t)\| < \infty, \tag{7.27}$$

$$\left\|\frac{1}{t}\int_0^t \varepsilon(\sigma)\varphi^T(\sigma)d\sigma\right\| \to 0 \quad \text{as} \quad t \to \infty \tag{7.28}$$

Condition (7.27), known as the persistent excitation condition,[2] requires the non-singularity of the matrix $\Gamma^{-1}(t) = \int_0^t \varphi(\sigma)\varphi^T(\sigma)d\sigma$. To avoid this restriction, introduce the term ρI where $0 < \rho \ll 1$ and I is the unit matrix, and redefine $\Gamma^{-1}(t)$ as

$$\Gamma^{-1}(t) = \int_0^t \left(\varphi(\sigma)\varphi^T(\sigma)d\sigma\right) + \rho I$$

In this case the value of $\Gamma^{-1}(t)$ is always nonsingular.

Notice that the introduction of the term ρI is equivalent to setting the initial conditions of Eq. (7.24), as

$$\Gamma(0) = \rho^{-1}I, \quad 0 < \rho\text{-small enough}$$

The introduction of the term ρ ensures the condition $\sup\|t\,\Gamma(t)\| < \infty$ but does not guarantee the convergence of the estimated parameters to their real values. The convergence of the estimated values to the real ones is ensured by the *persistent excitation condition*

$$\liminf_{t\to\infty}\frac{1}{t}\int_0^t \left(\varphi(\sigma)\varphi(\sigma)^T d\sigma\right) > 0$$

---

[2]See, for example, [173].

The condition in Eq. (7.28) relates to the filtering process, and it gives the convergence quality of the identification. How fast the term $\frac{1}{t}\int_0^t \varepsilon(\sigma)\varphi(\sigma)^T d\sigma$ converges to zero dictates how fast the parameters will be estimated. The above can be summarized in Theorem 7.1.

Theorem 7.1. *The algorithm in Eqs. (7.23), (7.24) ensures the convergence of $\hat{\Delta}_\theta \to \Delta_\theta$ if conditions (7.27), (7.28) are satisfied*

Remark 7.1. The effect of noise sensitivity in the suggested procedure can be easily seen from (7.28):

$$\frac{1}{t}\int_0^t \varepsilon(\sigma)\varphi^T(\sigma)\,d\sigma \to 0 \text{ when } t \to \infty$$

The term $\varepsilon(t)$ in (7.14) includes all error effects caused by observation noise (if there is any) and errors in the realization of the equivalent output injection. One can see that if $\varepsilon(t)$ and $\varphi(t)$ are uncorrelated and are "on average" equal to zero, i.e.,

$$\frac{1}{t}\int_0^t \varepsilon(\sigma)\,d\sigma \to 0, \quad \frac{1}{t}\int_0^t \varphi(\sigma)\,d\sigma \to 0$$

then the effect of noise vanishes.

The pendulum system is once again used to illustrate the previous algorithm.

Example 7.3. Consider the model of a pendulum from Example 7.1 with Coulomb friction given by the equation

$$\ddot{\theta} = \frac{1}{J}u - \frac{MgL}{2J}\sin(\theta) - \frac{V_s}{J}\dot{\theta} - \frac{P_s}{J}\text{sign}(\dot{\theta})$$

where $m = 1.1$ is the pendulum mass, $g = 9.815$ is the gravitational force, $L = 0.9$ is the arm length, $J = mL^2 = 0.891$ is the arm inertia, $V_S = 0.18$ is the viscous friction coefficient, and $P_s = 0.45$ models the Coulomb friction coefficient. Suppose that the angle θ is available for measurement. Introducing the variables $x_1 = \theta$, $x_2 = \dot{\theta}$, the state-space representation for the system becomes

$$\dot{x}_1 = x_2,$$
$$\dot{x}_2 = \frac{1}{J}u - \frac{mgL}{2J}\sin(x_1) - \frac{V_s}{J}x_2 - \frac{P_s}{J}\text{sign}(x_2),$$
$$y = x_1$$

where $a_1 = \frac{mgL}{2J} = 5.4528, a_2 = \frac{V_s}{J} = 0.2020$, and $a_3 = \frac{P_s}{J} = 0.5051$ are the unknown parameters. Design the super-twisting-based observer as

$$\dot{\hat{x}}_1 = \hat{x}_2 + \alpha_2|\tilde{x}_1|^{1/2}\,\text{sign}(\tilde{x}_1),$$
$$\dot{\hat{x}}_2 = \frac{1}{J}u - \bar{a}_1\sin(x_1) - \bar{a}_2\hat{x}_2 - \bar{a}_3\,\text{sign}(x_2) + \alpha_1\,\text{sign}(\tilde{x}_1),$$

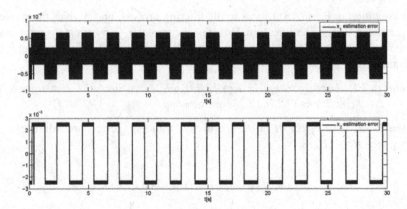

Fig. 7.6 x_1, x_2 estimation error LTI case

$$\tilde{x}_1 = y - \hat{x}_1$$

where $\bar{a}_1 = 2, \bar{a}_2 = \bar{a}_3 = 0.1$ are the nominal values of the unknown parameters. Let the control signal be generated by the twisting controller

$$u = -30\,\text{sign}(\theta - \theta_d) - 15\,\text{sign}(\dot{\theta} - \dot{\theta}_d), \tag{7.29}$$

where the reference signal is $\theta_d = 0.3\sin(3t + \pi/4) + 0.3\sin(1/2t + \pi)$. For a sampling time of $\Delta = 0.0001$ the state estimation error is shown in Fig. 7.6. In this case the identification variables are given by

$$z_{eq} = (\alpha_1\text{sign}(\tilde{x}_1))_{eq}$$
$$\Delta_\theta = [-a_1 + \bar{a}_1 \quad -a_2 + \bar{a}_2 \quad -a_3 + \bar{a}_3]$$
$$\Delta_\theta = [-3.4528 \quad -0.1020 \quad -0.4051]$$
$$\varphi = \begin{bmatrix} \sin(x_1) \\ x_2 \\ \text{sign}(x_2) \end{bmatrix}$$

It is now possible to use φ, the nonlinear regressor, to generate the dynamic adaptation gain $\Gamma(t)$ using Eq. (7.24). From Eq. (7.19), the value of $\bar{z}_{eq}$ is given by

$$\bar{z}_{eq} = z_{eq} = (\alpha_1\text{sign}(\tilde{x}_1))_{eq}$$

The dynamic estimate of the parameter error vector Δ_θ, which contains all the necessary information to retrieve the desired parameter vector θ, can be generated by implementing the algorithm in Eq. (7.23). Figure 7.7 shows the convergence of the estimated parameters to the real parameter values.

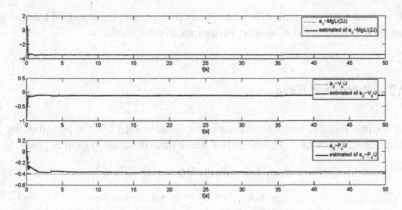

Fig. 7.7 Parameter identification for LTI case

7.2 Observation in Single-Output Linear Systems

The observer design problem for the general case of linear time-invariant systems will now be addressed in this section. The non-perturbed case will be revisited and then both unknown input and (exact) state estimation in the more complex perturbed case will be studied.

7.2.1 Non-perturbed Case

Consider a linear time-invariant system

$$\dot{x} = Ax + Bu$$
$$y = Cx \tag{7.30}$$

where $x \in \mathbb{R}^n$, $y \in \mathbb{R}$ are the system state and the output, $u \in \mathbb{R}^m$ is the known control, and the known matrices A, B, C have suitable dimensions. It is assumed that the pair (A, C) is observable. A standard Luenberger observer for this system is given by

$$\dot{\hat{x}} = A\hat{x} + Bu + L(y - \hat{y})$$
$$\hat{y} = C\hat{x} \tag{7.31}$$

where $L \in \mathbb{R}^{n \times 1}$ is a gain matrix chosen such that $(A - LC)$ is a Hurwitz matrix. Such a gain matrix L always exists because of the assumed observability of the system, and it ensures asymptotic convergence to zero of the estimation error $e = x - \hat{x}$.

It is important to remark that without any disturbance, the standard Luenberger observer is sufficient to reconstruct asymptotically the states.

7.2.2 Perturbed Case

Now assume that the linear time-invariant system in Eq. (7.30) is perturbed by an external disturbance $\zeta(t)$. The perturbed linear time-invariant system is given by

$$\dot{x} = Ax + Bu + D\zeta(t), \quad D \neq 0$$

$$y = Cx \tag{7.32}$$

where $x \in \mathbb{R}^n$, $y \in \mathbb{R}$, $u \in \mathbb{R}^m$, and $\zeta \in \mathbb{R}$ is an unknown input (disturbance). The corresponding matrices A, B, C, D have suitable dimensions. The unknown input $\zeta(t)$ is assumed to be a bounded Lebesgue-measurable function, $|\zeta(t)| \leq \zeta^+$, $\zeta^+ > 0$.

The equations are understood in the Filippov sense in order to provide for possibility to use discontinuous signals in observers. It is assumed also that all the inputs which are considered allow the existence and extension of the solution to the whole semi-axis $t \geq 0$.

Suppose that the Luenberger observer Eq. (7.31) is used to estimate the states. The dynamics of the estimation error $e = x - \hat{x}$ are given by

$$\dot{e} = (A - LC)e + D\zeta$$

$$y_e = Ce$$

In order to analyze the convergence properties of this observer, consider the Lyapunov-like function $V = e^T P_e e$, where $P_e = P_e^T > 0$ has suitable dimensions. Computing the first derivative of V we obtain

$$\dot{V} = e^T P_e ((A - LC)e + D\zeta] + [(A - LC)e + D\zeta)^T P_e e$$

Suppose the matrix P_e is the solution of the Lyapunov equation

$$P_e(A - LC) + (A - LC)^T P_e = -H \tag{7.33}$$

for some $H = H^T > 0$, then the first derivative of V becomes

$$\dot{V} = -e^T He + 2(D\zeta)^T P_e e$$

The condition $\dot{V} \leq 0$ is satisfied for all the estimation error satisfying the inequality

$$||e|| > \frac{2\zeta^+ ||D^T P_e||}{||H||}$$

This last inequality implies that the proposed Luenberger observer only can ensure the convergence of the estimation error to a bounded region around the origin. As a consequence, standard Luenberger observer cannot be applied for state reconstruction on perturbed systems.

The task in this section is to build an observer guaranteeing asymptotic (and preferably exact finite-time convergent) estimation of the states and the unknown input. Obviously, it can be assumed without loss of generality that the known input u is equal to zero (i.e., $u(t) = 0$).

It is very important to establish conditions when the unknown input can be reconstructed along with estate estimation. For this reason, several definitions will be introduced to study the state observation problem for perturbed linear systems. It is assumed in the following definitions that $u = 0$.

Definition 7.1 ([107]). System (7.32) is called strongly observable if for any initial state $x(0)$ and any input $\zeta(t)$, $y(t) \equiv 0$ with $\forall t \geq 0$ implies that also $x \equiv 0$.

Definition 7.2 ([107]). The system is strongly detectable , if for any $\zeta(t)$ and $x(0)$ it follows that $y(t) \equiv 0$ with $\forall t \geq 0$ implies $x \to 0$ with $t \to \infty$.

It is important to remark that these two definitions are not directly related to the structure of the system. However, important consequences on the system structure can be established.

Theorem 7.2. *The system (7.32) is strongly observable if and only if the output y has relative degree n with respect to the unknown input $\zeta(t)$, i.e., it has no invariant zeros.*

Proof. Let matrix P be defined by

$$P = \begin{bmatrix} C \\ CA \\ \vdots \\ CA^{n-1} \end{bmatrix}$$

Strong observability of the system requires observability, and therefore $rank\, P = n$. The observability implies the existence of a relative degree r associated with the output y with respect to the unknown input ζ. Indeed, otherwise $PD = 0$ and therefore $D = 0$. Then the coordinate transformation $x_O = Px$ turns system (7.32) into

$$\dot{x}_O = A_O x_O + B_O u(t) + D_O \zeta(t)$$
$$y(t) = C_O x_O \tag{7.34}$$

where

$$A_O = \begin{bmatrix} 0 & 1 & 0 & \cdots & 0 \\ 0 & 0 & 1 & \cdots & 0 \\ \vdots & \vdots & \vdots & \ddots & \vdots \\ 0 & 0 & 0 & \cdots & 1 \\ -a_1 & -a_2 & -a_3 & \cdots & -a_n \end{bmatrix} \tag{7.35}$$

$$D_O = [CD, \ldots, CA^{n-2}D, CA^{n-1}D]^T \tag{7.36}$$

$$C_O = [1, 0, \ldots, 0] \tag{7.37}$$

and the a_j, $j = 1, \ldots, n$ are some constants. The vector B_O does not have any specific form. Recall that u is assumed to be zero. When $r = n$ only the last component of D_O is not zero. It is obvious that in that case the identity $y \equiv 0$ implies $x_O \equiv 0$.

Assume now that $r < n$. That means that some nontrivial zero dynamics exists, which corresponds to nontrivial solutions satisfying $y \equiv 0$ and contradicts the conditions for strong observability. This ends the proof of the theorem. $\square$

7.2.3   Design of the Observer for Strongly Observable Systems

The importance of the property of strong observability for the type of linear systems described by Eq. (7.32) lies in the fact that it ensures the existence of the sliding mode state observer. This part of the chapter will explore the design of such an observer.

Assumption 7.1. *System (7.32) has the relative degree n with respect to the unknown input $\zeta(t)$ (i.e., the system is strongly observable).*

The observer is built in the form

$$\dot{z} = Az + Bu + L(y - Cz) \tag{7.38}$$

$$\hat{x} = z + Kv \tag{7.39}$$

$$\dot{v} = W(y - Cz, v) \tag{7.40}$$

where $z, \hat{x} \in \mathbb{R}^n$, $\hat{x}$ is the estimation of x and the matrix $L = [l_1, l_2, \ldots, l_n]^T \in \mathbb{R}^n$ is a correction factor chosen so that the eigenvalues of the matrix $A - LC$ have negative real parts. (Such an L exists due to Assumption 7.1 and Theorem 7.2.)

The proposed observer is actually composed of two parts. Equation (7.38) is a traditional Luenberger observer providing the boundedness of the difference $z - x$ in the presence of the unknown bounded input ζ. System (7.40) is based on high-order sliding modes and ensures the finite-time convergence of the resulting estimation error to zero.

Suppose that only the states are to be estimated and that Assumption 7.1 holds. Note that in the simplest case when $n = 1$ the only observable coordinate coincides with the measured output and, therefore, only the input estimation problem makes sense, requiring Assumption 7.1. Thus assume that $n > 1$.

Since the pair (C, A) is observable, arbitrary stable values can be assigned to the eigenvalues of the matrix $(A - LC)$, choosing an appropriate gain matrix L. Obviously the pair $(C, A - LC)$ is also observable and therefore its observability matrix

$$\tilde{P} = \begin{bmatrix} C \\ C(A - LC) \\ C(A - LC)^2 \\ \vdots \\ C(A - LC)^{n-1} \end{bmatrix} \tag{7.41}$$

is nonsingular. Set the gain matrix $K = \tilde{P}^{-1}$ and assign

$$\hat{x} = z + \tilde{P}^{-1} v \tag{7.42}$$

The nonlinear part of the observer Eq. (7.40) is chosen as

$$\dot{v}_1 = w_1 = -\alpha_n M^{1/n} |v_1 - y + Cz|^{(n-1)/n} \operatorname{sign}(v_1 - y + Cz) + v_2$$
$$\dot{v}_2 = w_2 = -\alpha_{n-1} M^{1/(n-1)} |v_2 - w_1|^{(n-2)/(n-1)} \operatorname{sign}(v_2 - w_1) + v_3$$

$$\vdots$$

$$\dot{v}_{n-1} = w_{n-1} = -\alpha_2 M^{1/2} |v_{n-1} - w_{n-2}|^{1/2} \operatorname{sign}(v_{n-1} - w_{n-2}) + v_n$$
$$\dot{v}_n = -\alpha_1 M \operatorname{sign}(v_n - w_{n-1}) \tag{7.43}$$

where v_i, z_i, and w_i are the components of the vectors $v, z \in \mathbb{R}^n$, and $w \in \mathbb{R}^{n-1}$, respectively. The parameter M must be chosen sufficiently large, and in particular $M > |d| \zeta^+$ must be satisfied, where $d = CA^{n-1}D$. The constants α_i are chosen recursively and must be sufficiently large (see Chap. 6 for a more detailed discussion). In particular, one of the possible choices is $\alpha_1 = 1.1, \alpha_2 = 1.5, \alpha_3 = 2$, $\alpha_4 = 3, \alpha_5 = 5, \alpha_6 = 8$, in a situation when $n \leq 6$. Note that Eq. (7.43) has a recursive form, useful for the parameter adjustment.

Recall that $x_O = Px$ is the vector of canonical observation coordinates and $e_O = P(\hat{x} - x)$ is the canonical observation error. With this in mind, the following theorem summarizes the exact finite-time convergence properties of the designed observer.

Theorem 7.3. *Let Assumption 7.1 be satisfied and the output be measured subject to noise, which is a Lebesgue-measurable function of time with maximal magnitude ε. Then with properly chosen α_j's, and a sufficiently large M, the state x of the system is estimated in finite time by the observer Eqs. (7.38), (7.41), (7.42)* and (7.43). With sufficiently small ε the observation errors $e_{0i} = \hat{x}_{0i} - \hat{x}_{0i} = CA^{i-1}(\hat{x}_i - x_i)$ are of the order of $\varepsilon^{(n-i+1)/n}$, i.e., they satisfy the inequalities $|e_{0i}| \leq \Gamma_i \varepsilon^{(n-i+1)/n}$ for some constants $\Gamma_i > 0$ depending only on the observer, the system parameters, and the input upper bound. A level of accuracy of the order of $\varepsilon^{1/n}$ is obtained in noncanonical coordinates due to the mix of coordinates. In particular, the state x is estimated **exactly** and in **finite time** in the absence of noises.

Remark 7.2. It is worth noting that using a Kalman filter instead of a Luenberger observers in this algorithm may be beneficial in the presence of measurement noise.

The finite-time convergence of the observation error, which is guaranteed using a Luenberger observer, allows us to address the problem of reconstructing ζ, the unknown input to the system.

Identification of the Unknown Input

Now let $v \in \mathbb{R}^{\bar{n}+1}$, where $\bar{n} = n + k$, satisfy the nonlinear differential equation (7.40) in the form

$$\dot{v}_1 = w_1 = -\alpha_{\bar{n}+1} M^{1/(\bar{n}+1)} |v_1 - y + Cz|^{(\bar{n})/(\bar{n}+1)} \operatorname{sign}(v_1 - y + Cz) + v_2$$

$$\dot{v}_2 = w_2 = -\alpha_{\bar{n}} M^{1/(\bar{n})} |v_2 - w_1|^{(\bar{n}-1)/(\bar{n})} \operatorname{sign}(v_2 - w_1) + v_3$$

$$\vdots$$

$$\dot{v}_n = -\alpha_{k+2} M^{1/(k+2)} |v_n - w_{n-1}|^{(k+1)/(k+2)} \operatorname{sign}(v_n - w_{n-1}) + v_{n+1} \quad (7.44)$$

$$\vdots$$

$$\dot{v}_{\bar{n}} = w_{\bar{n}} = -\alpha_2 M^{1/2} |v_{\bar{n}-1} - w_{\bar{n}-2}|^{1/2} \operatorname{sign}(v_{\bar{n}+1} - w_{\bar{n}-2}) + v_{\bar{n}}$$

$$\dot{v}_{\bar{n}+1} = -\alpha_1 M \operatorname{sign}(v_{\bar{n}+1} - w_{\bar{n}}) \quad (7.45)$$

where M is a sufficiently large constant. As described previously, the nonlinear differentiator has a recursive form, and the parameters α_i are chosen in the same way. In particular, one of the possible choices is $\alpha_1 = 1.1, \alpha_2 = 1.5, \alpha_3 = 2, \alpha_4 = 3, \alpha_5 = 5, \alpha_6 = 8$, in the situation when $n + k \leq 5$. In any computer realization one has to calculate the internal auxiliary variables w_j, $j = 1, \ldots, n + k$, using only the simultaneously sampled current values of y, z_1, and v_j. The equality $\bar{e} = \omega$ is established in finite time, where ω is the truncated vector

$$\omega = (v_1, \ldots, v_n)^T$$

Thus, in the case of unknown input reconstruction, the corresponding observer Eq. (7.39) is now modified and takes the form

$$\hat{x} = z + \tilde{P}^{-1} \omega \quad (7.46)$$

where $\tilde{P}$ is the observability matrix previously defined. The estimation of the input ζ is given as

$$\hat{\zeta} = \frac{1}{d}\left(v_{n+1} - (a_1 v_1 + a_2 v_2 + \cdots + a_n v_n)\right) \tag{7.47}$$

where $s^n - a_n s^{n-1} - \cdots - a_1 = (-1)^n \det(A - LC - sI)$ defines the characteristic polynomial of the matrix $A - LC$.

An example is now presented to illustrate the effectiveness of the proposed observer Eq. (7.46) for both state and unknown input reconstruction in linear systems.

Example 7.4. Consider system (7.32) with matrices

$$A = \begin{bmatrix} 0 & 1 & 0 & 0 \\ 0 & 0 & 1 & 0 \\ 0 & 0 & 0 & 1 \\ 6 & 5 & -5 & -5 \end{bmatrix}$$

$$B = D = \begin{bmatrix} 0 & 0 & 0 & 1 \end{bmatrix}^T, C = \begin{bmatrix} 1 & 0 & 0 & 0 \end{bmatrix}$$

and initial conditions $x(0) = \begin{bmatrix} 1 & 0 & 1 & 1 \end{bmatrix}$. Note that A is not stable since its eigenvalues are $-3, -2, -1, 1$. The relative degree r with respect to the unknown input equals 4. As a consequence, the system is strongly observable. The unknown input

$$\zeta = \cos(0.5t) + 0.5\sin(t) + 0.5$$

is used for pedagogical purposes. It is obviously a bounded smooth function with bounded derivatives. It is also assumed that $u = 0$. Furthermore, let the output of the system be affected by a deterministic noise of the form

$$\bar{w} = 0.1 \sin(1037|\cos(687t)|)$$

of amplitude $\epsilon = 0.1$. The correction factor $L = \begin{bmatrix} 5 & 5 & 5 & 5 \end{bmatrix}^T$ places the eigenvalues of $A - LC$ at $\{-1, -2, -3, -4\}$. The gain matrix $\tilde{P}^{-1}$ is thus given by

$$\tilde{P}^{-1} = \begin{bmatrix} 1 & 0 & 0 & 0 \\ 5 & 1 & 0 & 0 \\ 5 & 5 & 1 & 0 \\ 5 & 5 & 5 & 1 \end{bmatrix} \tag{7.48}$$

The parameters $\alpha_1 = 1.1, \alpha_2 = 1.5, \alpha_3, \alpha_4 = 3, M = 2$ are chosen, and the nonlinear part of the observer takes the form

$$\dot{v}_1 = w_1 = -3 \cdot 2^{1/4}|v_1 - y + Cz|^{(3)/4}\,\text{sign}(v_1 - y + Cz) + v_2$$

$$\dot{v}_2 = w_2 = -2 \cdot 2^{1/3}|v_2 - w_1|^{2/3}\,\text{sign}(v_2 - w_1) + v_3$$

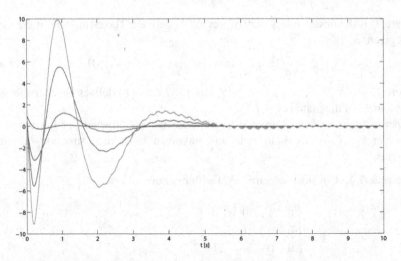

Fig. 7.8 State estimation errors in the presence of a deterministic noise of amplitude 10^{-1}

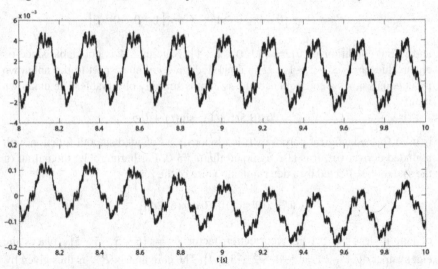

Fig. 7.9 Detail of observer error graphs. Estimation error of x_2 (*above*). Estimation error of x_4 (*below*)

$$\dot{v}_3 = w_{n-1} = -1.5 \cdot 2^{1/2} |v_3 - w_2|^{1/2} \operatorname{sign}(v_3 - w_2) + v_4$$

$$\dot{v}_4 = -1.1 \cdot 2 \operatorname{sign}(v_4 - w_3)$$

The observer performance and finite-time convergence for the sampling time interval $\tau = 0.001$ are depicted in Fig. 7.8. Figure 7.9 shows the details of the state convergence. Note that the estimation error associated with x_2 converges to a bounded region of order $5 \cdot 10^{-3}$, while the estimation error in x_4 converges to a

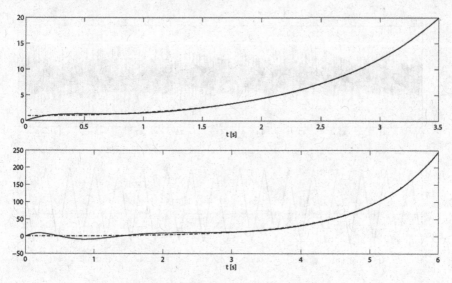

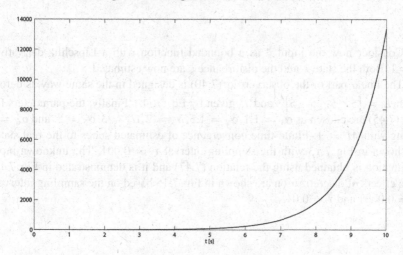

Fig. 7.10 Convergence of $\hat{x}_1, \hat{x}_4$ to x_1 and x_4

Fig. 7.11 System coordinates

bounded region of order $2 \cdot 10^{-1}$. The transient process is shown in Fig. 7.10 for
the states x_1 and x_4. It is seen from Fig. 7.11 that the system trajectories and their
derivatives of any order tend to infinity. Thus, the differentiator could not perform
the observation alone. Figure 7.12 shows the effect of discretization in observation.
The sampling time intervals $\tau = 0.0001$ and $\tau = 0.01$ were taken in the absence of
noises.

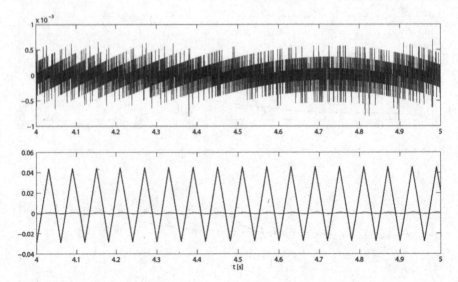

Fig. 7.12 Observer errors (detail) with sampling intervals $\tau = 0.0001$ (*above*) and $\tau = 0.01$ (*below*)

Consider now the input ζ as a bounded function with a Lipschitz derivative, $k = 1$. Both the state x and the disturbance ζ are now estimated.

The linear part of the observer Eq. (7.40) is designed in the same way as before with $L = \begin{bmatrix} 5 & 5 & 5 & 5 \end{bmatrix}^T$ and $\tilde{P}$ given by Eq. (7.48). Finally, the parameters for Eq. (7.45) are chosen as $\alpha_1 = 1.1, \alpha_2 = 1.5, \alpha_3 = 2, \alpha_4 = 3, \alpha_5 = 5$, and $\alpha_6 = 8$ along with $M = 1$. Finite-time convergence of estimated states to the real states is shown in Fig. 7.13 with the sampling interval $\tau = 0.001$. The unknown input estimation is obtained using the relation (7.47) and it is demonstrated in Fig. 7.14. The effects of discretization are shown in Fig. 7.15 based on the sampling intervals $\tau = 0.0001$ and $\tau = 0.01$.

7.3   Observers for Single-Output Nonlinear Systems

The sliding mode algorithms given in the last section can be extended to the unknown input reconstruction problem for a more general single-output nonlinear case. A differentiator-based scheme is used. This requires a system transformation into a canonical basis. The unknown inputs are expressed as a function of the transformed states and thus a diffeomorphism must be established to recover the unknown input in the original states.

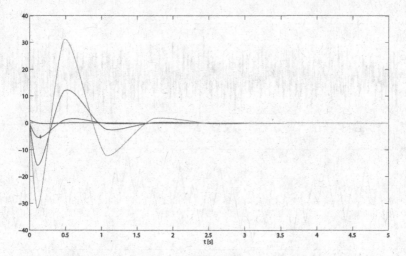

Fig. 7.13 Observer errors for the unknown input estimation case

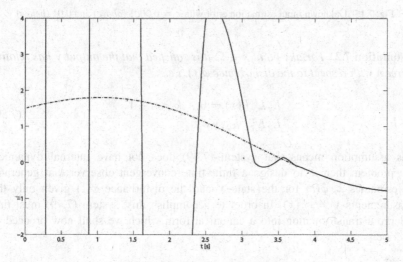

Fig. 7.14 Unknown input estimation

7.3.1   Differentiator-Based Observer

Consider a nonlinear system

$$\dot{x} = f(x) + g(x)\varphi(t)$$
$$y = h(x) \tag{7.49}$$

where $f(x) : \Omega \to \mathbb{R}^n$, $g(x) \in \mathbb{R}^n$, $h(x) : \mathbb{R}^n \to \mathbb{R}$ are smooth scalar and vector functions defined on an open set $\Omega \subset \mathbb{R}^n$. The states, outputs, and unknown inputs are given by $x \in \mathbb{R}^n$, $y \in \mathbb{R}$, and $\varphi(t) \in \mathbb{R}$.

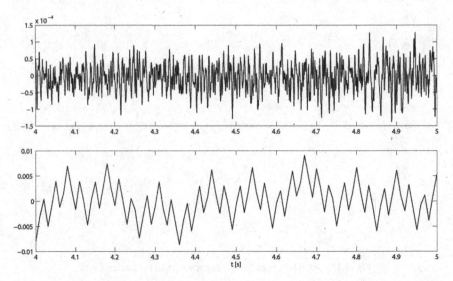

Fig. 7.15 Unknown input estimation error with $\tau = 0.0001$ (*above*) and 0.01 (*below*)

Assumption 7.2. *For any point $x \in \Omega$ it is satisfied that the output y has relative degree n with respect to the disturbance $\varphi(t)$, i.e.,*

$$
\begin{aligned}
L_g L_f^k h(x) &= 0, \quad k < n - 1 \\
L_g L_f^{n-1} h(x) &\neq 0
\end{aligned}
\tag{7.50}
$$

This assumption means that system (7.49) does not have internal dynamics. The problem then is to design a finite-time convergent observer that generates the estimates $\hat{x}$, $\hat{\varphi}(t)$ for the state x and the disturbance $\varphi(t)$ given only the measurements $y = h(x)$. In order to accomplish this, system (7.49) must first undergo a transformation into a canonical form which we shall now proceed to describe.

System Transformation

The system in Eq. (7.49) with relative degree n can be represented in a new basis that is introduced as follows:

$$
\xi = \begin{pmatrix} \xi_1 \\ \xi_2 \\ \vdots \\ \xi_n \end{pmatrix} = \begin{pmatrix} \phi_1(x) \\ \phi_2(x) \\ \vdots \\ \phi_n(x) \end{pmatrix} = \begin{pmatrix} h(x) \\ L_f h(x) \\ \vdots \\ L_f^{n-1} h(x) \end{pmatrix} \in \mathbb{R}^n
\tag{7.51}
$$

It is well known[3] that if Assumption 7.2 is satisfied, then the mapping

$$\Phi(x) = \begin{pmatrix} \phi_1(x) \\ \phi_2(x) \\ \vdots \\ \phi_n(x) \end{pmatrix} \tag{7.52}$$

defines a local diffeomorphism in a neighborhood of any point $x \in \bar{\Omega} \subset \Omega$, which means

$$x = \Phi^{-1}(\xi)$$

This is an important property since the signals obtained for the transformed system cannot be interpreted for the original system otherwise. Taking into account Eqs. (7.51) and (7.52), the system (7.49) with relative degree n can be written in the form

$$\dot{\xi} = \Lambda \xi + \psi(\xi) + \lambda(\xi, \varphi(t)) \tag{7.53}$$

where

$$\Lambda = \begin{bmatrix} 0 & 1 & 0 & \cdots & 0 \\ 0 & 0 & 1 & \cdots & 0 \\ \vdots & \vdots & \vdots & \cdots & \vdots \\ 0 & 0 & 0 & \cdots & 0 \end{bmatrix} \in \mathbb{R}^{n \times n} \tag{7.54}$$

and

$$\psi(\xi) = \begin{pmatrix} 0 \\ 0 \\ \vdots \\ L_f^n h(x) \end{pmatrix} = \begin{pmatrix} 0 \\ 0 \\ \vdots \\ L_f^n h(\Phi^{-1}(\xi)) \end{pmatrix} \tag{7.55}$$

and

$$\lambda(\xi, \varphi(t)) = \begin{pmatrix} 0 \\ 0 \\ \vdots \\ L_g L_f^{n-1} h(x) \varphi(t) \end{pmatrix} = \begin{pmatrix} 0 \\ 0 \\ \vdots \\ L_g L_f^{n-1} h(\Phi^{-1}(\xi)) \varphi(t) \end{pmatrix} \tag{7.56}$$

In order to obtain the necessary derivatives of the output, the following higher-order sliding mode observation/differentiation algorithm (from Chaps. 4 and 6) is used.

---

[3]See, for example, [112].

Higher-Order Sliding Mode Observer/Differentiator

The derivatives ξ_i, $i = 1, \ldots, n$ of the measured outputs $y = h(x)$ can be estimated in finite time by the higher-order sliding mode differentiator. This can be written in the form

$$
\begin{aligned}
\dot{z}_0 &= v_0 = z_1 - \kappa_0 |z_0 - y|^{\frac{n}{n+1}} \operatorname{sign}(z_0 - y) \\
\dot{z}_1 &= v_1 = z_2 - \kappa_1 |z_1 - v_0|^{\frac{n-1}{n}} \operatorname{sign}(z_1 - v_0) \\
&\cdots \\
\dot{z}_i &= v_i = z_i - \kappa_i |z_i - v_{i-1}|^{\frac{n-i}{n-i+1}} \operatorname{sign}(z_i - v_{i-1}) \\
&\cdots \\
\dot{z}_n &= -\kappa_n \operatorname{sign}(z_n - v_{n-1})
\end{aligned}
\tag{7.57}
$$

The choice of κ_i, $i = 0, \ldots, n$ is discussed in Chap. 6 (see, for instance, Eq. (6.29)). Therefore, the following estimates are available in finite time:

$$
\hat{\xi} = \begin{pmatrix} \hat{\xi}_1 \\ \hat{\xi}_2 \\ \vdots \\ \hat{\xi}_n \end{pmatrix} = \begin{pmatrix} \hat{\phi}_1(x) \\ \hat{\phi}_2(x) \\ \vdots \\ \hat{\phi}_n(x) \end{pmatrix} = \begin{pmatrix} z_0 \\ z_1 \\ \vdots \\ z_{n-1} \end{pmatrix} = \Phi(x) \tag{7.58}
$$

The finite-time estimation $\hat{x}$ of the state x can be easily obtained from Eqs. (7.52) and (7.58) as

$$
\hat{x} = \Phi^{-1}(\hat{\xi}) \tag{7.59}
$$

With a proper estimation of the system states achieved we can now concentrate on the other observation objective and proceed to identify the disturbance.

7.3.2   Disturbance Identification

Combining Eq. (7.59) and the last coordinate of the transformed system in Eq. (7.53), we obtain

$$
\dot{\xi}_n = L_f^n h(\Phi^{-1}(\xi)) + L_g L_f^{n-1} h(\Phi^{-1}(\xi)) \varphi(t) \tag{7.60}
$$

Since the exact finite-time estimate $\dot{\hat{\xi}}_n$ of $\dot{\xi}_n$ is available via the high-order sliding mode differentiator Eq. (7.57), and using the estimate $\hat{\xi}$ of ξ in Eq. (7.60), the finite-time estimate $\hat{\varphi}(t)$ of the disturbance can be obtained from

$$
\hat{\varphi}(t) = (L_g L_f^{n-1} h(\Phi^{-1}(\hat{\xi})))^{-1} \left[\dot{\hat{\xi}}_n - L_f^n h(\Phi^{-1}(\hat{\xi})) \right] \tag{7.61}
$$

Example 7.5. Consider a satellite system which is modeled as

$$\dot{\rho} = v$$

$$\dot{v} = \rho\omega^2 - \frac{k_g M}{\rho^2} + d$$

$$\dot{\omega} = -\frac{2v\omega}{\rho} - \frac{\theta\omega}{m}$$

In the equations above ρ is the distance between the satellite and the Earth's center, v is the radial speed of the satellite with respect to the Earth, Ω is the angular velocity of the satellite around the Earth, m and M are the mass of the satellite and the Earth, respectively, k_g represents the universal gravity coefficient, and θ is the damping coefficient. The quantity d which affects the radial velocity equation is assumed to be a disturbance which is to be reconstructed/estimated. Let $x := \mathrm{col}(x_1, x_2, x_3) := (\rho, v, \omega)$. The satellite system can then be rewritten as follows:

$$\dot{x} = \begin{pmatrix} x_2 \\ x_1 x_3^2 - \frac{k_1}{x_1^2} \\ -\frac{2x_2 x_3}{x_1} - k_2 x_3 \end{pmatrix} + \begin{pmatrix} 0 \\ 1 \\ 0 \end{pmatrix} d(t) \tag{7.62}$$

$$y = x_1 \tag{7.63}$$

where y is the system output, $k_1 = k_g M$ and $k_2 = \theta/m$. By direct computation, it follows that

$$L_g h(x) = 0, \qquad L_g L_f h(x) = 1$$

and thus the system in Eqs. (7.62) and (7.63) has global relative degree 2. Choose the coordinate transformation as $T : \xi_1 = x_1, \xi_2 = x_2, \eta = x_1^2 x_3$. Note that for $x_1 \neq 0$, this transformation is invertible and an analytic expression for the inverse can be obtained as $x_1 = \xi_1, x_2 = \xi_2, x_3 = \frac{\eta}{\xi_1^2}$. Since $x_1 = \rho$ is the distance of the satellite from the center of the Earth $x_1 \neq 0$. It follows that in the new coordinates $\mathrm{col}(\xi_1, \xi_2, \eta)$ the system from Eqs. (7.62) and (7.63) can be described by

$$\dot{\xi}_1 = \xi_2$$

$$\dot{\xi}_2 = \frac{\eta^2}{\xi_1^3} - \frac{k_1}{\xi_1^2} + d$$

$$\dot{\eta} = -k_2\eta$$

In this system the internal dynamics (given by the last equation) are linear and asymptotically stable, since $k_2 > 0$; and therefore the system itself is locally detectable. The higher-order sliding mode differentiator is described by

$$\dot{z}_0^1 = v_0^1$$
$$v_0^1 = -\lambda_0^1 |z_0^1 - y|^{2/3} \mathrm{sign}(z_0^1 - y) + z_1^1$$
$$\dot{z}_1^1 = v_1^1$$
$$v_1^1 = -\lambda_1^1 |z_1^1 - v_0^1|^{1/2} \mathrm{sign}(z_1^1 - v_0^1) + z_2^1$$
$$\dot{z}_2^1 = -\lambda_2^1 \mathrm{sign}(z_2^1 - v_1^1)$$

where $\mathrm{col}(z_0^1, z_1^2)$ or $\mathrm{col}(z_0^1, v_0^1)$ give an estimate of ξ and the estimate for η can be obtained from the equation $\dot{\hat{\eta}} = -k_2 \hat{\eta}$. Therefore, the estimate of the disturbance $d(t)$ is available online, and can be obtained from the expression

$$\hat{d} = \dot{\hat{\xi}}_2 - \frac{\hat{\eta}^2}{\hat{\xi}_1^3} + \frac{k_1}{\hat{\xi}_1^2}$$

In the simulations, the parameters have been chosen as follows: $m = 10, M = 5.98 \times 10^{24}, k_g = 6.67 \times 10^{-11}$, and $\theta = 2.5 \times 10^{-5}$. For simulation purposes, the disturbance $d(t) = \exp^{-0.002t} \sin(0.02t)$ has been introduced. The differentiator gains $\lambda_i^j j$ have been chosen as $\lambda_0^1 = 2$ and $\lambda_1^1 = \lambda_2^1 = 1$. In the following simulation, the initial values $x_0 = (10^7, 0, 6.3156 \times 10^{-4})$ are used for the plant states (in the original coordinates) while for the observer $z_0 = (1.001 \times 10^7, 0, 1)$ and $\hat{\eta}_0 = 6.3156 \times 10^{-4}$ (in the transformed coordinate system). Figures 7.16 and 7.17 show that the states and the disturbance signal $d(t)$ can be reconstructed faithfully.

7.4 Regulation and Tracking Controllers Driven by SM Observers

7.4.1 Motivation

The higher-order sliding mode observers presented in this chapter provide both theoretically exact observation and unknown input identification. This means that using such observers we cannot only observe the system states but also compensate matched uncertainties/disturbances (theoretically) exactly. Therefore, higher-order sliding mode observers create a situation in which two quite distinct control strategies can be used for the compensation of the matched uncertainties/disturbances:

- Sliding mode control based on observed system states
- The direct/continuous compensation of the uncertainties/disturbances based on its reconstructed/identified values

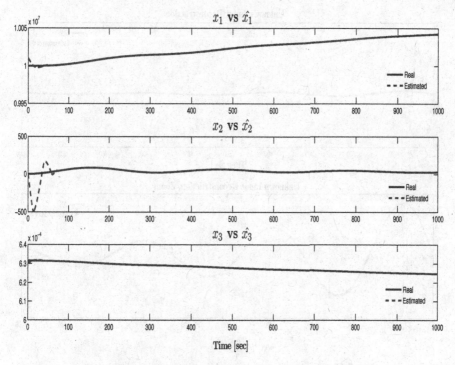

Fig. 7.16 The response of system states and their estimates

Both types of compensation are theoretically exact. In this section the control algorithms based on two proposed strategies are derived and compared. Finally, recommendations for the proper use of proposed algorithms are made which depend on the agility of the actuators, the parameters of discrete time implementation, and the measurement noise.

7.4.2 Problem Statement

Consider the linear time-invariant system with unknown inputs (LTISUI) of the form:

$$\begin{aligned} \dot{x}(t) &= Ax(t) + B(u(t) + w(t)) \\ y(t) &= Cx(t) \end{aligned}$$
(7.64)

where $x(t) \in \mathbb{R}^n$, $u(t) \in \mathbb{R}^m$, $y(t) \in \mathbb{R}^p$ $(1 \le p < n)$ are the state vector, the control, and the output of the system, respectively. The unknown inputs are represented by the function vector $w(t) \in \mathbb{R}^q$. Furthermore, $\text{rank}(C) = p$ and $\text{rank}(B) = m$. The following conditions are assumed to be fulfilled henceforth:

A1. The pair (A, B) is controllable.

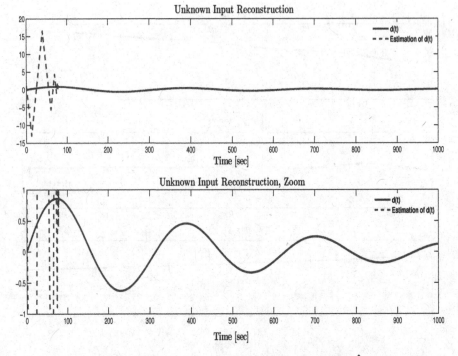

Fig. 7.17 The disturbance $d(t)$ and its reconstruction signal $\hat{d}(t)$

A2. For $u = 0$, the system is strongly observable (or the triple (A, C, B) has no invariant zeros).

A3. $w(t)$ is absolutely continuous, and there exists a constant w^+ such that $\|w(t)\| \le w^+$.

In this section we will firstly derive an observer-based robust output control for system (7.64) of the form

$$u(t) = u_0(t) + u_1(t) \qquad (7.65)$$

where $u_0(t)$ is a nominal control designed for the nominal system (i.e., $w(t) = 0$) and $u_1(t)$ is a compensator of the unknown input vector $w(t)$.

7.4.3 Theoretically Exact Output-Feedback Stabilization (EOFS)

Here, a compensation control law is designed based on the estimated states and the unknown input identification. Consider the nominal system

$$\dot{x}_0(t) = Ax_0(t) + Bu_0(t) \qquad (7.66)$$

The control signal $u_0(t)$ is a stabilizing state-feedback control for the nominal system,

$$u_0(t) = -Kx(t)$$

where the gain K can be designed using any control strategy.

Let us design the second part of the control input Eq. (7.65) as

$$u_1(t) = -\hat{w}(t)$$

where $\hat{w}(t)$ is the identified unknown input.

Theoretically, assuming exact observation and identification, the equalities $\hat{x}(t) \equiv x(t)$ and $\hat{w}(t) \equiv w(t)$ hold after a finite time T. When the EOFS control law is given by

$$u(t) = -Kx(t) - \hat{w}(t) \tag{7.67}$$

applied to system (7.64) it yields the following closed-loop dynamic equation: $\dot{x}(t) = (A - BK)x(t)$. Theoretically the continuous control $u_1(t)$ exactly compensates the matched perturbations and the solutions for systems (7.64) and (7.66) coincide.

7.4.4 Output Integral Sliding Mode Control

In this subsection, we propose applying the ISM method using the estimated states obtained by the HOSM observer. We will call it output integral sliding mode control (OISMC). Consider a control input of the form (7.65), where $u_0(t)$ is the nominal control for the system without uncertainties Eq. (7.66). Let the nominal control be $u_0(t) = -Kx(t)$. The compensator $u_1(t)$ should be designed to reject the disturbance $w(t)$ in the sliding mode on the manifold $\{x: s(x,t) = 0\}$, so that the equivalent control $u_{1eq} = -w(t)$. The switching function $s(x,t)$ is defined as $s(x,t) = s_0(x,t) + \zeta(x,t)$, where $s, s_0, \zeta \in \mathbb{R}^m$ and $s_0(x,t) = B^+ x(t)$ $(B^+ = (B^T B)^{-1} B^T)$ and the integral part ζ is selected such that $x(t) = x_0(t)$ for all $t \in [T, \infty)$. In other words, from $t_0 = T$ the system state belongs to the sliding surface, where the equivalent sliding mode control $u_{eq}(t)$ should compensate for the unknown input, that is, $u_{eq}(t) = -w(t)$. To achieve this purpose, ζ is determined from the equation $\dot{\zeta}(t) = -B^+ (Ax(t) + Bu_0(t))$, with $\zeta(t_0) = -B^+ s(x(t_0))$. The switching surface takes the form

$$s(x(t),t) = B^+ \left[x(t) - x(t_0) - \int_{t_0}^{t} [Ax(\tau) + Bu_0(x,\tau)] d\tau \right]$$

where $t_0 \geq T$. The compensator $u_1(t)$ is designed as a discontinuous unit-vector control $u_1(t) = -\rho \frac{s(x(t),t)}{\|s(x(t),t)\|}$. Thus, the sliding mode manifold $s(x,t)$ is attractive from t_0 if $\rho > w^+ \geq \|w(t)\|$. Finally, the control law Eq. (7.65) is designed as follows:

$$u(x,t) = -Kx(t) - \rho \frac{s(x,t)}{\|s(x,t)\|} \qquad (7.68)$$

Again, in the ideal case, system (7.64) with $u(x,t)$ given by Eq. (7.68) takes the form of Eq. (7.66).

7.4.5   Precision of the Observation and Identification Processes

Suppose that we would like to realize the observation with a sampling step Δ while considering that a deterministic noise signal $n(t)$ (a Lebesgue-measurable function of time with a maximal magnitude η) is present in the system output. Let

$$f(t) = f_0(t) + n(t), \quad \left\| f_0^{(i+1)}(t) \right\| < L, \quad \|n(t)\| \le \varepsilon \qquad (7.69)$$

From Theorem 6.2, the error caused by the sampling time τ in the absence of noise for an ith-order HOSM differentiator given by Equation (6.28) is

$$\left\| f_0^{(j)}(t) - z_j(t) \right\| \le O(\tau^{i-j+1}) \text{ for } j = 0,\dots,i \qquad (7.70)$$

and the differentiator error caused by a deterministic upper bounded noise will be

$$\left\| f_0^{(j)}(t) - z_j(t) \right\| \le O(\varepsilon^{\frac{i-j+1}{i+1}}) \text{ for } j = 0,\dots,i \qquad (7.71)$$

Here, we are dealing with an $(\alpha + k - 1)$th-order HOSM differentiator. To recover the estimated state, $(k - 1)$ differentiations are needed. From expressions (7.70) and (7.71) it follows that the observation error caused by the sampling time τ is $O(\tau^{\alpha+1})$, while the observation error caused by a deterministic upper bounded noise is $O(\varepsilon^{\frac{\alpha+1}{\alpha+k}})$.

It is clear that k differentiations are needed in order to recover the estimated unknown input. Therefore, from (7.70) the sampling step identification error will be $O(\tau^{\alpha})$, and the deterministic noise identification error Eq. (7.71) will be $O\left(\varepsilon^{\frac{\alpha}{\alpha+k}}\right)$.

The next proposition analyzes the total effect of both sampling step and deterministic noise errors:

Proposition 7.1. *Let us assume* $\delta \le k_\Delta \Delta$, *and* $\eta \le k_\eta \Gamma \Delta^{i+1}$ *with* k_Δ, k_η, Δ *some positive constants. Then after a finite-time, the HOSM observation and identification error will be* $O\left(\Delta^{\alpha+1}\right)$ *and* $O\left(\Delta^{\alpha}\right)$, *respectively.*

Remark 7.3. Table 7.1 summarizes the observation and identification errors when the unknown input $w(t)$ satisfies (7.69). Then, an $(\alpha + k - 1)$th-order differentiator is used to improve precision.

Table 7.1 Precision due to sampling step and bounded noise

Error	Sampling step δ	Bounded noise η	Total effect δ
Observation	$O\left(\delta^{\alpha+1}\right)$	$O\left(\eta^{\frac{\alpha+1}{\alpha+k}}\right)$	$O\left(\Delta^{\alpha+1}\right)$
Identification	$O\left(\delta^{\alpha}\right)$	$O\left(\eta^{\frac{\alpha}{\alpha+k}}\right)$	$O\left(\Delta^{\alpha}\right)$
HOSM differentiator	$O\left(\delta\right)$	$O\left(\eta^{\frac{1}{\alpha+k}}\right)$	$O\left(\Delta\right)$

EOFS Realization Error. Theoretically, the perturbations are exactly compensated in finite time. Nevertheless, in the previous section we discussed how the discretization and deterministic output noise present in the observation and identification processes affect the compensation accuracy. Furthermore, an additional error, due to the actuator time constant μ, will cause an error of order $O(\mu)$. Now, the EOFS controller stabilization error may be estimated by

$$\epsilon = O\left(\mu\right) + O\left(\Delta^{\alpha+1}\right) + O\left(\Delta^{\alpha}\right) \tag{7.72}$$

OISMC Realization Error. As we have seen, when the observation, identification, and control processes are free from nonidealities, both controllers, EOFS and OISMC, give identical results. However, in the practical case, the errors appearing in the complete control process should be taken into account. In the case of the OISMC the stabilization error is the sum of the observation error plus the control error, i.e.,

$$\epsilon = O(\mu) + O(\Delta^{\alpha+1}) \tag{7.73}$$

Now, we analyze the accuracy of the HOSM observer and the identification procedure, combined with both control methodologies. Recall that we are using a $(\alpha + k - 1)$th-order HOSM differentiator and that we need the $(k-1)$th and kth derivatives for the state observation and unknown input identification, respectively. Consider the following cases:

(a) $O\left(\Delta^{\alpha}\right) \ll O\left(\mu\right)$, i.e., the controller execution error is greater than the identification process error. In such a case, it would be suitable to use the EOFS strategy to avoid chattering.

(b) $O\left(\Delta^{\alpha+1}\right) \ll O\left(\mu\right) \ll O\left(\Delta^{\alpha}\right)$, i.e., the error related to the actuator time constant is less than the identification process error. Thus, the error in the EOFS control strategy is mainly determined by the identification error. In this case, OISMC strategy could be a better solution for systems tolerant to chattering with oscillation frequencies of order $O(\frac{1}{\mu})$.

(c) $O\left(\mu\right) \ll O\left(\Delta^{\alpha+1}\right)$, i.e., the error caused by the actuator time constant is less than the observation error. Once again, the precision of the EOFS controller is determined by the precision of the identification process $O(\Delta^{\alpha})$, and the precision of the OISMC controller is determined by the accuracy of the observation process $O(\Delta^{\alpha+1})$. However, it should be noted that in this case the use of the OISMC controller could amplify the observer noise.

7.5   Notes and References

The super-twisting-based observer for mechanical systems was first presented in [25,52].

The design of HOSM observers for strongly observable and detectable linear systems with unknown inputs is suggested in [96] and in [24]. For a proof of Theorem 7.4, see [94]. A step-by-step differentiator approach for linear systems with stable invariant zero is presented in [84].

The design of observers for nonlinear systems with unknown inputs for the case when the relative degree of unknown inputs with respect to measured outputs is well defined is given in [95]. The method in [95] does not require the system to be strongly observable, but the internal dynamics must be asymptotically stable. The first approach to state observation presented in Sect. 7.2 (but for a class of nonlinear systems) was given in [42]. The first paper in which the approach in Sect. 7.3.1 was presented for state observation is [41]. In [41] step-by-step differentiation was used. The work in [83] also uses step-by-step differentiation. Subsequently in [53, 54] HOSM differentiators were applied for the design of HOSM observers, which requires only the transformation of the observability Jacobian, but does not require the inversion of observability map. Parameter identification methods are also studied in [48].

The satellite system example is taken from [95], although the original model is from Marino and Tomei [180].

For details of the recursive least-squares algorithm see, for example, [173]. The proof of Theorem 7.1 is given in [173].

The comparison of effectiveness of HOSM-based uncertainty identification and compensation versus sliding mode based uncertainty compensation, is presented in [79]. HOSM observer based control for the compensation of unmatched uncertainties was developed in [80].

State estimation and input reconstruction in nonminimum phase causal nonlinear systems using higher-order sliding mode observers is studied in [166].

Automotive applications of sliding mode disturbance observer- based control can be found in the book [111].

7.6   Exercises

Exercise 7.1. The pendulum-cart system (see Fig. 7.18), when restricted to a two-dimensional motion, can be described by the following set of equations:

$$(M + m)\ddot{x} + ml\ddot{\theta} = u(I + ml^2)\ddot{\theta} + ml\ddot{x} = mgl \qquad (7.74)$$

where M and m are the mass of the cart and the pendulum, respectively, l is the pendulum length, $\theta(t)$ is its deviation from the vertical, and $x(t)$ represents the horizontal displacement of the cart. The system parameters are given as $M = 2[kg]$,

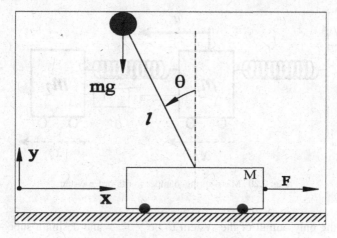

Fig. 7.18 The pendulum-cart system

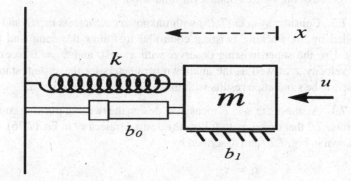

Fig. 7.19 Mass-spring-damper system

$m = 0.1$[kg], and $l = 0.5$[m]. Given the measured outputs θ and x, design a super-twisting observer for $\dot{\theta}$ and $\dot{x}$ for the uncontrolled case, $u = 0$. Assume the system's initial conditions are $\theta(0) = 0.3$[rad], $\dot{\theta}(0) = 0.03$[rad/s], $x(0) = 0$[m], and $\dot{x}(0) = 0.1$ [m/s]. Confirm the efficacy of the estimation algorithm via simulation if $\hat{\theta}(0) = 0$ [rad], $\dot{\hat{\theta}}(0) = 0$[rad/s], $\hat{x}(0) = 0$[m], and $\dot{\hat{x}} = 0$[m/s].

Exercise 7.2. Given the following mass-spring-damper system with friction (see Fig. 7.19)

$$m\ddot{x} + b_0\dot{x} + b_1 \text{sign}(\dot{x}) + kx = u \qquad (7.75)$$

where $m = 1$ [kg], $b_0 = 0.1$ [kg/s], $b_1 = 0.05$ [kg/s^2], and $k = 0.5$ [kg/s], design a feedback twisting control which achieves the tracking objective

$$x \to x_d = 0.7 \sin(2.3t) + 1.8 \sin(6.4t) \qquad (7.76)$$

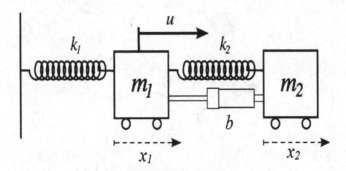

Fig. 7.20 Mass–spring–damper fourth-order system

Consider the only output of the system to be $y = x$ and design a super-twisting observer with $\hat{x} = 0$ and $\dot{\hat{x}} = 0$ to estimate $\dot{x}$ for the system starting at rest. Verify the convergence of the state estimates via simulation.

Exercise 7.3. Consider system (7.75) with unknown coefficients m, b_0, and k that is controlled by the tracking twisting controller to follow the command profile Eq. (7.76). Use the super-twisting observer with $\hat{x} = 0$ and $\dot{\hat{x}} = 0$ to estimate the mass velocity $\dot{x}$ as well as the unknown parameters for the system starting at rest. Compare the simulation results with those obtained in Exercise 7.2.

Exercise 7.4. Assuming that x_1 is measured, design the twisting tracking controller as in Exercise 7.2 that drives x_1 to follow the desired trajectory in Eq. (7.76), for the system shown in Fig. 7.20 and described by

$$\dot{x}_1 = x_2$$
$$\dot{x}_2 = -500x_1 + 150x_3 + 1.2u$$
$$\dot{x}_3 = x_4$$
$$\dot{x}_4 = 200x_1 - 600x_3 - 20x_4$$
$$y = x_1$$

Estimate x_2 via the super-twisting state observer and reconstruct the position of the second mass, x_3, treated as a disturbance in the first two equations. Use $x_0 = [0.4, -3.0, 0.28, 0]^T$ as initial conditions for the system and $\hat{x}_0 = [0, 0, 0, 0]^T$ for the observer in your simulations.

Exercise 7.5. Consider a flux-controlled DC motor whose dynamics are given by the following equations:

$$\dot{x}_1 = x_2$$
$$\dot{x}_2 = -0.1x_2 + 0.1x_3u + \zeta \qquad\qquad (7.77)$$
$$\dot{x}_3 = -2x_3 - 0.2x_2u + 200$$

where x_1 is the rotor position, x_2 is the angular velocity, x_3 is the current in the motor armature, and ξ is a disturbance. Assume that the system is affected by a perturbation signal $\zeta \neq 0$ and that both the position x_1 and the internal dynamics x_3 are measured. A zero-average random noise of amplitude $\varepsilon = 0.001$ is assumed to affect the measurement of x_1. Apply the concept of equivalent control to identify $\zeta = \cos(0.5t) + 0.5\sin(t) + 0.5$ and the discontinuous signal shown in Fig. 7.5, given the initial conditions $x_0 = [1000, 4\pi, 100]^T$ and $\hat{x}_0 = [0, 0, 0]^T$. Perform simulations to confirm the proper reconstruction of ζ.

Exercise 7.6. If a rotary spring is attached to the axis of the DC motor described in Exercise 7.5 then its dynamics may be represented by

$$
\begin{aligned}
\dot{x}_1 &= x_2 \\
\dot{x}_2 &= -a_1 x_1 - a_2 x_2 + 0.1 x_3 u \\
\dot{x}_3 &= -2x_3 - 0.2 x_2 u + 200
\end{aligned}
\tag{7.78}
$$

Design an appropriate controller u such that it satisfies the persistent excitation condition and estimate the unknown parameters a_1 and a_2 using the second-order sliding mode parameter estimator. Use $x_0 = [10, 4\pi, 100]^T$ and $\hat{x}_0 = [0, 0, 0]^T$ for simulation purposes.

Exercise 7.7. Consider the following LTI system:

$$
\dot{x} = \begin{bmatrix} 0 & 1 & 0 & 0 & 0 \\ 0 & 0 & 1 & 0 & 0 \\ 0 & 0 & 0 & 1 & 0 \\ 0 & 0 & 0 & 0 & 1 \\ 12 & -4 & -15 & 5 & 3 \end{bmatrix} x + \begin{bmatrix} 0 \\ 0 \\ 0 \\ 0 \\ 1 \end{bmatrix} u + \begin{bmatrix} 0 \\ 0 \\ 0 \\ 0 \\ 1 \end{bmatrix} \zeta(t),
$$

$$
y = \begin{bmatrix} 1 & 0 & 0 & 0 & 0 \end{bmatrix} x + \mu
$$

with initial conditions $x(0) = \begin{bmatrix} 1 & 0 & 1 & 0 & 1 \end{bmatrix}^T$. Assume that the output is affected by a deterministic noise

$$
\mu = 0.02 \cos\left(1024 \cdot |\sin(606t)|\right)
$$

Identify the constant unknown input $\zeta = 12\pi$ using the sliding mode observer for strongly observable systems assuming the control input is $u = 0$. Furthermore, apply feedback control to place the system poles at $-2, -4, -6, -1, -3.5$, and identify the sawtooth wave $\zeta(t) = 2\left(\frac{t}{0.5} - \text{floor}\left(\frac{t}{0.5} + \frac{1}{2}\right)\right)$. In both cases, use a Luenberger observer as the linear part of the observer and compare the results through simulations using a Kalman filter. The initial conditions for the observer should be $\hat{x}_0 = [0, 0, 0, 0, 0]^T$ in all cases.

Exercise 7.8. Consider the DC motor in Exercise 7.78 described by Eq. (7.77). Assume that a zero-average random noise of amplitude $\varepsilon = 0.001$ affects the measurement of x_1. Verify the strong observability of the system and apply the sliding mode observer to identify $\zeta = \cos(0.5t) + 0.5\sin(t) + 0.5$ and the discontinuous signal shown in Fig. 7.5 given the initial conditions $x_0 = [0, 4\pi, 100]^T$ and $\hat{x}_0 = [0, 0, 0]^T$. Compare the results with those obtained using the observer in Exercise 7.5 by means of simulation.

Exercise 7.9. Consider the following third-order nonlinear system:

$$\dot{x}_1 = -2x_1 - x_2 + x_3$$
$$\dot{x}_2 = x_1$$
$$\dot{x}_3 = -x_3{}^2 - 2x_3 \left(\frac{2x_1 + \sin(x_2)}{2 + \cos(x_3)} \right) + d(t) \qquad (7.79)$$
$$y = x_2$$

where $d(t) = \sin(3.18t) + 2\sin(7.32t) + 0.5\cos(0.79t)$ represents an unknown input. Design a differentiator-based observer that generates estimates of the state x and the unknown input $d(t)$, given the output y. Show the efficacy of the observer by means of simulations, considering the initial conditions $x(0) = [1, 2, -1]^T$ and $\hat{x}(0) = [0, 0, 0]^T$.

Exercise 7.10. The chaotic Chua's circuit can be described by the following state equations:

$$\dot{x}_1 = -acx_1 + ax_2 - ax_1{}^3 + d(t)$$
$$\dot{x}_2 = x_1 - x_2 + x_3$$
$$\dot{x}_3 = -bx_2 \qquad (7.80)$$
$$y = x_3$$

The parameters of the system are chosen as $a = 10, b = 16, c = -0.143$. Consider the unknown input term as $d(t) = 0.5\sin(t)$. Generate an estimate for the unknown input using the differentiator-based observer. Realize the corresponding simulations given the initial conditions $x(0) = [0.1\ 0.1\ 0.1]^T$ and $\hat{x}(0) = [0, 0, 0]^T$.

Exercise 7.11. Develop a differentiator-based observer when Assumption 7.2 is not satisfied, i.e., in the case when the output y has relative degree $r < n$ with respect to the unknown input $\varphi(t)$.

Exercise 7.12. Necessary and sufficient conditions for the strong observability of linear and nonlinear systems single-output systems were given in this chapter. Formulate the equivalent conditions for strong observability in multiple-input, multiple-output systems.

Chapter 8
Disturbance Observer Based Control: Aerospace Applications

The practical implementation of sliding mode controllers usually assumes knowledge of all system states. It also typically requires information (at least in terms of the boundaries) about the combined effect of drift terms, i.e., the internal and external disturbances of the system. In this chapter a feedback linearization-like technique is used for obtaining the input–output dynamics and reducing all disturbances to the matched ones. Then the sliding variables are introduced and their dynamics are derived. The higher-order sliding mode differentiator-based observer, which was discussed in Chap. 7, is used to the estimate system states, the derivatives of the sliding variables, as well as the drift terms. Therefore, in finite time, all information about the sliding variable dynamics becomes available. The estimated drift term is then used in the feedback loop to compensate the disturbances. The observed states are then used to design any (continuous) robust state-space controller while eliminating the chattering effect. Two case studies, launch vehicle and satellite formation control, illustrate the discussed robust control technique.

8.1 Problem Formulation

This section presents a continuous SMC technique based on a sliding mode disturbance observer (SMDO) that is applied to robust output tracking in feedback-linearizable perturbed/uncertain MIMO systems, with nonlinear feedback. Consider a nonlinear MIMO system

$$\dot{x} = f(x,t) + G(x,t)u \qquad (8.1)$$
$$y = h(x,t)$$

where $f(x,t), \in \mathbb{R}^n$, $h(x,t) = [h_1, h_2, \ldots, h_m]^T \in \mathbb{R}^m$, $G(x,t) = [g_1, g_2, \ldots, g_m] \in \mathbb{R}^{n \times m}$, and $g_i \in \mathbb{R}^n \ \forall i = 1, 2, \ldots, m$ are analytic vector and matrix functions. The system states, outputs, and inputs are given by $x \in \mathbb{R}^n$, $y \in \mathbb{R}^m$, and $u \in \mathbb{R}^m$,

Y. Shtessel et al., *Sliding Mode Control and Observation*, Control Engineering, 291
DOI 10.1007/978-0-8176-4893-0_8, © Springer Science+Business Media New York 2014

respectively. Assume that the system (8.1) is completely feedback linearizable in a reasonable compact domain $x \in \Omega$. Then the system can be transformed into a regular form

$$
\begin{bmatrix} y_1^{(r_1)} \\ y_2^{(r_2)} \\ \cdots \\ y_m^{(r_m)} \end{bmatrix} = \begin{bmatrix} L_f^{r_1} h_1(x,t) \\ L_f^{r_2} h_2(x,t) \\ \cdots \\ L_f^{r_m} h_m(x,t) \end{bmatrix} + E(x,t)u
\tag{8.2}
$$

where

$$
E(x,t) = \begin{bmatrix} L_{g_1}(L_f^{r_1-1} h_1) & L_{g_2}(L_f^{r_1-1} h_1) & \cdots & L_{g_m}(L_f^{r_1-1} h_1) \\ L_{g_1}(L_f^{r_2-1} h_2) & L_{g_2}(L_f^{r_2-1} h_2) & \cdots & L_{g_m}(L_f^{r_2-1} h_2) \\ \cdots & \cdots & \cdots & \cdots \\ L_{g_1}(L_f^{r_m-1} h_m) & L_{g_2}(L_f^{r_m-1} h_m) & \cdots & L_{g_m}(L_f^{r_m-1} h_m) \end{bmatrix}
\tag{8.3}
$$

and $\det E(x,t) \neq 0$. In the above $L_f^{r_i} h_i(x,t)$ and $L_{g_i}(L_f^{r_j-1} h_j) \; \forall i, j = 1, 2, \ldots, m$ are the corresponding Lie derivatives. The vector $\bar{r} = [r_1, r_2, \ldots, r_m]$ is the vector-relative degree, and $r_t = \sum r_i$ is the total relative degree, such that $r_t = n$.

The problem is in designing a continuous control u that provides robust decoupled output tracking in system (8.1):

$$
e_i = y_{ic} - y_i \to 0 \; \forall i = 1, 2, \ldots, m
\tag{8.4}
$$

where y_{ic} is the ith output command profile given online.

The formulated problem is addressed by enforcing the desired compensated dynamics. This is achieved by designing corresponding sliding variables, driving them to zero by means of continuous sliding mode controls u_i.

8.1.1 Asymptotic Compensated Dynamics

In this subsection the sliding variables is introduced as

$$
\sigma_i = e_i^{(r_i-1)} + c_{i,r_i-2} e_i^{(r_i-2)} + \ldots + c_{i,1} e_i^{(1)} + c_{i,0} e_i, \; i = 1, 2, \ldots m
\tag{8.5}
$$

The asymptotic decoupled output tracking compensated dynamics in system (8.2), such that $e_i \to 0 \; \forall i = 1, 2, \ldots m$ is achieved in the sliding mode

$$
e_i^{(r_i-1)} + c_{i,r_i-2} e_i^{(r_i-2)} + \ldots + c_{i,1} e_i^{(1)} + c_{i,0} e_i = 0 \; \forall i = 1, 2, \ldots m
\tag{8.6}
$$

if the coefficients $c_{i,j} > 0 \ \forall j = 0, 1, .., r_i - 2$ are chosen to provide the desired eigenvalue placement in the decoupled differential equations (8.6). The dynamics in Eq. (8.6) is enforced by designing control u that drives $\sigma_i \to 0$.

8.1.2 Finite-Time-Convergent Compensated Dynamics

In this subsection the sliding variables are chosen to guarantee the finite-time-convergent compensated dynamics in the sliding mode. They are:

$$e_i^{(r_i)} + k_{i,r_i} \left| e_i^{(r_i-1)} \right|^{\alpha_{i,r_i}} \text{sign} \left(e_i^{(r_i-1)} \right) + \ldots + k_{i,1} \left| e_i \right|^{\alpha_{i,1}} \text{sign} (e_i) = 0 \ \forall i = 1, 2, \ldots m$$

(8.7)

where the coefficients $k_{i,j} > 0 \ \forall i = 1, 2, \ldots m \ \forall j = 1, 2, \ldots r_i$, make the polynomials $P_i(\lambda) = (\lambda^{r_i} + k_{i,r_i} \lambda^{r_i-1} + \ldots + k_{i,1}$ Hurwitz, and $\alpha_{i,1}, \ldots, \alpha_{i,r_i}$ are defined as

$$\alpha_{i,j-1} = \frac{\alpha_{i,j} \alpha_{i,j+1}}{2\alpha_{i,j+1} - \alpha_{i,j}} \ \forall i = 1, 2, \ldots m \ \forall j = 2, 3, \ldots r_i$$

(8.8)

with $\alpha_{i,r_i+1} = 1$ and $\alpha_{i,r_i} = \alpha$ for some $\alpha \in (0, 1)$.

The compensated dynamics (8.7) and (8.8) can be enforced by designing an appropriate continuous control u.

Remark 8.1. In order to compute the sliding variable σ_i given by Eq. (8.5) or to enforce the finite-convergent-time asymptotic dynamics in Eqs. (8.7) and (8.8) the $r_i - 1$ consecutive derivatives of the tracking error e_i must be calculated/estimated. This can be achieved using the HOSM differentiator studied in Chaps. 4 and 6:

$$\dot{z}_{0i} = v_{0i} = z_{1i} - \kappa_{0i} |z_{0i} - e_i|^{\frac{r_i}{r_i+1}} \text{sign}(z_{0i} - e_i)$$
$$\dot{z}_{1i} = v_{1i} = z_{2i} - \kappa_{1i} |z_{1i} - v_{0i}|^{\frac{r_i-1}{r_i}} \text{sign}(z_{1i} - v_{0i})$$
$$\ldots$$
$$\dot{z}_{r_i,i} = -\kappa_{r_i,i} \text{sign}(z_{r_i,i} - v_{r_i-1,i})$$

(8.9)

where the κ_{ji} are defined as in Sect. 6.7. The finite-time estimation of the output tracking errors and their consecutive $r_i - 1$ derivatives $e_i, \ldots, e_i^{(r_i-1)}$ can be easily obtained from equations $z_{0i} = e_i, z_{1i} = \dot{e}_i, .., z_{r_i-1,i} = e_i^{(r_i-1)}$.
Denoting

$$v = E(x, t)u \Leftrightarrow u = E^{-1}(x, t)v, \ v = [v_1, v_2, .., v_m]^T$$

(8.10)

the sliding variable (8.5) dynamics can be presented in a decoupled form

$$\dot{\sigma}_i = \psi_i(.) - v_i$$

(8.11)

where $\psi_i(.) = y_{ic}^{(r_i)} - L_f^{r_i} h_i(x, t) + c_{i,r_i-2} e^{(r_i-1)} + \ldots + c_{i,0} e^{(1)} \ \forall i = 1, 2, \ldots m$.

Remark 8.2. In order to facilitate the computation of the matrix $E^{-1}(x,t)$ the state vector x must be estimated in finite time. In particular, it can be achieved using HOSM observers (see Chap. 7).

8.1.3   Sliding Variable Disturbed Dynamics

Assume that the drift terms $\psi_i(.)$ in Eq. (8.11) consist of a sum of known terms $\psi_i^0(.)$ and the unknown terms $\Delta\psi_i(.)$:

$$\psi_i(.) = \psi_i^0(.) + \Delta\psi_i(.) \tag{8.12}$$

The unknown terms that are due to external disturbances and model uncertainties are assumed to be bounded so that $|\Delta\psi_i(.)| \leq L_i$.

Then, Eq. (8.11) can be rewritten in a form

$$\dot\sigma_i = \psi_i^0(.) + \Delta\psi_i(.) - v_i \tag{8.13}$$

Now, the output tracking problem in system (8.2) is reduced to designing the continuous control laws v_i that drive the sliding variables σ_i to zero in the presence of the unknown bounded disturbances $\Delta\psi_i(.)$.

8.1.4   Output Tracking Error Disturbed Dynamics

In order to design the controller that drives the tracking errors to zero in finite time, the error dynamics (8.4) for system (8.2) are presented as

$$e_i^{(r_i)} = y_{ic}^{(r_i)} - L_f^{r_i} h_i(x,t) - v_i \tag{8.14}$$

where $\bar\psi_i(.) = y_{ic}^{(r_i)} - L_f^{r_i} h_i(x,t)$ can be presented as a sum of known $\bar\psi_i^0(.)$ and unknown bounded $|\Delta\bar\psi_i(.)| \leq \bar L_i$ terms:

$$\bar\psi_i(.) = \bar\psi_i^0(.) + \Delta\bar\psi_i(.) \tag{8.15}$$

Therefore, the error dynamics in Eq. (8.14) are reduced to the following equation:

$$e_i^{(r_i)} = \bar\psi_i^0(.) + \Delta\bar\psi_i(.) - v_i \tag{8.16}$$

Here the problem is reduced to designing the continuous control laws v_i that drive the output tracking errors in Eq. (8.16) to zero in *finite time* in the presence of the unknown bounded disturbances $\Delta\bar\psi_i(.)$.

8.2   Perturbation Term Reconstruction via a Disturbance Observer

In the first step of the continuous SMC/SMDO design methodology, the perturbation term must be estimated/reconstructed.

In order to design the SMDO for estimating the bounded disturbance $\Delta \psi_i(.)$ in Eq. (8.13), the auxiliary sliding variables s_i are introduced

$$s_i = \sigma_i + z_i \tag{8.17}$$

$$\dot{z}_i = -\psi_i^0(.) + v_i - \omega_i \tag{8.18}$$

whose dynamics are derived as

$$\frac{ds_i}{dt} = \Delta \psi_i(.) - \omega_i \tag{8.19}$$

where ω_i is the injection term of the SMDO in Eq. (8.19).

For estimating the bounded disturbance $\Delta \bar{\psi}_i(.)$ in Eq. (8.16), other auxiliary sliding variables $\bar{s}_i$ are introduced:

$$\bar{s}_i = e_i^{(r_i-1)} + \xi_i \tag{8.20}$$

$$\dot{\xi}_i = -\bar{\psi}_i^0(.) + v_i - \varpi_i \tag{8.21}$$

The $\bar{s}_i$ dynamics, taking into account Eq. (8.16), are presented as

$$\frac{d\bar{s}_i}{dt} = \Delta \bar{\psi}_i(.) - \varpi_i \tag{8.22}$$

where ϖ_i is the injection term of the SMDO in Eq. (8.22).

8.2.1   SMDO Based on Conventional SMC

The injection term given in a conventional SMC format is

$$\omega_i = \rho_i \, \text{sign}(s_i), \quad \rho_i = L_i + \varepsilon_i \tag{8.23}$$

which drives $s_i \to 0$ in finite time. Then, in the auxiliary sliding mode, $s_i = 0$; and the disturbance term $\Delta \psi_i(.)$ can be exactly estimated:

$$\Delta \psi_i(.) = \omega_{i_{eq}} \tag{8.24}$$

where $\omega_{i_{eq}}$ is the equivalent injection term. Note that it is practically impossible to compute the equivalent injection term $\omega_{i_{eq}}$ exactly, but only estimate it via low-pass filtering of control (8.23) in the sliding mode. For instance, the low-pass filter can be implemented as a first-order lag block

$$\bar{\tau}_i \frac{d\bar{\omega}_{i_{eq}}}{dt} = -\bar{\omega}_{i_{eq}} + \omega_i \tag{8.25}$$

where $\bar{\tau}_i > 0$ is a small-enough time constant of the low-pass filter (8.25). It is worth noting that the low-pass filter estimates the equivalent injection term $\hat{\omega}_{i_{eq}}(t)$ with the accuracy proportional to the time constant of the filter $\hat{\tau}_i$:

$$\left\| \bar{\omega}_{i_{eq}} - \omega_{i_{eq}} \right\| = O(\bar{\tau}_i) \tag{8.26}$$

Therefore, we obtain the estimation of the perturbation term $\Delta \psi_i(.)$ in a conventional sliding mode

$$\Delta \psi_i(.)_{est} \approx \bar{\omega}_{i_{eq}} \tag{8.27}$$

with the accuracy

$$\left\| \Delta \psi_i(.)_{est} - \Delta \psi_i(.) \right\| = O(\bar{\tau}_i) \tag{8.28}$$

Furthermore, the $\lim \left\| \Delta \psi_i(.)_{est} - \Delta \psi_i(.) \right\| = 0$ as $\bar{\tau}_i \to 0$.

Remark 8.3. Since the term $\Delta \bar{\psi}_i(.)$ can be estimated by the conventional SMC-based SMDO only asymptotically, some phase and amplitude distortions are expected (see, for instance, Eq. (8.28)). Apparently, it is better not to reconstruct the term $\Delta \bar{\psi}_i(.)$ using this asymptotic observer for enforcing the output error dynamics in Eqs. (8.7) and (8.8), since these output tracking dynamics are expected to be finite-time convergent. Consequently, it makes sense to reconstruct the disturbance terms $\Delta \bar{\psi}_i(.)$ also in finite time in order to retain overall finite-time convergence (dynamical collapse) in the output tracking.

8.2.2 SMDO Based on Super-Twisting Control

Assuming the derivatives of the perturbation terms $\Delta \psi_i(.)$ in Eq. (8.19) are bounded, i.e., $\left| \frac{d\Delta \psi_i(.)}{dt} \right| \leq \bar{L}_i$, the injection terms given in a super-twisting format are

$$\omega_i = \lambda_{0,i} |s_i|^{1/2} \text{sign}(s_i) + \eta_i \tag{8.29}$$
$$\dot{\eta}_i = \lambda_{1,i} \text{sign}(s_i)$$

with $\lambda_{0,i} = 1.5\sqrt{\bar{L}_i}$ and $\lambda_{1,i} = 1.1\bar{L}_i$ which drive $s_i, \dot{s}_i \to 0$ in finite time (see, for instance, Chaps. 4 and 7). Therefore, in the second-order sliding mode, an exact estimate of the perturbation term $\Delta \psi_i(.)$

$$\Delta \psi_i(.)_{est} = \omega_i \qquad (8.30)$$

is obtained. The super-twisting-based SMDO can be also designed for estimating/reconstructing the unknown disturbances $\Delta \bar{\psi}_i(.)$ in Eq. (8.16), assuming that the derivative of this term is also bounded: i.e., $\left| \frac{d\Delta\bar{\psi}_i(.)}{dt} \right| \leq \tilde{L}_i$. The injection terms ϖ_i are designed in the super-twist format as in Eq. (8.29) so that

$$\varpi_i = \beta_{0,i} |\bar{s}_i|^{1/2} \text{sign}(\bar{s}_i) + \chi_i \qquad (8.31)$$

$$\dot{\chi}_i = \beta_{1,i} \text{sign}(\bar{s}_i)$$

with $\beta_{0,i} = 1.5\sqrt{\tilde{L}_i}$ and $\beta_{1,i} = 1.1\tilde{L}_i$ and drive $\bar{s}_i, \frac{d\bar{s}_i}{dt} \to 0$ in finite time. Thus, the disturbances $\Delta \bar{\psi}_i(.)$ are exactly reconstructed in finite time in accordance with Eq. (8.22) and the super-twisting SMDO (8.31):

$$\Delta \bar{\psi}_i(.)_{est} = \varpi_i$$

8.2.3 Design of the SMC Driven by the SMDO

Asymptotic Continuous SMC/SMDO Design

The continuous sliding mode controllers v_i driven by the sliding mode disturbance observer (SMC/SMDO) that robustly asymptotically stabilize the sliding variable σ_i, whose dynamics are given by Eq. (8.13) in the presence of an unknown bounded disturbance $\Delta \psi_i(.)$, are designed as follows:

$$v_i = \psi_i^0(.) + \Delta \psi_i(.)_{est} + K_i \sigma_i \qquad (8.32)$$

The compensated sliding variable dynamics become

$$\dot{\sigma}_i = -K_i \sigma_i \qquad (8.33)$$

where $K_i > 0$ must be selected in order to provide the desired convergence rate $\sigma_i \to 0$.

Therefore, a continuous control law u that robustly drives the output tracking error of Eq. (8.1) $e_i \to 0 \ \forall i = 1, 2, \ldots m$ as time increases, in the presence of bounded disturbances, is designed given by

$$u = E^{-1}(x,t)v, \quad v = [v_1, v_2, .., v_m]^T \qquad (8.34)$$

$$v_i = \psi_i^0(.) + \Delta \psi_i(.)_{est} + K_i \sigma_i \ \forall i = 1, 2, \ldots m$$

where the $\Delta \psi_i(.)_{est}$ is estimated by either a conventional sliding mode or super-twisting SMDO. The sliding variables σ_i are computed using HOSM differentiators (8.9).

Remark 8.4. The SMC/SMDO design in Eqs. (8.27), (8.30), and (8.34) yields continuous control that is robust to the bounded disturbances and uncertainties. This SMC/SMDO controller can be interpreted as asymptotic SMC since $\sigma_i \to 0$ as time increases due to Eqs. (8.33) and (8.6).

Finite-Convergent-Time Continuous SMC/SMDO Design

The continuous sliding mode controllers v_i, driven by the sliding mode disturbance observer (SMC/SMDO), which robustly establish the error dynamics in Eq. (8.16) to zero in finite time are designed taking into account Eq. (8.7):

$$v_i = \bar{\psi}_i^0(.) + \Delta \bar{\psi}_i(.)_{est} + k_{i,r_i} \left| e_i^{(r_i-1)} \right|^{\alpha_{i,r_i}} \text{sign}\left(e_i^{(r_i-1)} \right) + \ldots + k_{i,1} \left| e_i \right|^{\alpha_{i,1}} \text{sign}(e_i)$$

(8.35)

The compensated finite-convergent-time output tracking error dynamics become as in Eq. (8.7).

Therefore, the continuous control law u that robustly drives the output tracking error e_i of Eq. (8.1) to zero in finite time in the presence of bounded disturbances is

$$u = E^{-1}(x,t)v$$

(8.36)

8.3   Case Study: Reusable Launch Vehicle Control

Flight control of both current and future reusable launch vehicles (RLV) in ascent and descent, as well as in approach and landing modes, involves attitude maneuvering through a wide range of flight conditions, wind disturbances, and plant uncertainties. The discussed SMC/SMDO becomes an attractive robust control algorithm for RLV maneuvers because, as well as being continuous, the SMC/SMDO controller is insensitive and robust to RLV uncertainties and external disturbances. In this section we study SMC/SMDO controller design for RLV flying in the terminal area energy management (TAEM) and approach/land (TAL) regions of flight (Fig. 8.1).

8.3.1   Mathematical Model of Reusable Launch Vehicle

The nonlinear Newton–Euler equations of motion for the rigid vehicle were chosen for the RLV flight control system design. The rotational equations of motion are

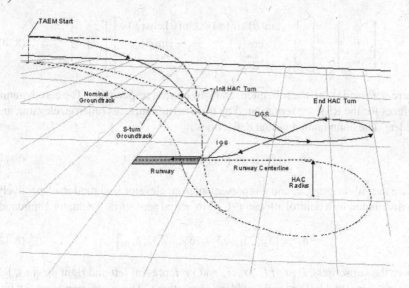

Fig. 8.1 Example TAEM and approach/land trajectory for a RLV [104]

$$\dot{p} = -L_{pq}pq - L_{qr}qr + L + f_L$$
$$\dot{q} = -M_{pr}pr - M_{r^2p^2}(r^2 - p^2) + M + f_M \qquad (8.37)$$
$$\dot{r} = -N_{pq}pq - N_{qr}qr + N + f_N$$

where p, q, and r are the roll, pitch, and yaw rates, respectively; L_{pq}, L_{qr}, M_{pr}, $M_{r^2p^2}$, N_{pq}, and N_{qr} are functions of the vehicle inertia, L, M, and N are roll, pitch, and yaw accelerations arising from the aerodynamics, and f_L, f_M, and f_N are disturbance accelerations. Translational accelerations $(\dot{u}, \dot{v}, \dot{w})$ in the vehicle body frame are presented as:

$$\dot{u} = rv - qw + X + G_x$$
$$\dot{v} = pw - ru + Y + G_y \qquad (8.38)$$
$$\dot{w} = qu - pv + Z + G_z$$

where the acceleration terms X, Y, and Z are from the aerodynamics and G_x, G_y, and G_z are gravity terms given by

$$G_x = -g\sin(\theta), \quad G_y = g\cos(\theta)\sin(\phi), \quad G_z = g\cos(\theta)\cos(\phi) \qquad (8.39)$$

The Euler angles that define the orientation of the RLV relative to an inertial frame are

$$\begin{bmatrix} \dot{\phi} \\ \dot{\theta} \\ \dot{\psi} \end{bmatrix} = \begin{bmatrix} 1 & \tan(\theta)\sin(\phi) & \tan(\theta)\cos(\phi) \\ 0 & \cos(\phi) & -\sin(\phi) \\ 0 & \frac{\sin(\phi)}{\cos(\theta)} & \frac{\cos(\phi)}{\cos(\theta)} \end{bmatrix} \begin{bmatrix} p \\ q \\ r \end{bmatrix} \qquad (8.40)$$

where ϕ, θ, and ψ are the roll, pitch, and yaw angles, respectively. The aerodynamic surfaces for the RLV are represented by three virtual surfaces (aileron, elevator, and rudder). The commanded virtual deflections

$$\delta_c = [\delta_{ac}, \delta_{ec}, \delta_{rc}]^T \qquad (8.41)$$

where the subscripts a, e, and r represent aileron, elevator, and rudder, respectively, are distributed by a control allocator (CA) to actual aero surface actuator commands

$$\bar{\delta}_c = [\delta_{elc}, \delta_{erc}, \delta_{flc}, \delta_{frc}, \delta_{rlc}, \delta_{rrc}]^T \qquad (8.42)$$

where the subscripts el, er, fl, fr, rl, and rr represent left and right elevons, left and right flaps, and left and right rudders, respectively. These commands are fed into the individual aero surface actuators, modeled here by first-order systems

$$\tau_i \dot{\delta}_i = \delta_{ic} - \delta_i, \quad i = el, er, fl, fr, rl, rr \qquad (8.43)$$

The actual deflection angles in Eq. (8.43) are δ_i, the actuator time constants are τ_i, and the actuator command inputs are δ_{ic}.

Remark 8.5. The aero surface actuator models in Eq. (8.43) are simplifications of the actual models, which are of higher order and nonlinear. Modeling them as simple, linear first order as in Eq. (8.43) simplifies the controller design. In the simulations used to verify the design, the actuators were modeled as second-order systems with the following position and rate limits:

$$-30° \le \delta_i \le 25°, \quad \left|\dot{\delta}_i\right| \le 30\,\text{deg}/\text{s}, \ i = el, er$$

$$-15° \le \delta_j \le 26°, \quad \left|\dot{\delta}_j\right| \le 10\,\text{deg}/\text{s}, \ j = fl, fr$$

$$-30° \le \delta_{rl} \le 60°, \quad \left|\dot{\delta}_{rl}\right| \le 30\,\text{deg}/\text{s}$$

$$-60° \le \delta_{rr} \le 30°, \quad \left|\dot{\delta}_{rr}\right| \le 30\,\text{deg}/\text{s}$$

8.3.2 Reusable Launch Vehicle Control Problem Formulation

The general problem for the flight control system design during TAL, is to determine aero surface deflection commands such that the guidance commands are robustly asymptotically followed, in the presence of external disturbances and uncertainties in the plant. Guidance commands in the TAL flight are N_{zc}, normal acceleration (or $A_{zc} = -N_{zc}$ in terms of the body frame acceleration command), ϕ_c, roll angle, and r_c, yaw rate. The RLV mathematical model (8.37)–(8.40) is presented in a quasi-cascade format:

$$\begin{bmatrix} \dot{A}_z \\ \dot{\phi} \end{bmatrix} = \begin{bmatrix} \dot{v} & -\dot{u} \\ 1 & \tan(\theta)\sin(\phi) \end{bmatrix} \begin{bmatrix} p \\ q \end{bmatrix} + \begin{bmatrix} f_{Az} \\ f_{\phi} \end{bmatrix} \tag{8.44}$$

$$\begin{bmatrix} \dot{p} \\ \dot{q} \end{bmatrix} = \begin{bmatrix} -L_{pq}pq - L_{qr}qr \\ -M_{pr}pr - M_{r^2p^2}(r^2 - p^2) \end{bmatrix} + \begin{bmatrix} L \\ M \end{bmatrix} + \begin{bmatrix} f_L \\ f_M \end{bmatrix} \tag{8.45}$$

$$\dot{r} = -N_{pq}pq - N_{qr}qr + N + f_N \tag{8.46}$$

where the f_{ϕ}, f_{Az} are bounded disturbances ($|f_{\phi}| \le L_{\phi}$, $|f_{Az}| \le L_{Az}$) in the normal acceleration and roll channels. The term f_{ϕ} is caused by modeling errors, while f_{Az} comprises modeling errors and external disturbances, and f_L, f_M, and f_N are the disturbances in the roll rate, the pitch rate channels, and the yaw rate channel. All disturbances are bounded in a reasonable flight domain.

The goal is to design a continuous SMC/SMDO controller that provides decoupled asymptotic output tracking, i.e., $A_z \to A_{zc}, \phi \to \phi_c$, and $r \to \bar{r}_c$ as time increases, in the presence of bounded disturbances, by means of roll, pitch, and yaw acceleration commands L_c, M_c, and N_c.

8.3.3   Multiple-Loop Asymptotic SMC/SMDO Design

The control problem is addressed using the continuous SMC–SMDO algorithm in a multiple-loop format. The controller is designed in the following steps:

Step 1   A control law for the outer loop that drives $A_z \to A_{zc}$ and $\phi \to \phi_c$ is designed in terms of virtual rate commands, p_c, q_c, that are fed into the inner loop.

Step 2   A control law for the inner loop that drives $e_L = p \to p_c, q \to q_c$, and $r \to \bar{r}_c r_c + r_{cSAS}$, where $\bar{r}_c = r_c + r_{cSAS}$ and r_{cSAS} is a special term generated by the stability augmentation system (SAS), is designed in terms of angular accelerations, L_c, M_c, and N_c.

Step 3   A control allocation matrix $\mathbf{B}_A$ is used to map the angular acceleration command vector to the actuator virtual deflection commands:

$$\delta_c = \mathbf{B}_A \mathbf{I} \begin{bmatrix} L_c \\ M_c \\ N_c \end{bmatrix}, \quad \delta_c = [\delta_{ac}, \delta_{lc}, \delta_{rc}] \in \mathbb{R}^3, \ \mathbf{B}_A \in \mathbb{R}^{3 \times 3}, \ \mathbf{I} \in \mathbb{R}^3 \tag{8.47}$$

CA is then used to distribute the virtual actuator commands δ_c to actual aero surface actuator commands $\bar{\delta}_c$, which are fed to the individual actuators (8.43): specifically

$$\bar{\delta}_c = CA(\delta_c), \quad \bar{\delta}_c = [\delta_{elc}, \delta_{erc}, \delta_{flc}, \delta_{frc}, \delta_{rlc}, \delta_{rrc}]^T \tag{8.48}$$

Outer-Loop SMC/SMDO Controller Design

The virtual controls, p_c and q_c, are designed in the following way:

$$\begin{bmatrix} p_c \\ q_c \end{bmatrix} = \begin{bmatrix} \dot{v} & -\dot{u} \\ 1 & \tan(\theta)\sin(\phi) \end{bmatrix}^{-1} \begin{bmatrix} \hat{d}_{Az} + K_{Az}\sigma_{Az} \\ \hat{d}_{\phi} + K_{\phi}\sigma_{\phi} \end{bmatrix} \tag{8.49}$$

where

$$\sigma_i = e_i, \ i = A_z, \phi, \tag{8.50}$$

$$e_{Az} = A_{zc} - A_z, e_{\phi} = \phi_c - \phi$$

and $\hat{d}_{Az}$ is the estimate of $d_{Az} = \dot{A} - f_{Az}$, and $\hat{d}_{\phi}$ is the estimate of $d_{\phi} = \dot{\phi}_{Ac} - f_{\phi}$. Assuming $|d_{Az}| \leq \bar{L}_{Az}$, and $|d_{\phi}| \leq \bar{L}_{\phi}$ in a reasonable flight envelope, and denoting

$$\begin{bmatrix} \lambda_{Az} \\ \lambda_{\phi} \end{bmatrix} = \begin{bmatrix} \dot{v} & -\dot{u} \\ 1 & \tan(\theta)\sin(\phi) \end{bmatrix} \begin{bmatrix} p \\ q \end{bmatrix} \tag{8.51}$$

the finite-convergence-time SMDO based on a super-twisting injection term is designed, in accordance with Eqs. (8.17), (8.19), (8.23), (8.24), (8.25), and (8.27).

Firstly, auxiliary sliding variables are introduced as

$$s_i = \sigma_i + z_i, \ \dot{z}_i = \lambda_i - v_i, \ i = A_z, \phi \tag{8.52}$$

and the auxiliary sliding variable dynamics are derived as

$$\dot{s}_i = d_i - v_i, \ i = A_z, \phi \tag{8.53}$$

Secondly, the super-twisting injection terms that drive the auxiliary sliding variables $s_i, i = A_z, \phi$ to zero in finite time are designed (see Sect. 8.2.2)

$$v_i = \lambda_{0,i} |s_i|^{1/2} \text{sign}(s_i) + \eta_i \tag{8.54}$$

$$\dot{\eta}_i = \lambda_{1,i} \text{sign}(s_i)$$

with $\lambda_{0,i} = 1.5\sqrt{\bar{L}_i}, \lambda_{1,i} = 1.1\bar{L}_i$. The auxiliary injection terms v_{Az} and $v_{A\phi}$ exactly estimate the disturbances d_{A_z} and $d_{A\phi}$ in Eq. (8.53) in finite times, i.e.,

$$d_{A_z} = v_{A_z}, \ d_{A\phi} = v_{A\phi} \tag{8.55}$$

The obtained disturbance estimates are substituted into the virtual control laws in Eq. (8.49).

Inner-Loop SMC/SMDO Controller Design

The inner-loop input–output dynamics are taken as the lower part of Eq. (8.46). The inner-loop SMC/SMDO takes the body rate virtual control commands, p_c and q_c, generated by the outer-loop controller (8.51)–(8.55), and the yaw rate command r_c that comes from the guidance function. The inner-loop continuous SMC/SMDO controller is supposed to enforce following the body rate virtual control commands in the presence of bounded disturbances and by generating pitch, roll, and yaw acceleration commands L_c, M_c, and N_c. In addition to tracking virtual body rate commands, the inner loop causes the system to exhibit a linear decoupled motion in the sliding mode. No timescale (separation) between the inner- and outer-loop compensated dynamics is required (but it is welcome). Following the continuous SMC/SMDO controller design algorithm presented above, the components of the vector-sliding variable are introduced as

$$\sigma_i = e_i, \ i = L, M, N$$
$$e_L = p_c - p, \ e_M = q_c - q, \ e_N = e_{SAS} \tag{8.56}$$

where, in order to provide turn coordination, a SAS generates a yaw rate command

$$r_{cSAS} = \frac{g}{U} \sin(\phi) \tag{8.57}$$

where g is the gravitational constant, U is true air speed, and the error signal used in the yaw rate loop SMC/SMDO is

$$e_{sas} = (r_c + r_{cSAS} - r) \tag{8.58}$$

In the sliding mode $\sigma_L = \sigma_M = \sigma_N = 0$, and the inner-loop compensated error dynamics are decoupled. The objective of the inner-loop control is to generate continuous commands L_c, M_c, and N_c that provide asymptotic convergence of $\sigma_L, \sigma_M, \sigma_N \to 0$. The pitch, roll, and yaw acceleration control commands L_c, M_c, and N_c are designed:

$$\begin{bmatrix} L_c \\ M_c \\ N_c \end{bmatrix} = \begin{bmatrix} \tilde{f}_L \\ \tilde{f}_M \\ \tilde{f}_N \end{bmatrix} + \begin{bmatrix} L_{pq} + L_{qr}qr \\ M_{pr} + M_{r^2p^2}(r^2 - p^2) \\ N_{pq}pq + N_{qr}qr \end{bmatrix}$$

$$+ \begin{bmatrix} K_L & 0 & 0 \\ 0 & K_M & 0 \\ 0 & 0 & K_N \end{bmatrix} \begin{bmatrix} \sigma_L \\ \sigma_M \\ \sigma_N \end{bmatrix} \tag{8.59}$$

where $\begin{bmatrix} \tilde{f}_L & \tilde{f}_M & \tilde{f}_N \end{bmatrix}^T$ is the estimate vector of the bounded accumulated disturbance

$$
\begin{bmatrix} \bar{f}_L \\ \bar{f}_M \\ \bar{f}_N \end{bmatrix} = \begin{bmatrix} \dot{p}_c - f_L \\ \dot{q}_c - f_M \\ \dot{r}_c - f_N \end{bmatrix}
$$

Assuming $\left| \bar{f}_L \right| \le \bar{L}_L$, $\left| \bar{f}_M \right| \le \bar{L}_{Mo}$, and $\left| \bar{f}_N \right| \le \bar{L}_N$ in a reasonable flight envelope, the SMC/SMDO is designed in accordance with Eqs. (8.17)–(8.27) in the following steps:

Firstly, the components of the auxiliary vector-sliding variable $[s_L, s_M, s_N]^T$ are introduced

$$
s_i = \sigma_i + z_i, \quad \dot{z}_i = \lambda_i - v_i, \quad i = L, M, N, \; \lambda_i = i \qquad (8.60)
$$

and the auxiliary vector-sliding variable dynamics are derived as

$$
\dot{s}_i = \bar{f}_i - v_i, \quad i = L, M, N \qquad (8.61)
$$

Secondly, the super-twisting injection terms are defined as in Eq. (8.54), and drive the auxiliary sliding variables s_i to zero in finite times, $i = L, M, N$. Therefore, the auxiliary injection terms v_L, v_M, and v_N exactly estimate the disturbances $\bar{f}_L$, $\bar{f}_M$, and $\bar{f}_N$ in Eq. (8.61) in finite times, i.e.,

$$
\bar{f}_L = v_L, \; \bar{f}_M = v_M, \; \bar{f}_N = v_N \qquad (8.62)
$$

The disturbance estimates are fed into the virtual control laws in Eq. (8.49). Then, the obtained disturbance estimates in Eq. (8.62) are fed into the torque control laws in Eq. (8.59).

Control System Parameterization

The following choice of the controller (8.49), (8.51), (8.52), (8.53), (8.55), (8.59), (8.60), (8.61), and (8.62) parameters is recommended:

(a) $K_i > 0 \; \forall i = A_z, \phi, L, M, N$
(b) $K_i > K_j \; \forall i = L, M, N \; \forall j = A_z, \phi$
(c) $\frac{1}{\tau_i} > K_j \; \forall i = A_z, \phi, L, M, N \; \forall j = A_z, \phi$

Condition (a) is necessary for providing the asymptotic stability of sliding variable dynamics. Condition (b) enforces decoupling (time-scaling) between the inner- (σ_L, σ_M, σ_N) and the outer-loop (σ_{Az}, σ_ϕ) sliding variable dynamics given by Eqs. (8.33). Condition (c) provides for the transient times in the SMDO, which basically are controlled by the time constants of the filters in Eqs. (8.55) and (8.62). These transients are supposed to be faster than the transients in the control loops.

Meeting this condition helps to reduce the effect of the SMDO dynamics on the compensated dynamics of the tracking errors. All these conditions are not very restrictive and can be easily met in the RLV control system design.

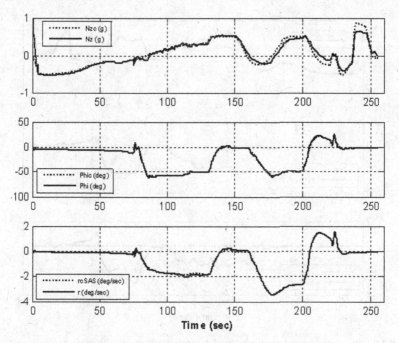

Fig. 8.2 Guidance command tracking: normal acceleration (*top*), roll angle (*center*), yaw rate (*bottom*), SMC/SMDO simulation, and nominal wind [104]

8.3.4 Flight Simulation Results and Analysis

The designed SMC/SMDO was coded and implemented in MATLAB/SIMULINK and the simulation results were compared with the baseline classical PID with gain-scheduling controller design. Only the autopilots were different between the two simulations; all other models remained unchanged, including a second-order actuator model with position and rate limiting and a wind model.

The simulations began at the TAEM interface and terminated at main gear touchdown on the runway. The control functions (8.59) in terms of roll, pitch, and yaw acceleration commands are transformed into virtual deflection commands by Eq. (8.47), allocated into actual actuator commands by means of the control allocator (8.48), and these are fed to the actuators (8.43) with deflection and rate limits being taken into account. The proposed continuous SMC/SMDO and the baseline controllers were compared in two simulation cases: one with a nominal wind and one with a severe headwind gust. Simulation results from the nominal wind case are shown in Fig. 8.2 through 8.5.

In comparing Figs. 8.2 (SMC/SMDO) and 8.3 (baseline), it can be seen that normal acceleration and roll angle tracking is far better with the SMC/SMDO (in particular, the roll angle command is followed almost perfectly). The yaw rate command is also followed very closely by the SMC/SMDO. The baseline controller

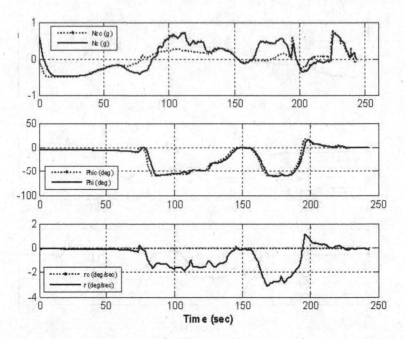

Fig. 8.3 Guidance command tracking: normal acceleration (*top*), roll angle (*center*), yaw rate (*bottom*), baseline simulation, and nominal wind [104]

does not follow the yaw rate command (which is always equal to zero, generated by the guidance loop) very well, due to the fact that it too uses turn coordination through interchannel cross-feeds. It is worth noting that the guidance command profiles are slightly different between the two simulations due to the fact that closed-loop guidance shapes its commands as a function of the flight condition and the distance from the landing site. The flight conditions differ because the two controllers produce different aero surface commands resulting in different aerodynamic accelerations on the vehicle. The aero surface deflections are shown in Figs. 8.4 and 8.5 for the SMC/SMDO and for the baseline controller respectively. The average deflections are similar in magnitude; however, in order to provide more accurate guidance command tracking, the SMC commands are more aggressive due to the use of the SMDO in response to maneuvers and other disturbances. This results in large transient deflections that quickly decrease once the tracking error is again close to zero.

To provide a more challenging case for the control system, a severe head-wind gust was simulated by artificially imposing a 25° angle of attack for a duration of 3 s, at 185 s flight time. The simulation results for this more stressing case are shown in Fig. 8.8 through 8.10. The gust appears as a step increase in angle of attack. Guidance command tracking is shown in Figs. 8.6 and 8.7 for the SMC/SMDO and baseline controller, respectively. At the time of the wind gust, the SMC/SMDO allows roughly 4g-error in normal acceleration, while roll and yaw errors are about

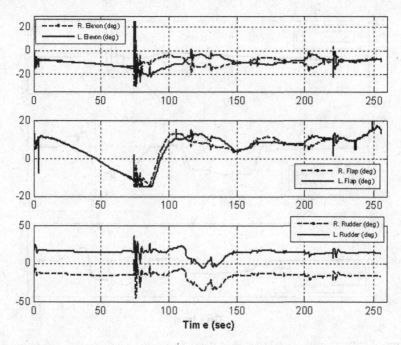

Fig. 8.4 Elevon (*top*), flap (*center*), and rudder (*bottom*) deflections, SMC/SMDO simulation, and nominal wind [104]

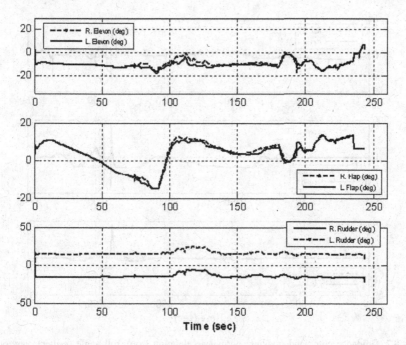

Fig. 8.5 Elevon (*top*), flap (*center*), and rudder (*bottom*) deflections, baseline simulation, and nominal wind [104]

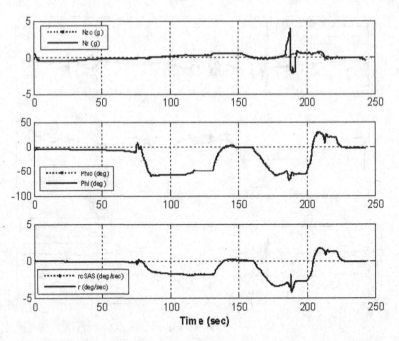

Fig. 8.6 Guidance command tracking: normal acceleration (*top*), roll angle (*center*), yaw rate (*bottom*), and SMC simulation with severe wind gust [104]

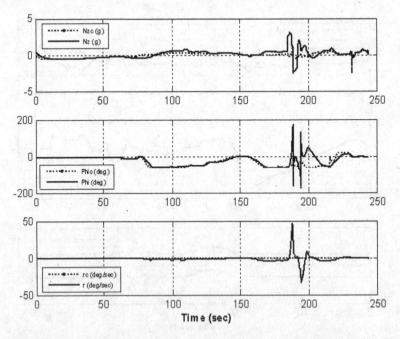

Fig. 8.7 Guidance command tracking: normal acceleration (*top*), roll angle (*center*), yaw rate (*bottom*), and baseline controller simulation with severe wind gust [104]

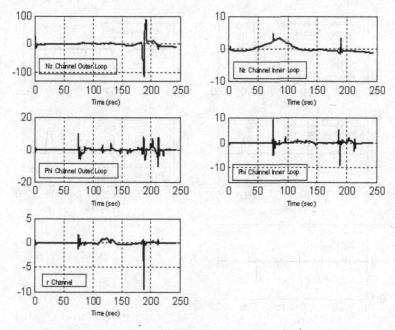

Fig. 8.8 Normal acceleration channel (Nz), roll channel (Phi), and yaw rate channel (r) SMC sliding variables and severe wind gust simulation [104]

the same as in the nominal simulation. In the baseline controller simulation, the vehicle rolls through 360° and the yaw rate climbs to a maximum of 50 degree per second. Figure 8.8 shows the sliding variables for the SMC/SMDO outer loop (8.50) and inner loop (8.56). Figure 8.9 shows the sliding variables for the SMDO outer loop (8.52) and SMDO inner loop (8.60). The sliding variables diverge occasionally due to the guidance maneuvers and the wind disturbance but then re-converge to a region around zero. The outer- and inner-loop SMDO disturbance estimates are shown in Fig. 8.10. The disturbance estimates are a function of maneuver rates, wind disturbances, and other unmodeled dynamics.

Remark 8.6.   The proposed SMC/SMDO algorithm for the reusable launch vehicle flight control system was coded and implemented in the X-33 MAVERIC simulation. In its implementation, problems faced in real-world applications (like filtering and multi-rate subsystems) were successfully addressed.

8.4   Case Study: Satellite Formation Control

In this section we consider controlling the satellite formation, in particular, controlling the motion of one satellite as it follows a defined path around another satellite (that is orbiting the Earth). This tracking must be robust to model uncertainties

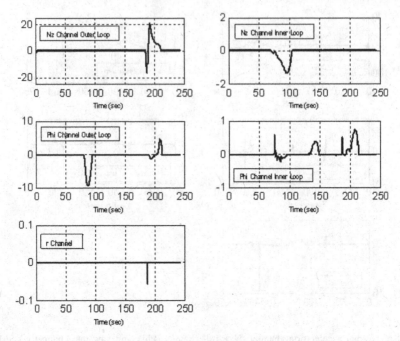

Fig. 8.9 Normal acceleration channel (Nz), roll channel (Phi), and yaw rate channel (r) SMDO sliding variables and severe wind gust simulation [104]

and external disturbances. The problem with any satellite formation control is that all orbiting bodies are subject to forces, which include gravitational perturbations, atmospheric drag, and solar radiation pressure, that tend to force the satellites out of their stable Keplerian orbits . Since the satellites are usually controlled by on–off thrusters, a conventional SMC with a dead-band is often used. The width of the dead-band usually is selected to minimize the fuel consumption. At the same time, simulations show that the jet-thruster duty cycle depends on a board computer step size. It means that using a dead-band controller does not permit controlling the frequency of switching. This could create a problem in using SMC in satellite formation control actuated by jet-thrusters that can be switched " on" and " off" only with a required duty cycle.

8.4.1   Satellite Formation Mathematical Model

The formation control problem studied in this section is a two-satellite formation. The coordinate system used is illustrated in Fig. 8.11.

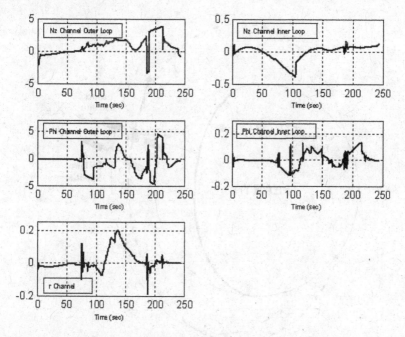

Fig. 8.10 Normal acceleration channel (Nz), roll channel (Phi), and yaw rate channel (r) SMDO disturbance estimates and severe wind gust simulation [104]

The first satellite, the leader, is in a circular orbit around a spherical Earth and for simplicity it is assumed to maintain that orbit in free flight. Assuming the second satellite is in a slightly elliptical orbit but remains close to the first when compared to the overall radii of their orbits around the Earth, we use the linearized Hill's equations to describe the relative motion between a leader and follower satellite. Since the available thrust levels are very low in many satellite applications, initial misalignments as well as transients resulting from disturbances may take many hours and even days to settle out. So, scaling the natural time t, a new time is defined such that $\tau = wt$, where w is the mean motion of the leader around the Earth (the value of the constant angular velocity that is required for a satellite to complete one revolution). The normalized Hill equations in the new time τ are

$$\ddot{x} - 2\dot{y} - 3x = u_x + d_x \qquad (8.63)$$
$$\ddot{y} + 2\dot{x} = u_y + d_y$$
$$\ddot{z} + z = u_z + d_z$$

In Eq. (8.63) u_x, u_y, and u_z are the net (difference between leader and follower)-specific control forces acting on the follower satellite, and d_x, d_y, and d_z encompass the net-specific disturbances experienced by the two-satellite system, including linearization residuals. The specific control forces and disturbances are defined as

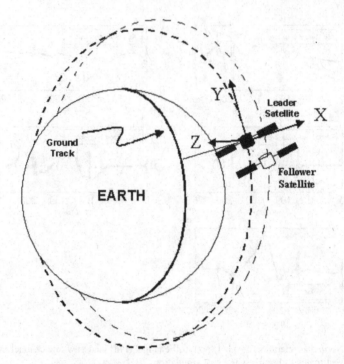

Fig. 8.11 Satellite formation coordinate system [139]

forces per unit mass per mean motion squared. All terms in Eq. (8.63) have a length dimension. The coordinates x, y, and z describe the position of the follower satellite relative to the leader satellite. Parameterizing Equation (8.63) as a function of τ with all external forces (control and disturbances) is equal to zero provides the reference trajectories for the simulation

$$x_c(\tau) = r\sin(\tau + \theta) \tag{8.64}$$

$$y_c(\tau) = 2r\cos(\tau + \theta) \tag{8.65}$$

$$z_c(\tau) = mr\sin(\tau + \theta) + 2nr\cos(\tau + \theta) \tag{8.66}$$

where r, m, n, and θ are arbitrary constants that define the elliptical desired path that the follower satellite maintains, relative to the leader in a force-free environment. In particular, r gives the size of the ellipse, θ is the initial angular position of the follower satellite on the path, and m and n are the slopes of the plane in which the path is located. For any given θ the follower satellite must enter the path (8.66) at initial conditions that are derived by setting $\tau = 0$ in Eq. (8.66), as well as in the derivative of Eq. (8.66), yielding

$$x_c(0) = r\sin(\theta), \quad y_c(0) = 2r\cos(\theta), \quad z_c(0) = mr\sin(\theta) + 2nr\cos(\theta) \tag{8.67}$$

$$\dot{x}_c(0) = r\cos(\theta), \quad \dot{y}_c(0) = 2r\sin(\theta), \quad \dot{z}_c(0) = mr\cos(\theta) - 2nr\sin(\theta)$$

The value of θ defines the angle at which the follower satellite is making with a reference coordinate of the leader. Physically Equation (8.66) provides a circular or elliptical (depending on the choice of constants) reference trajectory around a leader satellite starting with the initial conditions defined in Eq. (8.67).

8.4.2 Satellite Formation Control in SMC/SMDO

If there is a mismatch between the follower satellite and the reference trajectory, then the starting point for the satellite will not be on one of the desired trajectories. Then a control system will be required to force the follower satellite back on track. The continuous SMC/SMDO controllers are supposed to achieve this goal while completely compensating for the bounded disturbances d_x, d_y, and d_z. Otherwise, the disturbances d_x, d_y, and d_z, which are due to the oblateness of the Earth, atmospheric drag, tesseral resonance, solar radiation pressure (so-called J_2 effects), and bulges of the Earth (so-called J_3 effects), will gradually disperse an uncontrolled formation. The J_2 effect is by far the dominant perturbation force compared to J_3. For this study, the representative d_x, d_y, and d_z disturbances are sine waves with a frequency of $0.6\,\text{rad/s}(\tau)$ and are kept the same for all of the controllers studied. Also we assume that the leader satellite is maintained on the orbit by its own control.

Assuming the relative positions in Eq. (8.63) are measured, the vector-relative degree for this system is identified as $\bar{r} = [2, 2, 2]^T$ and the sliding variables (8.5) are formed as

$$\sigma = \left[\sigma_x, \sigma_y, \sigma_z\right]^T, \ \sigma_i = \dot{e}_i + c_{i,0}e_i + c_{i,-1}\int e_i d\tau \ \forall i = x, y, z \quad (8.68)$$

$$e_x = x_c - x, \ y_y = y_c - y, \ e_z - z_c - z$$

The integral terms in Eq. (8.68) are added in order to compensate for constant bias that could occur in the sliding modes at the implementation step. The following, optimized for fuel consumption, compensated dynamics of the output tracking errors in the sliding mode in a normalized time τ are

$$\sigma_i = \dot{e}_i + 1.8e_i + 1.0\int e_i d\tau = 0 \rightarrow \ddot{e}_i + 1.8\dot{e}_i + e_i = 0 \ \forall i = x, y, z \quad (8.69)$$

The SMC/SMDO of (8.17)–(8.25), (8.32), and (8.34) is designed ($u_i = v_i$ since $E(x,t) = I$ in Eq. (8.63)) for controlling the satellite formation model (8.63) and (8.66):

$$u_i = 10\sigma_i + \hat{v}_{ieq} \ \forall i = x, y, z \quad (8.70)$$

$$0.05\frac{d\hat{v}_{ieq}}{d\tau} = -\hat{v}_{ieq} + 2\text{sign}(s_i)$$

$$s_i = \sigma_i + \int (u_i - 2\text{sign}(s_i))\, d\tau$$

It is worth noting that the injection term v_i in the SMC/SMDO (8.70) can be also designed in a form of the super-twisting control as in Sect. 8.2.2.

8.5 Simulation Study

A Simulink model of the follower satellite motion around the leader was developed to test the effectiveness of the continuous SMC/SMDO algorithm in Eq. (8.70) against the set of disturbances. For this study, the representative disturbance is a sine wave with amplitude 0.5 km and a frequency of 0.6 rad/sec(τ).

The SMC/SMDO in Eq. (8.70) is compared with the conventional (discontinuous with dead-band) SMC. The same orbital period is used, i.e., 100.7 m. Recall that τ is defined as the product of the mean motion and time. Therefore, for a 100.7 m orbit, the actual time required to reach $\tau = 1$ is 961.6 s. For the simulations an Euler first-order integration algorithm is used, and the time step is fixed at 0.01 ms.

In Eq. (8.66) the values of the parameters are taken as $r = 0.5$ km for the reference trajectory, whereas $r = 0.6$ km for the follower satellite in order to create the initial condition offset. Also $n = 0$ and $m = \sqrt{3}$ are defined to set the follower satellite in a circular orbit around the leader at $\pm \frac{2\pi}{3} rad$ inclination with the x-axis. The angular position $\theta = \frac{\pi}{2}$ is the same for both the reference trajectory and the simulated satellite's trajectory. A pulse-width-modulation (PWM) technique is used to turn the continuous SMC/SMDO signals into a pulse train that could be realizable by a satellite on–off jet-thruster. It is assumed that the thruster can cycle at a frequency of 10 Hz. Due to the 961.6 s(τ) conversion factor, 10 Hz signal in the t time domain would need to be approximately a $10000\, cycles/\tau$ signal in the τ domain. Therefore, a dither signal, a sine wave of 10000 cycles/τ with amplitude of $0.25 \cdot 10^{-5}$ km/s^2, is combined with the control signal to model the required duty cycle of the thrusters. In order to further optimize weekly fuel consumption, the dead-zone is used in the relay elements of PWM. The recommended dead-band optimal value is taken as $\delta_i = 0.5 \cdot 10^{-5}$ km/s^2 and the corresponding PWM with a dead-band is shown in Fig. 8.12.

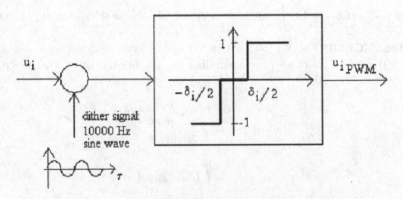

Fig. 8.12 PWM with dead band [139]

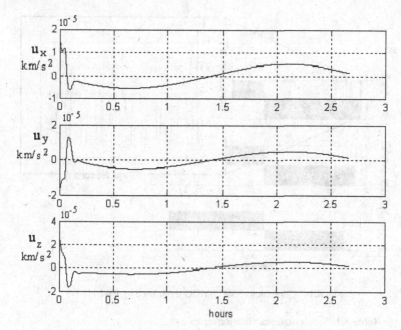

Fig. 8.13 Continuous SMC/SMDO (km/s^2) [139]

Figure 8.13 shows the control inputs to the follower satellite while using the continuous SMC driven by the SMDO. After a few hours, this control drives the sliding variables and the errors to near zero. However, since this algorithm is supposed to be controlling a satellite thruster that fires pulses of thrust of constant magnitude, the continuous control input shown in Fig. 8.13 versus the original time is not very useful.

Figure 8.14 shows the control inputs with the PWM technique implemented in the original time t.

Here, the 10 Hz duty cycle is clearly illustrated. With the PWM technique, all continuous SMC reduce the errors between reference trajectory and the follower satellite's trajectory to near zero. The upper boundaries of steady-state errors are compared in Table 8.1.

It can be seen that the SMC/SMDO provides more accurate stabilization compared to conventional SMC. Figure 8.15 shows a 3-dimensional view of the relative motion between the leader and follower satellite using the continuous SMC/SMDO with the PWM technique incorporated.

The corresponding weekly fuel consumptions in terms of ΔV, which are low for both methods, are shown in Table 8.2

Remark 8.7. The studied SMC/SMDO easily drives the control on–off thrusters with given duty cycle using the PWM technique, which is an improvement over a conventional (discontinuous with dead-band) SMC. The simulation analysis shows

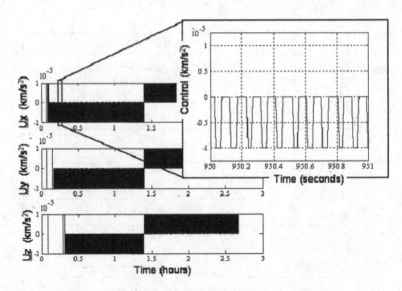

Fig. 8.14 Continuous SMC/SMDO modulated by PWM [139]

Table 8.1 Steady-state stabilization errors

Method	Conventional SMC	SMC/SMDO
$|e_x|$ (km)	3.610^{-4}	1.510^{-5}
$|e_y|$ (km)	3.410^{-4}	0.410^{-5}
$|e_z|$ (km)	3.710^{-4}	2.810^{-5}

that the continuous SMC/HOSM algorithm is effective in reducing the tracking errors of the follower satellite in the presence of a bounded disturbance function while keeping fuel consumption at a low level. It also appears that the PWM technique introduced does not degrade the tracking performance to any significant degree retaining fixed duty cycle.

8.6 Notes and References

For details about vector and total relative degree, see, for example, [112]. It is worth noting that the problem formulated in Sect. 8.1 can be also addressed for systems with stable zero dynamics [112].

The equation for finite-time-convergent compensated dynamics given in Sect. 8.1.2 is taken form [23].

The model of the RLV is from [106] and [190]. The multiple-loop asymptotic SMC design algorithm used for the RLV is from [104]. The overall stability analysis of the double-loop SMC/SMDO is also from [104].

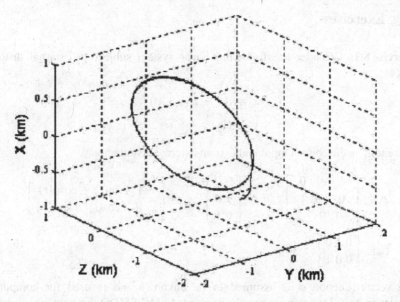

Fig. 8.15 Three-dimensional trajectory: continuous SMC/SMDO modulated by PWM (km) [139]

Table 8.2 The weekly fuel consumption

Method	Conventional SMC	SMC/SMDO
$\Delta V_x \ (m/s)$	0.212	0.210
$\Delta V_y \ (m/s)$	0.203	0.202
$\Delta V_z \ (m/s)$	0.203	0.202

The description of the Keplerian orbits and the satellite fuel consumption problems are from [194]. The continuous SMC/SMDO robust controller, realized by PWM to provide a required duty cycle in the control command to the jet-thrusters and used to robustly control the satellite formation while providing low weekly fuel consumption, is reported in [139]. An analysis of the application of a broader variety of SMC/SMDO techniques to the satellite formation control problem, including a super-twisting disturbance observer and a continuous robust SMC based on integral sliding surface design, can be found in [139].

The challenging task of output tracking in systems with transmission zeros located in the right-hand side of the complex plane or, generally speaking, in systems with unstable zero dynamics, is called a nonminimum phase output tracking problem [112]. A solution to the nonminimum phase output tracking problem using sliding modes, for an aerospace example, can be found in [160]. Output tracking in nonminimum phase systems with arbitrary relative degree, using higher-order sliding mode control, is studied in [102, 164, 167].

The 2-SM controllers that are applied to the launch vehicle and missile system are discussed in [163, 165, 176]. Other aerospace applications of SMC/HOSM control and disturbance observation techniques appear in a special issue of the Journal of Franklin Institute [169].

8.7 Exercises

Exercise 8.1. Consider the following linear system subject to external disturbances:

$$\dot{x} = Ax + B(u + \varphi)$$
$$y = Cx$$

where $u \in^2$ is the control input and φ is an external disturbance

$$A = \begin{bmatrix} 1 & -1 & 0 \\ 0 & -3 & 1 \\ -1 & 0 & -2 \end{bmatrix}, \ B = \begin{bmatrix} 0 & 0 \\ 1 & 0 \\ 0 & 1 \end{bmatrix}, \ \varphi = \begin{bmatrix} 3 + 4\sin(t) + \cos(4t) \\ x_1 + 4\cos(t) + \sin(3t) \end{bmatrix},$$

$$C = \begin{bmatrix} 0 & 1 & 0 \\ 0 & 0 & 1 \end{bmatrix}$$

The vector-function φ is assumed to be unknown and is used for computer simulation only. Design a continuous decoupled SMC/SMDO that achieves asymptotic tracking $y \rightarrow y_c$ with a given settling time $t_s = 0.5$ s, when $y_c = \begin{bmatrix} \cos(t), \sin^2(t) \end{bmatrix}^T$, and $x(0) = [1, -2, 3]^T$. Support the simulation results by plotting the corresponding variables.

Exercise 8.2. Repeat Exercise **8.1** and design a finite-time SMC/SMDO that achieves tracking $y \rightarrow y_c$ with an overall given convergence time $t_r = 0.5$ s.

Exercise 8.3. Consider Chua's circuit (Example 2.7) with an output $y = x_1$. The unknown function $f(x_1) = -x_1(1 - x_1^2)$ is used for simulation purposes only. Design a continuous SMC/SMDO that achieves asymptotic tracking $y \rightarrow y_c$ with a given settling time of $t_s = 1.5$ s. Simulate the control system with

$$y_c(t) = 10 + \sin^2 t \ (V), \ x(0) = [0.5, 0.4, 0.2]^T$$

Support the simulation results by plotting the corresponding variables.

Exercise 8.4. Repeat Exercise **8.3** and design a finite-time SMC/SMDO that achieves tracking $y \rightarrow y_c$ with a given convergence time of $t_r = 1.5$ s.

Exercise 8.5. The dynamics of an armature-controlled DC motor are given by

$$\begin{cases} J\frac{d\omega}{dt} = k_m i - T_L \\ L\frac{di}{dt} = -iR - k_b\omega + u \\ y = \omega \end{cases}$$

where J is the moment of inertia, i is armature current, L and R are armature inductance and resistance, respectively, ω is the motor angular speed, k_b is a constant relating to back electromotive force, k_m is a motor torque constant, T_L

is load torque, and u is a control function in terms of the armature voltage. Design a sliding mode control u that drives $y = \omega \rightarrow y_c$ as time increases, assuming that $y = \omega$ is measurable. Assume all the parameters are known except for the load torque T_L that is assumed to be bounded together with its derivative: $|T_L| \leq L_m$, $|\dot{T}_L| \leq \bar{L}_m$ Design a continuous SMC/SMDO that achieves asymptotic tracking $y \rightarrow y_c$ with a given settling time t_s. Simulate the control system with $R = 1\,\text{Ohm}$, $L = 0.5\,\text{H}$, $k_m = 5 \cdot 10^{-2}\,\text{N} \cdot \text{m/A}$, $k_b = k_m$, $J = 10^{-3}\,\text{N} \cdot \text{m} \cdot \text{s}^2/\text{rad}$, $T_L = 10\cos(3t)\,\text{N} \cdot \text{m}$, $y_c(t) = 60 + 50\sin(2t)$, $t_s = 1.2\,\text{s}$, $\omega(0) = 10\,\text{rad/s}$, and $i(0) = 0$. Support the simulation results by plotting the corresponding variables.

Exercise 8.6. Repeat Exercise **8.5** and design a finite-convergent-time SMC/SMDO that achieves tracking $y \rightarrow y_c$ with an overall given convergence time $t_r = 1.2\,\text{s}$.

Exercise 8.7. A simplified model of the dynamics of the longitudinal motion of an underwater vehicle is given by

$$m\ddot{x} + k\dot{x}|\dot{x}| = u + T_d$$

$$y = x$$

where x is the position of the vehicle, u is the control function (a force that is provided by a propeller), m is the mass of the vehicle, $k > 0$ is the drag coefficient, and T_d is an unknown smooth disturbance bounded together with its derivative. Assuming the value of m is known exactly, the drag coefficient is bounded $k_1 \leq k \leq k_2$ and the position and its derivative (velocity) x, $\dot{x}$ are measured, design a continuous SMC/SMDO that drives $y \rightarrow y_c$ as time increases, with given settling time t_s. Simulate the control system for $x_1(0) = 2\,\text{m}$, $x_2(0) = 0.5\,\text{m/s}$, $m = 4\,\text{kg}$, $t_s = 5\,\text{s}$, $k = 1.5 + 0.4\sin(2t)\left[\frac{\text{kg}}{\text{m} \cdot \text{s}}\right]$, $T_d = 10^4 + 10^3\sin(t)$, and

$$y_c(t) = \begin{cases} 0.1t^2, & \text{if } 0 \leq t \leq 10\,\text{s} \\ 10 + 2(t-10), & \text{if } t > 10\,\text{s} \end{cases}$$

Support the simulation results by plotting the corresponding variables.

Exercise 8.8. Repeat Exercise **8.7** and a finite-convergent-time SMC/SMDO that achieves tracking $y \rightarrow y_c$ with an overall given convergence time $t_c = 3s$.

Exercise 8.9. The evolution of the angle of attack of a missile controlled by fin deflection can be described by a simplified system of differential equations

$$\begin{cases} \dot{\alpha} = q - 2.7\alpha + d(t) \\ \dot{q} = -5.5\alpha - 0.4q - 19\delta \\ \dot{\delta} = -20\delta + 20u + 20h(t) \\ y = \alpha \end{cases}$$

where α, δ are angle of attack and fin deflection in rad, respectively; q is pitch rate in rad/s, u is the control function in rad, and $d(t)$ and $h(t)$ are smooth bounded disturbances in rad/s. The angle of attack and pitch rate are measured. Design a continuous double-loop SMC/SMDO that drives $y \rightarrow y_c$ as time increases with given settling time t_s. Take pitch rate q as a "virtual" outer-loop control and u as the inner-loop control. Simulate the control system for $\alpha(0) = 0$, $q(0) = 0$, $\delta(0) = 0$, $h(t) = 0.1 \sin(t) \, \text{rad/s}$, $h(t) = 0.05 \cos(2t) \, \text{rad/s}$, $t_s = 0.3 \, \text{s}$, and

$$\alpha_c(t) = \begin{cases} 0.5 \sin(0.5t), & \text{if } 0 \leq t \leq \pi \\ 0.5, & \text{if } t > \pi \end{cases}$$

Support the simulation results by plotting the corresponding variables.

Exercise 8.10. Repeat Exercise **8.9** and design a single-loop finite-convergent-time SMC/SMDO that achieves tracking $y \rightarrow y_c$ with an overall given convergence time $t_c = 0.25s$.

Appendix A
Mathematical Preliminaries

A.1 Linear Algebra

A.1.1 Rank and Determinant

If A is an arbitrary $m \times n$ matrix over the field of real numbers $\mathbb{R}$, then the *row space* of A is the subspace of $\mathbb{R}^n$ generated by the rows of A, and the *column space* of A is the subspace of $\mathbb{R}^m$ generated by the columns of A. The dimensions of the row space and of the column space are called the *row rank* of A and the *column rank* of A, respectively. The following results hold linking the row space and the column space:

Theorem A.1 *The row rank and column rank of the matrix A are equal.*

Row rank and column rank (since they are equal) are usually called the *rank* of A and written rank(A). For arbitrary matrices A and B of appropriate dimension that the product AB exists has the property that

$$\text{rank}(AB) \leq \min\{\text{rank}(A),\ \text{rank}(B)\}. \tag{A.1}$$

For every square matrix A (i.e., a matrix with the same number of rows and columns) there can be assigned a unique scalar, called its *determinant*, and denoted $\det(A)$. This value can be assigned recursively as follows: first consider a scalar as matrix with one row and one column. The determinant of such a matrix is equal to the element itself. For a matrix of dimension 2 given by

$$A = \begin{bmatrix} a_{11} & a_{12} \\ a_{21} & a_{22} \end{bmatrix} \tag{A.2}$$

then the determinant of A is given (by definition) as

Y. Shtessel et al., *Sliding Mode Control and Observation*, Control Engineering,
DOI 10.1007/978-0-8176-4893-0, © Springer Science+Business Media New York 2014

$$\det(A) = a_{11}a_{22} - a_{12}a_{21}. \tag{A.3}$$

The determinant of a matrix A of dimension 3 may be defined by

$$\det \begin{bmatrix} a_{11} & a_{12} & a_{13} \\ a_{21} & a_{22} & a_{23} \\ a_{31} & a_{32} & a_{33} \end{bmatrix} = a_{11} \det \begin{bmatrix} a_{22} & a_{23} \\ a_{32} & a_{33} \end{bmatrix}$$

$$-a_{12} \det \begin{bmatrix} a_{21} & a_{23} \\ a_{31} & a_{33} \end{bmatrix} + a_{13} \det \begin{bmatrix} a_{21} & a_{22} \\ a_{31} & a_{32} \end{bmatrix}.$$

The representation above demonstrates how determinants can be defined in terms of determinants of matrices of lower order. In order to develop a specific generic formula, first define the *minor* M_{rs} of any element a_{rs} of a general matrix A of dimension n is the determinant of the matrix obtained by omitting the row and column containing a_{rs}. The *cofactor* of a_{rs}, here denoted by A_{rs}, is given by

$$A_{rs} = (-1)^{r+s} M_{rs}$$

With these definitions, the determinant of a general square matrix A can then be expressed (in terms of minors and cofactors) as

$$\det(A) = a_{r1}A_{r1} + a_{r2}A_{r2} + \ldots + a_{rn}A_{rn}, \qquad r = 1, 2, \ldots, n.$$

Some useful properties of determinants are:

- If A has two identical rows (columns), then $\det(A) = 0$.
- If A is invertible then $\det(A)$ is nonzero.
- The determinant of a product of two matrices A and B is equal to the product of their determinants, $\det(AB) = \det(A)\det(B)$.

Define the *identity matrix* or order n which will be denoted as I_n as the matrix which has ones on the diagonal elements and zeros elsewhere. It is easy to verify that the determinant of such a matrix is unity.

The *transpose* of a matrix A, denoted by A^T, is defined by interchanging the rows and columns, i.e., $\{A^T\}_{ij} = A_{ji}$. A square matrix is said to be *orthogonal* if $A^T A = AA^T = I_n$.

A matrix A is a *symmetric matrix* if $A^T = A$.

A.1.2   Eigenvalues and Eigenvectors

Definition A.1. If $A \in \mathbb{R}^{n \times n}$ is a square matrix, then $\lambda \in \mathbb{C}$ is an *eigenvalue* of A, if for some nonzero vector $v \in \mathbb{C}^n$,

$$Av = \lambda v$$

In this case v is said to be the *eigenvector* corresponding to the eigenvalue λ.

The matrix $\lambda I_n - A$ is called the *characteristic matrix* of A. The determinant of the characteristic matrix is called the *characteristic polynomial* of A and

$$\det(\lambda I_n - A) = 0 \tag{A.4}$$

is called the *characteristic equation* of A. This is an nth-order polynomial in λ. Note that the values of λ which constitute roots of the characteristic equation determine the eigenvalues of A.

If A is a square matrix partitioned as

$$A = \begin{bmatrix} A_{11} & A_{12} \\ 0 & A_{22} \end{bmatrix}, \tag{A.5}$$

where the top left and bottom right sub-blocks (A.5) are square, then the following properties hold:

Proposition A.1 *The determinant and the eigenvalues of A satisfy:*

1. $det(A) = det(A_{11})\, det(A_{22})$
2. $\lambda(A) = \lambda(A_{11}) \cup \lambda(A_{22})$

A.1.3 QR Decomposition

A frequently used decomposition is the so-called QR reduction, whereby an arbitrary matrix $X \in \mathbb{R}^{n \times m}$ can be expressed as

$$X = QR, \tag{A.6}$$

where R is an upper triangular matrix of the same dimension as X and Q is an orthogonal matrix.

A.1.4 Norms

Definition A.2. A norm is a function which assigns to every vector x in a vector space a real number $\|x\|$ such that:

1. $\|x\| \geq 0$
2. $\|x\| = 0$ if and only if $x = 0$
3. $\|kx\| = |k|\, \|x\|$ where k is a scalar and $|k|$ is the absolute value of k
4. $\|x + y\| \leq \|x\| + \|y\|$

Property 4 is usually called the *triangle inequality*.
Three common vector norms are:

1. The 1-norm

$$\|x\|_1 = |x_1| + |x_2| + \ldots + |x_n|. \tag{A.7}$$

2. The Euclidean norm (or 2-norm)

$$\|x\|_2 = (|x_1|^2 + |x_2|^2 + \cdots + |x_n|^2)^{1/2}. \tag{A.8}$$

3. The ∞ norm

$$\|x\|_\infty = \max |x|_i \quad (i = 1, \ldots, n). \tag{A.9}$$

The Euclidean norm corresponds exactly to the usual notion of distance, i.e., the length of the straight line between two points.

Corresponding to the vector norms above are three *induced matrix norms*:

1. The 1-norm

$$\|A\|_1 = \max_j (\sum_i |a_{ij}|). \tag{A.10}$$

2. The 2-norm is given by the maximum eigenvalue of $A^T A$.
3. The ∞ norm

$$\|A\|_\infty = \max_i (\sum_j |a_{ij}|). \tag{A.11}$$

For any two vectors x and y the equality

$$y^T x = \|x\| \|y\| \cos(\theta) \tag{A.12}$$

holds where θ is the angle (or more formally the direction cosine) between the two vectors and $\| \cdot \|$ denotes the Euclidean norm. From the properties of the cosine function it follows

$$y^T x \leq \|x\| \|y\| \tag{A.13}$$

for all x and y. The inequality in Eq. (A.13) is called the *Cauchy–Schwarz inequality*.

A.1.5   Quadratic Forms

A quadratic form is a function Q of n real variables $x_1, x_2, \ldots, x_n$ such that

$$Q(x_1, x_2, \ldots, x_n) = \sum_{i,j=1}^{n} q_{ij} x_i x_j \qquad q_{ij} = q_{ji}. \tag{A.14}$$

Without loss of generality, the q_{ij} can be thought of as the entries of a symmetric matrix Q and the x_i can be considered the components of the vector x. The quadratic form (A.14) can more conveniently be represented as

$$Q(x_1, x_2, \ldots, x_n) = x^T Q x, \qquad (A.15)$$

where the matrix $Q \in \mathbb{R}^{ntimesn}$ is a symmetric matrix.

Proposition A.2 *Let A be a symmetric matrix, then:*

1. *The eigenvalues of A are all real.*
2. *There exists an orthogonal matrix Q such that*

$$A = Q \Lambda Q^T, \qquad (A.16)$$

where Λ is a diagonal matrix formed from the eigenvalues of A and the orthogonal matrix Q is formed from the associated eigenvectors.

Quadratic forms always satisfy the Rayleigh principle, namely,

$$\lambda_{min}(Q)\|x\|^2 \leq x^T Q x \leq \lambda_{max}(Q)\|x\|^2. \qquad (A.17)$$

In particular, if $\lambda_{min}(Q) \geq 0$ then it follows that $x^T Q x \geq 0$ for all x.

Definition A.3. The quadratic form $x^T Q x$ where Q is a real symmetric matrix is said to be positive semidefinite if

$$x^T Q x \geq 0 \qquad \text{for all } x$$

In particular if

$$x^T Q x > 0 \qquad x \neq 0$$

then the quadratic form is said to be positive definite

If $x^T Q x$ is positive definite then the matrix Q is said to be a *positive definite matrix* and this will be written $Q > 0$. It can be seen from the modal decomposition expression in Eq. (A.16) that a symmetric matrix Q will be positive definite if all its eigenvalues are positive. It also follows that a positive definite matrix Q is invertible since from Eq. (A.16)

$$\det(Q) = \det(\Lambda) = \prod_{i=1}^{n} \lambda_i > 0.$$

Given any square matrix Q let a sequence of matrices $M_1 \ldots M_n$, termed the *principal minors*, be recursively defined so that $M_1 = Q$ and M_{i+1} is obtained from deleting the first row and first column from M_i.

Theorem A.2 *A necessary and sufficient condition for the quadratic form* $x^T Q x$, *where Q is a square symmetric matrix, to be positive definite is that the determinant of Q be positive and the successive principal minors of the determinant of Q be positive.*

Another propertythat will be exploited is that, given a symmetric matrix P and a nonsingular matrix T of the same dimension,

$$P > 0 \quad \Leftrightarrow \quad T^T P T > 0$$

A symmetric matrix P is said to be *negative definite* if $-P$ is positive definite.

Appendix B
Describing Functions

B.1 Describing Function Fundamentals

The main use of *describing function (DF) technique* [11, 100, 171] is in studying
the stability of the nonlinear systems, in particular predicting of limit cycles and
their stability in nonlinear systems. A system with zero input that contains only one
combined nonlinearity is depicted in Fig. B.1. The transfer function $G(s)$ is denoted
as the plant. In general case the transfer function $G(s)$ can include also a transfer
function of the controller and the sensor. It is assumed that there exists a *periodic
solution*

$$e(t) = A \sin \omega t, \tag{B.1}$$

where ω is a frequency of *self-sustained oscillations* (*limit cycle*). The output v of
the nonlinear block will be given by the Fourier series

$$v = \frac{\bar{a}_0}{2} + \sum_{n=1}^{\infty} \left[\bar{a}_n \cos(n\omega t) + \bar{b}_n \sin(n\omega t) \right], \tag{B.2}$$

where $\bar{a}_0 = 0$ for odd nonlinearities and

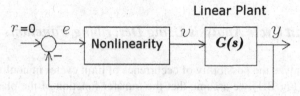

Fig. B.1 Block diagram of nonlinear system

Y. Shtessel et al., *Sliding Mode Control and Observation*, Control Engineering,
DOI 10.1007/978-0-8176-4893-0, © Springer Science+Business Media New York 2014

$$\bar{a}_n = \frac{\omega}{\pi} \int_0^{2\pi/\omega} v(t) \cos(n\omega t)\, dt$$

$$\bar{b}_n = \frac{\omega}{\pi} \int_0^{2\pi/\omega} v(t) \sin(n\omega t)\, dt. \tag{B.3}$$

B.1.1 Low-Pass Filter Hypothesis and Describing Function

Assume that the transfer function of the plant, $G(s)$, has low-pass filter charac-
teristics with respect to higher harmonics in the signal $e(t)$. It means that $|G(s)|$
is assumed to be small with respect to the higher harmonic components for $n = 2, 3, \ldots$ of the signal $v(t)$ given by Eq. (B.2) compared to the value $|G(s)|$ with
respect to a fundamental component for $n = 1$. In this case the fundamental output
$v(t)$ of the nonlinearity can be computed as

$$v = \bar{a}_1 \cos(\omega t) + \bar{b}_1 \sin(\omega t) = M \sin(\omega t + \varphi). \tag{B.4}$$

This assumption is the foundation of the DF technique.

The DF of the nonlinearity is defined as the fundamental output (presented in
a complex function or phasor format) divided by the input amplitude. This is in a
Cartesian format

$$N(A, \omega) = \frac{\bar{a}_1 + j\bar{b}_1}{A} \tag{B.5}$$

or in polar coordinates

$$N(A, \omega) = \frac{M(A, \omega)}{A} e^{j\varphi(A,\omega)}, \tag{B.6}$$

where

$$M(A, \omega) = \sqrt{\bar{a}_1^2 + \bar{b}_1^2}$$
$$\varphi(A, \omega) = -\tan^{-1}\frac{\bar{b}_1}{\bar{a}_1} \tag{B.7}$$

B.1.2 Limit Cycle Analysis Using Describing Functions

In order to analyze the possibility of occurrence of limit cycles in nonlinear closed-
loop system (Fig. B.1) we assume that the transfer function of the plant $G(s)$ has
low-pass filter characteristics and the input $e(t)$ of the nonlinearity is sinusoidal
and given by Eq. (B.1). Then the nonlinearity in Fig. B.1 is replaced by its DF
$N(A, \omega)$ given by Eqs. (B.5) or (B.6) and (B.7), and the sinusoidal signals $e(t)$

and $v(t)$ are presented in a phasor format. Resulting system with $M = \sqrt{\bar{a}_1^2 + \bar{b}_1^2}$, $\varphi = -\tan^{-1} \frac{\bar{b}_1}{\bar{a}_1}$ is presented in Fig. B.2.

A steady state sinusoidal analysis is performed next. The nonlinear system will have a limit cycle if there exists a harmonic balance in this system. It means that the sinusoidal signal at the input of the nonlinearity will propagate through the cascade of the nonlinear block and the linear plant and being negated at a negative feedback will regenerate itself. Therefore, we obtain in the frequency domain:

$$
\begin{aligned}
Ae^{j\omega t} N(A, \omega) &= Me^{j(\omega t + \varphi)} \\
Me^{j(\omega t + \varphi)} G(j\omega) &= -Ae^{j\omega t}
\end{aligned}
\tag{B.8}
$$

Equation (B.8) implies

$$
1 + G(j\omega)N(A, \omega) = 0 \;\Rightarrow\; G(j\omega) = -\frac{1}{N(A, \omega)} \tag{B.9}
$$

that is known as the harmonic balance equation. If Eq. (B.9) is satisfied for $A = A_c$, and $\omega = \omega_c$, then these values give the amplitude and frequency of the *predicted limit cycle*. If Eq. (B.9) does not have a real-valued positive solution, the limit cycle *is not predicted*. We have to acknowledge that due to the previous assumptions, including low-pass filter hypothesis, and due to taking into account only the fundamental harmonic when performing the harmonic balance, we can only *predict* the limit cycle. The prediction is supposed to be verified via simulations. It is worth noting the stability of the predicted limit cycle must be studied separately.

B.1.3 Stability Analysis of the Limit Cycle

The DF technique can be used to analyze the stability of the limit cycles. The Nyquist criterion of stability will be also employed. Let's assume that a solution (A_c, ω_c) of Eq. (B.9) exists. Figure B.3 illustrates the procedure of getting the solution graphically.

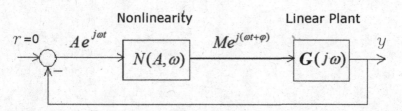

Fig. B.2 Block diagram of nonlinear system with the describing function in a frequency domain

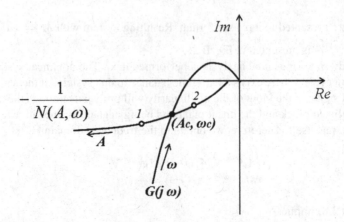

Fig. B.3 Graphical solution of the harmonic balance equation

Now, assume that the perturbation of the limit cycle with the parameters (A_c, ω_c) occurs such that the amplitude A_c of the limit cycle increases slightly. Then the operating point moves slightly outside of the zone encircled by the Nyquist plot (point 1 on the plot Fig. B.3). Assuming the linear plant does not have poles in the right-hand side of the complex plane, then in accordance with the Nyquist criterion of stability (that can be applied to the linearized system in the vicinity of the operating point (A_c, ω_c)), the system is stable, and the amplitude A_c starts decreasing moving the operating point to its original position (A_c, ω_c). Let's perturb the operating point (limit cycle with the parameters (A_c, ω_c)) such that the amplitude of oscillations A_c decreases slightly. Then the operating point moves insight the Nyquist plot (point 2 on the plot Fig. B.3). Applying the Nyquist criterion of stability to the system linearized in the operating point (A_c, ω_c) we can conclude that the linearized system now is unstable and the amplitude starts increasing moving the operating point to its original position (A_c, ω_c). Therefore, for the considered case the limit cycle is stable.

Appendix C
Linear Systems Theory

C.1 Introduction

This appendix provides the necessary background in terms of systems theory that is required for the sliding mode control developments. For further details see [47,155].

C.1.1 Linear Time-Invariant Systems

Consider the linear system

$$\dot{x}(t) = Ax(t), \tag{C.1}$$

where $A \in \mathbb{R}^{n \times n}$ and the state vector $x \in \mathbb{R}^n$. A necessary and sufficient condition for asymptotic stability is that the eigenvalues of A have negative real parts.

Now consider the system

$$\dot{x}(t) = Ax(t) + Bu(t). \tag{C.2}$$

Essentially a forcing function or control input $u(t)$ has been introduced into the system. The general solution to this state equation may be expressed in the form

$$x(t) = Ve^{\Lambda t}W^T x(0) + V \int_0^t e^{\Lambda(t-\tau)} W^T Bu(\tau)\, d\tau, \tag{C.3}$$

where V is a matrix whose columns consist of the linearly independent right eigenvectors corresponding to the distinct eigenvalues λ_i, $i = 1, \ldots, n$ of the matrix A, $\Lambda = \text{diag}(\lambda_1, \ldots, \lambda_n)$, and the rows of the matrix W^T consist of the corresponding left eigenvectors, i.e., $V^{-1} = W^T$. Here $x(0)$ represents an arbitrary initial condition. Again the time domain characteristics of the system are determined by the eigenvalues of A. In addition, the associated right eigenvectors determine the "shape" of a given mode. As the solution depends upon a linear combination of

Y. Shtessel et al., *Sliding Mode Control and Observation*, Control Engineering,
DOI 10.1007/978-0-8176-4893-0, © Springer Science+Business Media New York 2014

functions of the form $v_i e^{\lambda_i t}$, appropriate eigenvector entries enable the transient $e^{\lambda_i t}$ to contribute, or not, to a particular state variable. In this way it is seen that the entire eigenstructure, and not just the eigenvalues, are effective in determining the time response of a system.

C.1.2   Controllability and Observability

Consider the linear time-invariant system

$$\dot{x}(t) = Ax(t) + Bu(t) \tag{C.4}$$

$$y(t) = Cx(t), \tag{C.5}$$

where $A \in \mathbb{R}^{n \times n}$, $B \in \mathbb{R}^{n \times m}$, and $C \in \mathbb{R}^{p \times n}$. The variables $u(t)$ and $y(t)$ will be referred to as the *inputs* and *outputs*, respectively. The matrices A, B, and C will be termed the *system*, *input distribution*, and *output distribution* matrices, respectively. For convenience the system in Eqs. (C.4) and (C.5) will be referred to as a system triple (A, B, C).

If $T \in \mathbb{R}^{n \times n}$ is nonsingular, then the change of coordinates $x \mapsto Tx$ induces a new system representation with system matrix TAT^{-1}, input distribution matrix TB, and output distribution matrix CT^{-1}, i.e., the triple (TAT^{-1}, TB, CT^{-1}).

Definition C.1. The system is said to be completely controllable if given any initial condition $x(t_0)$ there exists an input function on the finite interval $[t_0, t_1]$ such that $x(t_1) = 0$.

From this definition the following theorem can be proved.

Theorem C.1 *Given any pair (A, B) the following conditions are all equivalent:*

- *(A, B) is completely controllable.*
- *The controllability matrix $[B \ \ AB \ \ A^2 B \ \ \ldots \ \ A^{n-1}B]$ has full rank.*
- *The matrix $[sI - A \ \ B]$ has full rank for all $s \in \mathbb{C}$.*
- *The spectrum of $(A + BF)$ can be assigned arbitrarily by choice of $F \in \mathbb{R}^{m \times n}$.*

The third condition, which is often the most convenient method of establishing controllability, is often referred to as the Popov–Belevitch–Hautus rank test or PBH test.

Definition C.2. The linear system is said to be completely observable if the output function $y(t)$ over some time interval $[t_0, t_1]$ uniquely determines the initial condition $x(t_0)$.

An important duality exists between the notions of controllability and observability which can be summarized as follows.

Theorem C.2 *The pair (A, C) is completely observable if and only if the pair (A^T, C^T) is completely controllable.*

From the theorem above, the results of Theorem C.1 can be modified to provide a list of equivalent statements for observability.

In addition to these concepts there exist two slightly weaker notions: *stabilizability* and *detectability*.

Theorem C.3 *Given any pair (A, B) the following conditions are equivalent:*

- *(A, B) is stabilizable.*
- *The matrix $[sI - A \quad B]$ has full rank for all $s \in \mathbb{C}_+$.*
- *There exists an $F \in \mathbb{R}^{m \times n}$ such that the eigenvalues of $A + BF$ belong to $\mathbb{C}_-$.*

The notion of detectability can be defined as the dual of stabilizability.

If the pair (A, B) in Eq. (C.4) is not controllable then there exists a change of coordinates $x \mapsto Tx$ so that in the new coordinate system the new system and input distribution matrices have the form

$$
TAT^{-1} = \begin{bmatrix} A_{11} & 0 \\ A_{21} & A_{22} \end{bmatrix} \quad TB = \begin{bmatrix} 0 \\ B_2 \end{bmatrix}, \tag{C.6}
$$

where the pair (A_{22}, B_2) is completely controllable. Note that because of the special structure of the canonical form in Eq. (C.6) it follows

$$
\lambda(A_{11}) \subset \lambda(A + BF)
$$

for any $F \in \mathbb{R}^{m \times n}$. Consequently, the pair (A, B) is stabilizable if and only if A_{11} is stable.

By duality, a similar canonical form exists for pairs (A, C) which are not observable.

C.1.3 Invariant Zeros

Consider the linear time-invariant system

$$
\dot{x}(t) = Ax(t) + Bu(t) \tag{C.7}
$$

$$
y(t) = Cx(t). \tag{C.8}
$$

Assume the initial state is given by $x(0)$. Taking Laplace transforms of the system representation yields

$$
\begin{bmatrix} sI - A & -B \\ C & 0 \end{bmatrix} \begin{bmatrix} X(s) \\ U(s) \end{bmatrix} = \begin{bmatrix} x(0) \\ Y(s) \end{bmatrix}. \tag{C.9}
$$

The polynomial system matrix

$$P(s) = \begin{bmatrix} sI - A & -B \\ C & 0 \end{bmatrix} \tag{C.10}$$

is usually referred to as Rosenbrock's system matrix. A necessary and sufficient condition for an input

$$u(t) = u(0)e^{zt} \tag{C.11}$$

to yield rectilinear motion in the state space of the form

$$x(t) = x(0)e^{zt} \tag{C.12}$$

such that the output of the system is identically zero for all time is that z, $x(0)$, and $u(0)$ satisfy

$$P(z) \begin{bmatrix} x(0) \\ u(0) \end{bmatrix} = 0. \tag{C.13}$$

This result defines a set of complex frequencies z which are associated with specific directions $x(0)$ and $u(0)$ in the state and input spaces for which the output of the system is zero. These elements are called *invariant zeros*. It is clear that information regarding the existence of invariant zeros comes from examining the rank of $P(z)$ in Eq. (C.13). For example, in the case of a square system—i.e., systems with equal numbers of inputs and outputs—in order for Eq. (C.13) to have a nonzero solution for $x(0)$ and $u(0)$, $\det(P(z))$ must be zero.

C.1.4   State Feedback Control

For the controllable state-space system represented by

$$\dot{x}(t) = Ax(t) + Bu(t) \tag{C.14}$$

a state feedback controller is defined by

$$u(t) = -Kx(t), \tag{C.15}$$

where $K \in \mathbb{R}^{m \times n}$. The state equation of the closed-loop system is given by

$$\dot{x}(t) = (A - BK)x(t). \tag{C.16}$$

As the system is controllable, from Theorem C.1, the closed-loop poles can be allocated to any desired location by appropriate choice of K.

The linear quadratic regulator (LQR) is a particular formulation of the state feedback control problem. Given the state-space system (C.14) with given initial condition $x(0)$, the input signal $u(t)$ is sought which regulates the system state to the origin by minimizing the cost function

$$J = \frac{1}{2} \int_0^\infty x(t)^T Q x(t) + u(t)^T R u(t) \, dt, \qquad (C.17)$$

where Q and R are positive definite symmetric matrices which penalize the deviation of the state from the origin and the magnitude of the control signal, respectively. The optimal solution, for any initial state, is given by

$$u(t) = -Kx(t) = -R^{-1} B^T X x(t), \qquad (C.18)$$

where X is the unique positive semidefinite solution ($X \geq 0$), of the algebraic Riccati equation

$$A^T X + X A - X B R^{-1} B^T X + Q = 0. \qquad (C.19)$$

For further details see [8].

C.1.5 Static Output Feedback Control

It has been demonstrated in the previous section that it is possible to readily determine a state feedback-based control strategy for a controllable linear system. In practice it may not be feasible to measure all the state variables for a given system. If only a subset of state information is available, the feedback control problem must now consider the system

$$\dot{x}(t) = Ax(t) + Bu(t) \qquad (C.20)$$

$$y(t) = Cx(t), \qquad (C.21)$$

where $y \in \mathbb{R}^p$ denotes an available vector of output measurements. The development of a control law

$$u(t) = -KCx(t) = -Ky(t), \qquad (C.22)$$

where $K \in \mathbb{R}^{m \times p}$ is needed. It is important to note that this problem has associated with it further constraints. A sufficient condition to ensure that the eigenvalues of $A - BKC$ may be placed arbitrarily close to desired locations is that the pair (A, B) is controllable, the pair (A, C) is observable, and the dimensionality requirement

$$m + p + 1 \geq n \qquad (C.23)$$

is satisfied. These requirements are referred to as the *Kimura–Davison* conditions.

Appendix D
Lyapunov Stability

Earlier the notion of stability was considered using the concept of phase portraits for second-order systems. Now a more general (and more abstract) approach will be taken. For the general nonlinear system

$$\dot{x}(t) = f(t, x), \tag{D.1}$$

where $x \in \mathbb{R}^n$, an *equilibrium point* in the state space is given by vectors x_e satisfying

$$f(t, x_e) = 0 \text{ for all } t > 0. \tag{D.2}$$

In general there may be many values of x_e which satisfy this condition. In terms of analyzing the stability of a particular equilibrium point it is useful to assume the point is at the origin. This can be done without loss of generality since given an equilibrium point x_e the change of variables (actually just a simple translation of the origin) $x \mapsto \tilde{x}$ where $\tilde{x} = x - x_e$ implies the origin is an equilibrium point of

$$\dot{\tilde{x}}(t) = f(t, \tilde{x}) \tag{D.3}$$

since $f(t, \tilde{x} = 0) = f(t, x_e) = 0$.

The solution to Eq. (D.1) will be written as $x(t, x_0)$ where t signifies the evolution of the state with respect to time and x_0 represents the initial conditions—usually the value of the state at $t = 0$. Note, it is usually very difficult if not impossible to obtain an analytic expression for $x(t, x_0)$ for general nonlinear systems.

Definition D.1. The origin of Eq. (D.1) is said to be stable if given any $\epsilon > 0$ there exists a $\delta > 0$ such that if $\|x_0\| < \delta$ then $\|x(t, x_0)\| < \epsilon$ for all $t > 0$.

Less formally the definition means that by starting close enough to the equilibrium point, the solution will always remain arbitrarily close to it.

Definition D.2. The origin of Eq. (D.1) is said to be asymptotically stable if it is stable and the solution $x(t, x_0) \to 0$ as $t \to \infty$.

Y. Shtessel et al., *Sliding Mode Control and Observation*, Control Engineering,
DOI 10.1007/978-0-8176-4893-0, © Springer Science+Business Media New York 2014

Less formally this definition means that by starting close enough to the equilibrium point, the solution will always remain arbitrarily close, but in addition, the trajectory will always move towards the equilibrium point.

D.1 Local Results

Consider a domain $\mathcal{D}$ in the state space which contains a neighborhood of the origin $\{x \in \mathbb{R}^n : \|x\| < r\}$ where r is a positive scalar. In two dimensions this represents a circle centered at the origin of radius r; in three dimensions $\mathcal{D}$ represents a sphere of radius r.

Definition D.3. A function $V : \mathcal{D} \mapsto \mathbb{R}$ is positive definite in the domain $\mathcal{D}$ if:

- $V(x) \geq 0$ for all $x \in \mathcal{D}$.
- $V(x) = 0$ implies $x = 0$.

The function is said to be positive semidefinite in the domain $\mathcal{D}$ if only the first condition holds. If $\mathcal{D} = \mathbb{R}^n$ the function is said to be positive definite or positive semidefinite, respectively.

The function $V : \mathbb{R}^2 \mapsto \mathbb{R}$ given by $V(x_1, x_2) = x_1^2 + x_2^2$ is positive definite. This is a special case of a generic class of functions called *quadratic forms*.

One version of Lyapunov's theorem is given below:

Theorem D.1 *Consider the nonlinear system in Eq. (D.1). Suppose in the domain $\mathcal{D}$ there exists a differentiable positive definite function $V : \mathbb{R} \times \mathcal{D} \mapsto \mathbb{R}$ such that*

$$\dot{V} = \frac{\partial V}{\partial t} + \frac{\partial V}{\partial x} f(t, x) \leq 0$$

then the origin is stable. Furthermore if $\dot{V} < 0$ for $x \neq 0$ then the origin is asymptotically stable.

D.2 Global Results

To extend these local results to be global, intuitively all that needs to be done is to ensure the conditions of Theorem D.1 hold for $\mathcal{D} = \mathbb{R}^n$. It turns out this is not quite sufficient. A further constraint needs to be included, namely, $V(x) \to \infty$ as $\|x\| \to \infty$. A function V with such a property is termed *radially unbounded*.

Theorem D.2 *Consider the nonlinear system in Eq. (D.1). Suppose there exists a differentiable function $V : \mathbb{R} \times \mathbb{R}^n \mapsto \mathbb{R}$ which is radially unbounded and positive definite such that*

$$\dot{V} = \frac{\partial V}{\partial t} + \frac{\partial V}{\partial x} f(t, x) \leq 0$$

then the origin is globally stable. Furthermore if $\dot{V} < 0$ for $x \neq 0$ then the origin is asymptotically stable.

Quadratic forms are functions of the type $x^T P x$ where $P \in \mathbb{R}^{n \times n}$ is symmetric, i.e., $P^T = P$. Thought of in terms of the components x_i of x, a quadratic form is a weighted sum of all possible products $x_i x_j$: specifically

$$x^T P x = \sum_{i=1}^{n} \sum_{j=i}^{n} p_{ij} x_i x_j$$

Consider a general linear system described by

$$\dot{x}(t) = Ax(t), \tag{D.4}$$

where $A \in \mathbb{R}^{n \times n}$. Consider a Lyapunov candidate of the form

$$V(x) = x^T P x, \tag{D.5}$$

where $P \in \mathbb{R}^{n \times n}$ is some symmetric positive definite matrix. Differentiating Eq. (D.5) with respect to time gives

$$\dot{V} = \dot{x}^T P x + x^T P \dot{x}$$
$$= x^T (PA + A^T P) x. \tag{D.6}$$

In order to prove that the system is asymptotically stable the symmetric matrix in Eq. (D.6) must be negative definite. Consider the matrix equation

$$PA + A^T P = -Q, \tag{D.7}$$

where $Q \in \mathbb{R}^{n \times n}$ is a symmetric positive definite matrix. Given A and Q, if a symmetric positive definite matrix P exists, solving the so-called *Lyapunov equation* given in Eq. (D.7), then the linear system in Eq. (D.4) will be asymptotically stable. In fact the following can be shown:

Theorem D.3 *There exists a unique s.p.d matrix P satisfying Eq. (D.7) if and only if the matrix A is stable, i.e., the eigenvalues of A lie in the open left half plane.*

D.2.1 Quadratic Stability

The previous section considered the special case of linear systems. This enabled an analytic expression for the solution to be obtained. For general nonlinear systems this is usually impossible. An approach for studying the stability of differential equations, without the need to obtain an explicit solution, is the method of

Lyapunov. Loosely speaking, if a differentiable function $V : \mathbb{R}^n \mapsto \mathbb{R}$ can be found which is positive except at an equilibrium point and whose total time derivative decreases along the system trajectories, then the equilibrium point is stable. The key point is that this approach obviates the need to solve the nonlinear differential equation when assessing its stability properties.

Unfortunately, no systematic way exists to synthesize Lyapunov functions for nonlinear systems. This section considers the special case when the scalar function $V : \mathbb{R}^n \to \mathbb{R}$ is the quadratic form

$$V(x) = x^T P x, \tag{D.8}$$

where $P \in \mathbb{R}^{n \times n}$ is some symmetric positive definite matrix. By construction the function is nonzero except at the origin. Next, form the function of time

$$V(t) = x(t)^T P x(t), \tag{D.9}$$

where $x(t)$ represents the solution of the differential equation (D.1). Differentiating Eq. (D.9) with respect to time gives

$$\dot{V}(t) = \dot{x}(t)^T P x(t) + x(t)^T P \dot{x}(t)$$
$$= 2x(t)^T P \dot{x}(t)$$
$$= 2x(t)^T P f(x,t),$$

where the second equality follows because the quantities are scalars and hence

$$\dot{x}(t)^T P x(t) = \left(\dot{x}(t)^T P x(t) \right)^T = x(t)^T P \dot{x}(t)$$

Definition D.4. The origin of the system (D.1) is said to be quadratically stable if there exist symmetric positive definite matrices $P, Q \in \mathbb{R}^{n \times n}$ such that the total time derivative satisfies

$$\dot{V}(x) = 2x^T P f(x,t) \le -x^T Q x$$

The inequality above implies $\|x(t)\| < e^{-\alpha t}$ where $\alpha = \lambda_{min}(P^{-1}Q)$ and hence the origin is asymptotically stable. If $f(x,t) = Ax(t)$ then it is well known that A has stable eigenvalues if and only if, given any symmetric positive definite matrix Q, there exists a unique symmetric positive definite matrix P satisfying the *Lyapunov equation*

$$PA + A^T P = -Q. \tag{D.10}$$

Consequently, any stable linear system is quadratically stable. A symmetric positive definite matrix P satisfying Eq. (D.10) will be referred to as a *Lyapunov matrix* for the matrix A.

Lyapunov theory may also be used as a means of examining the *robustness* of a given linear system; suppose

$$\dot{x}(t) = Ax(t) + \xi(t, x), \tag{D.11}$$

where the matrix A is stable and $\xi(\cdot)$ is an imprecisely known function which represents uncertainty in the system. Let the pair of positive definite matrices (P, Q) satisfy the Lyapunov Eq. (D.10) and define

$$\mu = \frac{\lambda_{min}(Q)}{\lambda_{max}(P)} \tag{D.12}$$

and suppose that the uncertain function satisfies

$$\|\xi(t, x)\| \leq \tfrac{1}{2}\mu\|x(t)\| \tag{D.13}$$

then the system in Eq. (D.11) is stable. This can be established by using $V = x^T P x$ as a Lyapunov function: the derivative along the trajectories satisfies

$$\begin{aligned} \dot{V} &= x(t)^T PAx(t) + x(t)^T A^T Px(t) + 2x(t)^T P\xi(t, x) \\ &= -x(t)^T Qx(t) + 2x(t)^T P\xi(t, x) \\ &\leq -x(t)^T Qx(t) + 2\|Px(t)\|\|\xi(t, x)\|, \end{aligned} \tag{D.14}$$

where the Cauchy–Schwarz inequality (see Appendix A.1.4) has been used to obtain the last inequality. Now

$$\|Px\| = \sqrt{x^T P^2 x} \leq \sqrt{\lambda_{max}(P^2)\|x\|^2} = \lambda_{max}(P)\|x\|, \tag{D.15}$$

where the Rayleigh principle has been used to obtain the middle inequality. Also directly from the Rayleigh principle

$$-x^T Qx \leq -\lambda_{min}(Q)\|x\|^2. \tag{D.16}$$

Thus from the inequality in Eq. (D.14) and using Eqs. (D.15) and (D.16) it follows that

$$\begin{aligned} \dot{V}(t) &\leq -\lambda_{min}(Q)\|x(t)\|^2 + 2\lambda_{max}(P)\|x(t)\|\|\xi(t, x)\| \\ &= -\lambda_{max}(P)\|x(t)\|\,(\mu\|x(t)\| - 2\|\xi(t, x)\|) \end{aligned}$$

and therefore if $\xi(\cdot)$ satisfies Eq. (D.13) the Lyapunov derivative is always negative and stability is proved.

In view of the condition (D.13), it is natural to attempt to choose Q in an effort to maximize Eq. (D.12). It can be shown that the maximum is given by

$$\hat{\mu} = \frac{1}{\lambda_{max}(P)} \tag{D.17}$$

when $Q = I$. Furthermore, it can be shown that

$$\hat{\mu} \leq -2 \max [\text{Re } \lambda(A)] \qquad \qquad (\text{D.18})$$

with equality if the matrix A is *normal*, i.e., if it has n orthonormal eigenvectors.

When dealing with uncertain systems, it may not be possible to guarantee asymptotic stability. Consider the nonlinear system (D.1) and suppose it is subject to an imprecisely known exogenous signal $\xi(\cdot)$ so that

$$\dot{x}(t) = f(x, t, \xi). \qquad \qquad (\text{D.19})$$

Let $\mathcal{E} \subset \mathbb{R}^n$ be a bounded set, then the following definition can be made. For further details see [116].

Definition D.5. The solution $x(\cdot)$ to the uncertain system (D.19) is said to be ultimately bounded with respect to the set $\mathcal{E}$ if:

- On any finite interval the solution remains bounded, i.e., if $\|x(t_0)\| < \delta$ then $\|x(t)\| < d(\delta)$ for any $t \in [t_0, t_1)$.
- In finite time the solution $x(t)$ enters the bounded set $\mathcal{E}$ and remains there for all subsequent time.

The set $\mathcal{E}$ is usually an acceptably small neighborhood of the origin and the concept is often termed *practical stability*.

For further details see [116].

Bibliography

[1] Abe, M.: Vehicle dynamics and control for improving handling and active safety: from four-wheel steering to direct yaw moment control. Proc. Inst. Mech. Eng. 213(2), 87–101 (1999)

[2] Ackermann, J., Utkin V.I.: Slidingmode control design based on Ackermann's formula. IEEE Trans. Automat. Contr. 43(2), 234–237 (1998)

[3] Acary, V., Brogliato, B., Orlov, Y.: Chattering-free digital sliding-mode control with state observer and disturbance rejection. IEEE Trans. Automat. Contr. 57(5), 1087–1101 (2012)

[4] Aguilar, L., Boiko, I., Fridman, L., Iriarte, R.: Generating self-excited oscillations for underactuated mechanical systems via two-relay controller. Int. J. Control 82(9), 1678–1691 (2009)

[5] Aguilar, L., Boiko, I., Fridman, L., Iriarte, R.: Oscillations via two-relay controller. IEEE Trans. Automat. Contr. 54(2), 416–420 (2009)

[6] Alwi, H., Edwards, C.: Fault tolerant control using sliding modes with online control allocation. Automatica 44, 1859–1866 (2008)

[7] Alwi, H., Edwards, C., Tan, C.P.: Fault Detection and Fault-tolerant Control Using Sliding Modes. Advances in Industrial Control Series. Spring, Berlin (2011)

[8] Anderson, B., Moore, J.: Optimal Control. Series Information and Systems Science. Prentice Hall, London (1990)

[9] Anosov, D.V.: On stability of equilibrium points of relay systems. Automat. Rem. Contr. 20(2), 135–149 (1959)

[10] Atassi, A.N., Khalil, H.K.: Separation results for the stabilization of nonlinear systems using different high-gain observer designs. Syst. Control Lett. 39(3), 183–191 (2000)

[11] Atherton, D.P.: Nonlinear Control Engineering – Describing Function Analysis and Design. Van Nostrand Company Limited, Workingham, Berkshire (1975)

[12] Baccioti, A., Rosier, L.: Lyapunov Functions and Stability in Control Theory, 2nd edn. Springer, New York (2005)

[13] Bag, S.K., Spurgeon, S.K., Edwards, C.: Output feedback sliding mode design for linear uncertain systems. Proc. IEE, Part D 144, 209–216 (1997)

[14] Bandyopadhyay, B., Janardhanan, S.: Discrete-time Sliding Mode Control: A Multi-rate Output Feedback Approach. Lecture Notes in Control and Information Sciences, vol. 323. Springer, Berlin (2006).

[15] Bartolini, G., Pisano, A., Usai, E.: First and second derivative estimation by sliding mode technique. J. Signal Proc. 4(2), 167–176 (2000)

[16] Bartolini, G., Pydynowski, P.: An improved, chattering free, V.S.C. scheme for uncertain dynamical systems. IEEE Trans. Automat. Contr. 41(8), 1221–1226 (1996)

[17] Bartolini, G., Ferrara, A., Usai, E.: Output tracking control of uncertain nonlinear second-order systems. Automatica 33(12), 2203–2212 (1997)

[18] Bartolini, G., Ferrara, A., Usai, E.: Chattering avoidance by second-order sliding mode control. IEEE Trans. Automat. Contr. **43**(2), 241–246 (1998)

[19] Bartolini, G., Ferrara A., Levant, A., Usai, E.: On second order sliding mode controllers. In: Young, K.D., Ozguner, U. (eds.) Variable Structure Systems, Sliding Mode and Nonlinear Control. Lecture Notes in Control and Information Sciences, vol. 247, pp. 329–350. Springer, Berlin (1999)

[20] Bartolini, G., Pisano, A., Punta, E., Usai, E.: A survey of applications of second-order sliding mode control to mechanical systems. Int. J. Control **76**(9/10), 875–892 (2003)

[21] Basin, M., Rodriguez, J., Fridman, L.: Optimal and robust control for linear state-delay systems. J. Franklin Inst. **344**(6), 830–845 (2007)

[22] Bhat, S.P., Bernstein, D.S.: Finite time stability of continuous autonomous systems. SIAM J. Control Optim. **38**(3), 751–766 (2000)

[23] Bhat, S.P., Bernstein, D.S.: Geometric homogeneity with applications to finite time stability. Math. Control Signal. **17**, 101–127 (2005)

[24] Bejarano, J., Fridman, L.: High order sliding mode observer for linear systems with unbounded unknown inputs. Int. J. Control **83**(9), 1920–1929 (2010)

[25] Bejarano, J., Fridman, L., Poznyak, A.: Exact state estimation for linear systems with unknown inputs based on hierarchical super-twisting algorithm. Int. J. Robust. Nonlin. **17**(18), 1734–1753 (2007)

[26] Bejarano, J., Fridman, L., Poznyak, A.: Output integral sliding mode control based on algebraic hierarchical observer. Int. J. Control **80**(3), 443–453 (2007)

[27] Bejarano, J., Fridman, L., Poznyak, A.: Output integral sliding mode for min-max optimization of multi-plant linear uncertain systems. IEEE Trans. Automat. Contr. **54**(11), 2611–2620 (2009)

[28] Boiko, I.: Frequency domain analysis of fast and slow motions in sliding modes. Asian J. Control **5**(4), 445–453 (2003)

[29] Boiko, I.: Discontinuous Control Systems: Frequency-Domain Analysis and Design. Birkhauser, Boston (2009)

[30] Boiko, I.: On frequency-domain criterion of finite-time convergence of second-order sliding mode control algorithms. Automatica **47**(9), 1969–1973 (2011)

[31] Boiko, I., Fridman, L.: Analysis of chattering in continuous sliding-mode controllers. IEEE Trans. Automat. Contr. **50**(9), 1442–1446 (2005)

[32] Boiko, I., Fridman, L., Castellanos, M.I.: Analysis of second order sliding mode algorithms in the frequency domain. IEEE Trans. on Automat. Contr. **49**(6), 946–950 (2004)

[33] Boiko,I., Iriarte, R., Pisano, A., Usai, E.: Parameter tuning of second-order sliding mode controllers for linear plants with dynamic actuators. Automatica **42**(5), 833–839 (2006)

[34] Boiko, I., Fridman, L., Pisano, A., Usai, E.: Analysis of chattering in systems with second order sliding modes. IEEE Trans. Automat. Contr. **52**(11), 2085–2102 (2007)

[35] Boiko, I., Fridman, L., Pisano, A., Usai, E.: Performance analysis of second-order sliding-mode control systems with fast actuators. IEEE Trans. Automat. Contr. **52**(6), 1053–1059 (2007)

[36] Boiko, I., Fridman, L., Pisano, A., Usai, E.: On the transfer properties of the generalized suboptimal second-order sliding mode control algorithm. IEEE Trans. Automat. Contr. **54**(2), 399–403 (2009)

[37] Bondarev, A.G., Bondarev, S.A., Kostylyeva, N.Y., Utkin, V.I.: Sliding modes in systems with asymptotic state observers. Automatica i telemechanica (Automat. Rem. Contr.) **46**(5), 679–684 (1985)

[38] Boyd, S.P., El Ghaoui, L., Feron, E. Balakrishnan V.: Linear Matrix Inequalities in Systems and Control Theory. SIAM, Philadelphia (1994).

[39] Burlington, R.S.: Handbook of Mathematical Tables and Formulas. McGraw-Hill, New York (1973)

[40] Burton, J.A., Zinober, A.S.I.: Continuous approximation of variable structure control. Int. J. Syst. Sci. **17**(6), 876–885 (1986)

[41] Cannas, B., Cincotti, S., Usai, E.: An algebraic observability approach to chaos synchro-nisation by sliding differentiators. IEEE Trans. Circ. Systems-I: Fundamental Theory and Applications **49**(7), 1000–1006 (2002)

[42] Cannas, B., Cincotti, S., Usai, E.: A chaotic modulation scheme based on algebraic observability and sliding mode differentiators. Chaos Soliton. Fract. **26**(2), 363–377 (2005)

[43] Castanos, F., Fridman, L.: Analysis and design of integral sliding manifolds for systems with unmatched perturbations. IEEE Trans. Automat. Contr. **55**(5), 853–858 (2006)

[44] Castanos, F., Fridman, L.: Dynamic switching surfaces for output sliding mode control: an $\mathcal{H}_\infty$ approach. Automatica **47**(7), 1957–1961 (2011)

[45] Castanos, F., Xu, J.X., Fridman, L.: Integral sliding modes for systems with matched and unmatched uncertainties. In: Edwards, C., Colet, E.F., Fridman, L. (eds.) Advances in Variable Structure and Sliding Mode Control. Lecture Notes in Control and Information Sciences, vol. 334, pp. 227–246. Springer, Berlin (2006)

[46] Castellanos, M.I., Boiko, I. Fridman, L.: Parameter identification via modified twisting algorithm. Int. J. Control **81**(5), 788–796 (2008)

[47] Chen, C.: Linear Systems: Theory and Design. Oxford University Press, New York (1999)

[48] Chen, B.M.: Robust and $\mathcal{H}_\infty$ control. Series: Communication and Control Engineering. Springer, Berlin (2000)

[49] Chen, M.-S., Hwang, Y.-R., Tomizuka, M.: A State-Dependent Boundary Layer Design for Sliding Mode Control. IEEE Trans. Automat. Contr. **47**(10), 1677–1681 (2002)

[50] Craig, J.: Introduction to Robotics: Mechanics and Control. Addison-Wesley Publishing, Boston (1989)

[51] Cruz-Zavala, E., Moreno, J., Fridman, L.: Uniform robust exact differentiator. IEEE Trans. Automat. Contr. **56**(11), 2727–2733 (2011)

[52] Davila, J., Fridman, L., Levant, A.: Second-order sliding-modes observer for mechanical systems. IEEE Trans. Automat. Contr. **50**(11), 1785–1789 (2005)

[53] Davila, J., Fridman, L., Pisano A., Usai, E.: Finite-time state observation for nonlinear uncertain systems via higher order sliding modes. Int. J. Control **82**(8), 1564–1574 (2009)

[54] Davila, J., Rios, H., Fridman, L.: State observation for nonlinear switched systems using non-homogeneous higher order sliding mode observers. Asian J. Control **14**(4), 911–923 (2012)

[55] DCCT: The Diabetes Control and Complications Trial Research Group, The effect of inten-sive treatment of diabetes on the development and progression of long-term complications in insulin-dependent diabetes mellitus. New Engl. J. Med. **329**, 977–986 (1993)

[56] DeCarlo, R.A., Zak, S.H., Matthews, G.P.: Variable structure control of nonlinear multivari-able systems: a tutorial. Proc. IEEE **76**(3), 212–232 (1988)

[57] DeCarlo, R.A., Drakunov, S., Li, X.: A unifying characterization of sliding mode control: a Lyapunov approach. ASME J. Dyn. Syst. Meas. Control **122**(4), 708–718 (2000)

[58] DeCarlo, R.A., Zak, S.H., Drakunov, S.V.: Variable structure,sliding mode controller design. In: The Control Handbook, 2nd edn. Electrical Engineering Handbook Series, pp. 50.1–50.21. CRC Press, Boca Raton, USA (2010)

[59] Defoort, M., Floquet, T., Kokosy, A., Perruquetti, W.: Integral sliding mode control for trajectory tracking of a unicycle type mobile robot. Integr. Comput. Aid. E. **13**(3), 277–288 (2006)

[60] DeJager, B.: Comparison of methods to eliminate chattering and avoid steady state errors in sliding mode digital control. In: Proceedings of the IEEE VSC and Lyapunov Workshop, pp. 37–42, Sheffield, UK (1992)

[61] Drakunov, S.V., Utkin, V.: Sliding mode control in dynamic systems. Int. J. Control **55**(4), 1029–1037 (1992)

[62] Drakunov, S., Utkin, V.I.: Sliding mode observers: tutorial. In: Proceedings of the 34th IEEE Conference of Decision and Control, vol. 4, pp. 3376–3378, New Orleans, LA (1995)

[63] Drakunov, S., Su, W., Ozguner, U.: Constructing discontinuity surfaces for variable structure systems: a Lyapunov approach. Automatica **32**(6), 925–928 (1996)

[64] Drazenivic, B.: The invariance conditions for variable structure systems. Automatica **5**(3), 287–295 (1969)

[65] Edwards, C., Spurgeon, S.K.: On the development of discontinuous observers. Int. J. Control **59**(5), 1211–1229 (1994)

[66] Edwards, C., Spurgeon, S.K.: Sliding mode stabilisation of uncertain systems using only output information. Int. J. Control **62**(5), 1129–1144 (1995)

[67] Edwards, C., Spurgeon, S.: Sliding Mode Control: Theory and Applications. Taylor and Francis, London (1998)

[68] Edwards, C., Spurgeon, S.: On the limitations of some variable structure output feedback controller designs. Automatica **36**, 743–748 (2000)

[69] Edwards, C., Tan, C.P.: A comparison of sliding mode and unknown input observers for fault reconstruction. Eur. J. Control **16**, 245–260 (2006)

[70] Edwards, C., Spurgeon, S.K, Patton, R.J.: Sliding mode observers for fault detection and isolation. Automatica **36**(4), 541–553 (2000)

[71] Edwards, C., Spurgeon, S.K., Akoachere, A.: Sliding mode output feedback controller design using linear matrix inequalities. IEEE Trans. Automat. Contr. **46**(1), 115–119 (2001)

[72] Edwards, C., Spurgeon, S.K., Hebden, R.G.: On the design of sliding mode output feedback controllers. Int. J. Control **76**, 893–905 (2003)

[73] El-Khazali R., DeCarlo, R.A.: Output feedback variable structure control design. Automatica **31**(6), 805–816 (1995)

[74] Emelyanov, S.V., Korovin, S.K.: Applying the principle of control by deviation to extend the set of possible feedback types. Sov. Phys. Dokl. **26**(6), 562–564 (1981)

[75] Emelyanov, S.V., Korovin, S.K. Levantovsky, L.V.: Higher order sliding modes in the binary control systems. Sov. Phys. Dokl. **31**(4), 291–293 (1986)

[76] Estrada, A., Fridman, L.: Integral HOSM semiglobal controller for finite-time exact compensation of unmatched perturbations. IEEE Trans. Automat. Contr. **55**(11), 2644–2649 (2010)

[77] Estrada, A., Fridman, L.: Quasi-continuous HOSM control for systems with unmatched perturbations. Automatica **46**, 1916–1919 (2010)

[78] Fan, X. Arcak, M.: Observer design for systems with multivariable monotone nonlinearities. Syst. Control Lett. **50**, 319–330 (2003)

[79] Ferreira, A., Bejarano, F.J., Fridman, L.: Robust control with exact uncertainties compensation: with or without chattering? IEEE Trans. Contr. Syst. Tech. **19**(5), 969–975 (2011)

[80] Ferreira, A., Bejarano, F.J., Fridman, L.: Unmatched uncertainties compensation based on high-order sliding mode observation. Int. J. Robust. Nonlin. (2012). doi: 10.1002/rnc.2795

[81] Filippov, A.: Differential Equations with Discontinuous Right-hand Sides. Kluwer Academic Publishers, Dordrecht (1988)

[82] Fisher, M.E.: A semi closed-loop algorithm for the control of blood glucose levels in diabetics. IEEE Trans. Biomed. Eng. **38**(1), 57–61 (1991)

[83] Floquet, T., Barbot, J.P.: Super twisting algorithm-based step-by-step sliding mode observers for nonlinear systems with unknown inputs. Int. J. Syst. Sci. **38**(10): 803–815 (2007)

[84] Floquet, T., Edwards, C., Spurgeon, S.K.: On sliding mode observers for systems with unknown inputs. Int J. Adapt. Control Signal Process. **21**, 638–656 (2007)

[85] Flugge-Lotz, I.: Discontinuous Automatic Control. Princeton University Press, New Jersey (1953)

[86] Franklin, G.F., Powell, J.D., Emami-Naeini, A.: Feedback Control of Dynamic Systems. Prentice Hall, New Jersey (2002)

[87] Fridman, L.: The problem of chattering: an averaging approach. In: Young, K., Ozguner, U. (eds.) Variable Structure, Sliding Mode and Nonlinear Control. Lecture Notes in Control and Information Science, vol. 247, pp. 363–386. Springer, London (1999)

[88] Fridman, L.: An averaging approach to chattering. IEEE Trans. Automat. Contr. **46**(8), 1260–1264 (2001)

[89] Fridman, L.: Singularly perturbed analysis of chattering in relay control systems. IEEE Trans. Automat. Contr. **47**(12), 2079–2084 (2002)

[90] Fridman, L.: Chattering analysis in sliding mode systems with inertial sensors. Int. J. Control **76**(9/10), 906–912 (2003)

[91] Fridman, L., Levant, A.: Higher order sliding modes as the natural phenomena of control theory. In: Garafalo, F., Glielmo, G. (eds.) Robust Control Variable Structure and Lyapunov Techniques, pp. 107–133. Springer, Berlin (1996)

[92] Fridman, L., Levant, A.: Higher order sliding modes. In: Barbot, J.P., Perruguetti, W. (eds.) Sliding Mode Control in Engineering, pp. 53–102. Dekker, New York (2002)

[93] Fridman, L., Poznyak, A., Bejarano, J.: Decompositionof the min-max multimodel problem via integral sliding mode. Int. J. Robust. Nonlin. **15**(13), 559–574 (2005)

[94] Fridman, L., Levant, A., Davila, J.: Observation of linear systems with unknown inputs via high-order sliding-modes, Int. J. Syst. Sci. **38**(10), 773–791 (2007)

[95] Fridman, L., Shtessel, Y., Edwards, C., Yan, X.G.: Higher-order sliding-mode observer for state estimation and input reconstruction in nonlinear systems. Int. J. Robust. Nonlin. **18**(4/5), 399–413 (2008)

[96] Fridman, L., Davila, J., Levant, A.: High-order sliding-mode observation for linear systems with unknown inputs. Nonlinear Anal. Hybrid Syst. **5**(2), 174–188 (2011)

[97] Furuta, K.: Sliding mode control of a discrete system. Syst. Control Lett. **14**(2), 145–152 (1990)

[98] Furuta, K., Pan, Y.: Variable structure control with sliding sector. Automatica **36**(2), 211–228 (2000)

[99] Gahinet, P., Nemirovski, A., Laub, A., Chilali, M.: LMI Control Toolbox, User Guide. MathWorks Inc., Natick (1995)

[100] Gelb, A., Vander Velde, W.E.: Multiple-Input Describing Functions and Nonlinear System Design. McGraw-Hill, New York (1968)

[101] Gonzalez, T., Moreno, J., Fridman, L.: Variable Gain Super-Twisting Sliding Mode Control. IEEE Trans. Automat. Contr. **57**(8), 2100–2105 (2012)

[102] Gopalswamy S., Hedrick, J.K.: Tracking nonlinear non-minimum phase systems using sliding control. Int. J. Control **57**(5), 1141–1158 (1993)

[103] Gutman, S.: Uncertain dynamic Systems – a Lyapunov min-max approach. IEEE Trans. Automat. Contr. **24**(3), 437–449 (1979)

[104] Hall, C., Shtessel, Y.: Sliding mode disturbance observers-based control for a reusable launch vehicle. AIAA J. Guid. Control Dynam. **29**(6), 1315–1329 (2006)

[105] Hamayun, M.T., Edwards, C. Alwi, H.: Design and analysis of an integral sliding mode fault-tolerant control scheme. IEEE Trans. Automat. Contr. **57**, 1783–1789 (2012)

[106] Hanson, J., Jones, R., Krupp, D.: Advanced guidance and control methods for reusable launch vehicles: test results. AIAA Paper, pp. 2002–4561 (2002)

[107] Hautus, M.: Strong detectability and observers. Linear Algebra. Appl. **50**(4), 353–368 (1983)

[108] Hebden, R.G., Edwards, C., Spurgeon, S.K.: Automotive stability in a split-mu manoeuvre using an observer based sliding mode controller. Department of Engineering Report 02–4, Leicester University (2002)

[109] Heck, B.S., Ferri, A.A.: Application of output feedback to variable structure systems. J. Guid. Control Dynam. **12**, 932–935 (1989)

[110] Heck, B.S., Yallapragada, S.V., Fan, M.K.H.: Numerical methods to design the reaching phase of output feedback variable structure control. Automatica **31**(2), 275–279 (1995)

[111] Imine, H., Fridman, L., Shraim, H., Djemai, M.: Sliding mode based analysis and identification of vehicle dynamics. Lecture Notes in Control and Information Sciences, vol. 414. Springer, Berlin (2011)

[112] Isidori, A.: Nonlinear Control Systems. Springer, New York (1995)

[113] Itkis, Y.: Control Systems of Variable Structure. Wiley, New York (1976)

[114] Jaremco J., Rorstad, O.: Advances toward the implantable artificial pancreas for treatment of diabetes. Diabetes Care **21**(3), 444–450 (1998)

[115] Kaveh, P., Shtessel, Y.: Blood glucose regulation using higher order sliding mode control. Int. J. Robust. Nonlin. Special Issue on Advances in Higher Order Sliding Mode Control **18**(4–5), 557–569 (2008)

[116] Khalil, H.: Nonlinear Systems, 3d edn. Prentice Hall, New Jersy (2002)
[117] Kobayashi, S., Furuta, K.: Frequency characteristics of Levant's differentiator and adaptive sliding mode differentiator. Int. J. Syst. Sci. **38**(10), 825–832 (2007)
[118] Kochalummoottil, J., Shtessel, Y., Moreno, J.A., Fridman, L.: Adaptive twist sliding mode control: a Lyapunov design. In: Proceedings of the 50th Conference on Decision and Control, pp. 7623–7628, Orlando, FL (2011)
[119] Kochalummoottil, J., Shtessel, Y., Moreno, J.A., Fridman, L.: Output feedback adaptive twisting control: a Lyapunov design. In: Proceedings of the American Control Conference, pp. 6172–6177, Montreal, Canada (2012)
[120] Kolmogoroff, A. N.: On inequalities between upper bounds of consecutive derivatives of an arbitrary function defined on an infinite interval. Amer. Math. Soc. Transl. **2**, 233–242 (1962)
[121] Krstic, J.M., Kanellakopoulos, I., Kokotovic, P.: Nonlinear and Adaptive Control Design. Wiley, New York (1995)
[122] Krupp, D.R., Shkolnikov, I.A., Shtessel, Y.B.: 2-sliding mode control for nonlinear plants with parametric and dynamic uncertainties. In: Proceedings of Conference on Guidance Navigation and Control, AIAA paper, pp. 2000–3965, Denver, CO (2000)
[123] Levant, A.: Robust exact differentiation via sliding mode technique. Automatica **34**(3), 379–384 (1998)
[124] Levant, A.: Universal SISO sliding-mode controllers with finite-time convergence. IEEE Trans. Automat. Control **46**(9), 1447–1451 (2001)
[125] Levant, A.: Higher-order sliding modes, differentiation and output-feedback control. Int. J. Control **76**(9/10), 924–941 (2003).
[126] Levant, A.: Homogeneity approach to high-order sliding mode design. Automatica **41**(5), 823–830 (2005)
[127] Levant, A.: Quasi-continuous high-order sliding-mode controllers. IEEE Trans. Automat. Contr. **50**(11) 1812–1816 (2006)
[128] Levant, A.: Construction principles of 2-sliding mode design. Automatica **43**(4), 576–586 (2007)
[129] Levant, A.: Finite differences in homogeneous discontinuous control. IEEE Trans. Automat. Contr. **52**7, 1208–1217 (2007)
[130] Levant, A.: Chattering analysis. IEEE Trans. Automat. Contr. **55**(6), 1380–1389 (2010)
[131] Levant, A., Fridman, L.: Accuracy of Homogeneous Sliding Modes in the Presence of Fast Actuators. IEEE Trans. Automat. Contr. **55**(3), 810–814 (2010)
[132] Levant, A., Levantovsky, L.V.: Sliding order and sliding accuracy in sliding mode control. Int. J. Control **58**6, 1247–1263 (1993)
[133] Levant, A., Michael, M.: Adjustment of high-order sliding-mode controllers. Int. J. Robust. Nonlin. **19**(15), 1657–1672 (2009)
[134] Luenberger, D.G.: Observing the state of a linear system. IEEE Trans. Mil. Electron. **8**(2), 74–80 (1964)
[135] Luenberger, D.G.: An introduction to observers. IEEE Trans. Automat. Contr. **16**(6), 96–602 (1971)
[136] Lukyanov, A.G.: Reducing dynamic systems: regular form. Automat. Rem. Contr. **41**(3), 5–13 (1981)
[137] Lukyanov, A.G., Utkin, V.I.: Methods for reducing equations for dynamic system to a regular form. Automat. Rem. Contr. **4**, 14–18 (1981)
[138] Man Z., Paplinski, A.P., Wu, H.R.: A robust MIMO terminal sliding mode control for rigid robotic manipulators. IEEE Trans. Automat. Contr. **39**(12), 2464–2468 (1994)
[139] Massey, T., Shtessel, Y.: Continuous traditional and high order sliding modes for satellite formation control. AIAA J. Guid. Control Dynam. **28**(4), 826–831 (2005)
[140] Matthews, G.P., DeCarlo, R.A.: Decentralized tracking for a class of interconnected nonlinear systems using variable structure control. Automatica **24**(2), 187–193 (1988)
[141] Milosavljevic, C.: Discrete-time VSS. In: Sabanovich, E., Spurgeon, S., Fridman, L. (eds.) Variable Structure systems: From Principles to Implementation. IEE Control Series, vol. 66, pp. 99–128. IEE-publisher Stevenage, UK (2004)

[142] Moreno, J.A.: Lyapunov approach to analysis and design of second order sliding mode algorithms. In: Fridman, L., Moerno, J., Iriarte, R. (eds.) Sliding Modes after the first Decade of the 21st Century. Lecture Notes in Control and Information Science, vol. 412, pp. 115–149. Springer, Berlin (2011)

[143] Moreno, J., Osorio, M.: Strict lyapunov functions for the super-twisting algorithm. IEEE Trans. Automat. Contr. **57**(4), 1035–1040 (2012)

[144] Neatpisarnvanit, C., Boston, J.R.: Estimation of plasma insulin from plasma glucose. IEEE Trans. Biomed. Eng. **49**(11), 1253–1259 (2002)

[145] Orlov, Y.: Finite time stability and robust control synthesis of uncertain switched systems. SIAM J. Cont. Optim. **43**(4), 1253–1271 (2005)

[146] Orlov, Y.: Discontinuous Control. Springer, Berlin (2009)

[147] Orlov, Y., Aguilar, L., Cadiou, J.C.: Switched chattering control vs. backlash/friction phenomena in electrical servo-motors. Int. J. Control **76**(9–10), 959–967 (2003)

[148] Patel, N., Edwards, C., Spurgeon, S.K.: Tyre/road friction estimation - a comparative study. Proc. Inst. Mech. Eng. Part D: J. Automob. Eng. **22**, 2337–2351 (2008)

[149] Petersen, I.R.: A stabilization algorithm for a class of uncertain linear systems. Syst. Control Lett. **8**(4), 351–357 (1987)

[150] Pisano, A., Usai, E.: Contact force regulation in wire-actuated pantographs via variable structure control and frequency-domain techniques. Int. J. Control **81**(11), 1747–1762 (2008)

[151] Plestan, F., Grizzle, J.W., Westervelt, E.R., Abba, G.: Stable walking of a 7-DOF biped robot. IEEE Trans. Robotic. Autom. **19**(4), 653–668 (2009)

[152] Polyakov, A.: Nonlinear feedback design for fixed-time stabilization of linear control systems. IEEE Trans. Automat. Contr. **57**(8), 2106–2110 (2012)

[153] Polyakov, A., Poznyak, A.: Reaching time estimation for super-twisting second order sliding mode controller via Lyapunov function designing. IEEE Trans. Automat. Contr. **54**(8), 1951–1955 (2009)

[154] Poznyak, A., Fridman, L., Bejarano, F.: Mini-max integral sliding mode control for multimodel linear uncertain systems. IEEE Trans. Automat. Contr. **49**(1), 97–102 (2004)

[155] Rosenbrock, H.H.: State Space and Multivariable Theory. Wiley, New York (1970)

[156] Rubagotti, M., Estrada, A., Castanos, F., Ferrara, A., Fridman, L.: Integral sliding mode control for nonlinear systems with matched and unmatched perturbations. IEEE Trans. Automat. Contr. **56**(11), 2699–2704 (2011)

[157] Ryan, E.P., Corless, M.: Ultimate boundedness and asymptotic stability of a class of uncertain dynamical systems via continuous and discontinuous control. IMA J. Math. Control Inform. **1**(3), 223–242 (1984)

[158] Saks, S.: Theory of the integral. Dover Publ. Inc., New York (1964)

[159] Sastry, S., Bodson, M.: Adaptive Control: Stability, Convergence, and Robustness. Prentice-Hall Advanced Reference Series (Engineering). Prentice-Hall, New Jersey (1994)

[160] Shkolnikov, I.A., Shtessel, Y.B.: Aircraft Nonminimum Phase Control in Dynamic Sliding Manifolds. AIAA J. Guid. Control Dynam. **24**(3), 566–572 (2001)

[161] Shtessel, Y.B., Lee, Y.J.: New approach to chattering analysis in systems with sliding modes. In: Proceedings of 35th IEEE Conference on Decision and Control, pp. 4014–4019, Kobe, Japan (1996)

[162] Shtessel, Y., Buffington, J., Banda, S.: Tailless aircraft flight control using multiple time scale re-configurable sliding modes. IEEE Trans. Contr. Syst. Tech. **10**, 288–296 (2002)

[163] Shtessel, Y., Shkolnikov, I., Levant, A.: Smooth second order sliding modes: Missile guidance application. Automatica **43**(8), 1470–1476 (2007)

[164] Shtessel, Y., Baev, S., Shkolnikov, I.: Nonminimum-phase output tracking in causal systems using higher order sliding modes. Int. J. Robust. Nonlin. Special Issue on Advances in Higher Order Sliding Mode Control **18**(4–5), 454–467 (2008)

[165] Shtessel, Y., Shkolnikov, I., Levant, A.: Guidance and control of missile interceptor using second order sliding modes. IEEE Trans. Aero. Elec. Syst. **45**(1), 110–124 (2009)

[166] Shtessel, Y., Baev, S., Edwards, C., Spurgeon, S.: HOSM observer for a class of non-minimum phase causal nonlinear systems. IEEE Trans. Automat. Contr. **55**(2),543–548(2010)

[167] Shtessel, Y., Baev, S., Edwards, C., Spurgeon, S.: Output feedback tracking in causal nonminimum-phase nonlinear systems using higher order sliding modes. Int. J. Robust. Nonlin. **20**(16), 1866–1878 (2010)

[168] Shtessel, Y., Taleb, M., Plestan, F.: A novel adaptive-gain super-twisting sliding mode controller: methodology and application. Automatica **48**(5), 759–769 (2012)

[169] Shtessel, Y., Tournes, C., Fridman, L.: Advances in guidance and control of aerospace vehicles using sliding mode control and observation techniques. J. Fraklin Inst. **349**(2), 391–396 (2012)

[170] Shtessel. Y., Kochalummoottil, J., Edwards, C., Spurgeon, S.: Continuous adaptive finite reaching time control and second order sliding modes. IMA J. Math. Control Inform. (2012). doi: 10.1093/imamci/dns013

[171] Slotine, J.-J., Li, W.: Applied Nonlinear Control. Prentice Hall, New Jersey (1991)

[172] Slotine, J.J.E., Hedrick, J.K., Misawa, E.A.: On sliding observers for nonlinear systems. Trans. ASME: J. Dyn. Syst. Meas. Control **109**, 245–252 (1987)

[173] Soderstrom, T., Stoica, P.: System Identification. Prentice Hall International, Cambridge (1989)

[174] Spurgeon, S.: Sliding mode observers - a survey. Int. J. Syst. Sci. **39**(8), 751–764 (2008)

[175] Steinberg, A., Corless, M.J.: Output feedback stabilisation of uncertain dynamical systems. IEEE Trans. Automat. Contr. **30**(10), 1025–1027 (1985)

[176] Stott, J., Shtessel, Y.: Launch vehicle attitude control using sliding mode control and observation techniques. J. Franklin Inst. Special Issue on Advances in Guidance and Control of Aerospace Vehicles using Sliding Mode Control and Observation Techniques **349**(2), 397–412 (2012)

[177] Strang, G.: Linear Algebra and its Applications. Harcourt Brace Jovanovich, London (1988)

[178] Tan, C.P. Edwards, C.: Sliding mode observers for robust detection and reconstruction of actuator and sensor faults. Int. J. Robust. Nonlin. **13**, 443–463 (2003)

[179] Tan, C.P. Edwards, C.: Robust fault reconstruction in linear uncertain systems using multiple sliding mode observers in cascade. IEEE Trans. Automat. Contr. **55**, 855–867 (2010)

[180] Tomei, P. Marino, R.: Nonlinear Control Design: Geometric, Adaptive and Robust. Prentice Hall, London (1995)

[181] Tsypkin, Y.Z.: Relay Control Systems. Cambridge University press, Cambridge (1984)

[182] Utkin, V.I.: Sliding Modes in Optimization and Control Problems. Springer, New York (1992)

[183] Utkin, V.I.: Variable structure systems with sliding modes. IEEE Trans. Automat. Contr. **22**(2), 212–222 (1977)

[184] Utkin, V.I.: First stage of VSS: peolpe and events. In: Yu, X., Xu, J. (eds.) Variable Structure Systems: Towards the 21st Century. Lecture Notes in Control and Information Sciences, vol. 247, pp. 1–33. Springer, Berlin (2002)

[185] Utkin, V.I., Shi, J.: Integral sliding mode in systems operating under uncertainty conditions. In: Proceedings of the 35th IEEE Conference on Decision and Control, pp. 4591–4596, Kobe, Japan (1996)

[186] Utkin, V., Guldner, J., Shi, J.: Sliding Mode Control in Electromechanical Systems. Taylor and Francis, London (1999)

[187] Utkin, V.I, Guldner, J., Shi, J.: Sliding Mode Control in Electro-Mechanical Systems, 2nd edn. CRC Press, Boca Raton (2009)

[188] Walcott, B.L., Zak, S.H.: State observation of nonlinear uncertain dynamical systems. IEEE Trans. Automat. Contr. **32**(2), 166–170 (1987)

[189] Walcott, B.L., Corless, M.J., Zak, S.H.: Comparative study of nonlinear state observation techniques. Int. J. Control **45**(6), 2109–2132 (1987)

[190] Wise, K.A., Broy, D.J.: Agile missile dynamics and control. AIAA J. Guid. Control Dynam. **21**(3), 441–449 (1998)

[191] Xu, J.X., Pan, Y.J., Lee, T.H., Fridman, L.: On nonlinear hinfty sliding mode control for a class of nonlinear cascade systems. Int. J. Syst. Sci. **36**(15), 983–992 (2005)

[192] Xu, J.X., Abidi, K. On the discrete-time integralsliding-mode control. IEEE Trans. Automat. Contr. **52**(4), 709–715 (2007)

[193] Yan, X.G., Edwards, C.: Nonlinear robust fault reconstruction and estimation using a sliding mode observer. Automatica **43**(9), 1605–1614 (2007)

[194] Yeh, H., Nelson, E., Sparks, A.: Nonlinear tracking control for satellite formations. J. Guid. Control Dynam. **25**(2), 376–386 (2002)

[195] Young, K.D., Utkin, V.I., Ozguner, U.: A control engineer's guide to sliding mode control. IEEE Trans. Contr. Syst. Tech. **7**(3), 328–342 (1999)

[196] Zubov, V.I.: Methods of A. M. Lyapunov and Their Applications. Noordhoff International, Groningen (1964)

Index

Systems
  Linear time invariant (LTI), 50, 91
  matrix, 332
  Multi input multi output (MIMO), 48, 67
  Single input single output (SISO), 55, 61,
        62
  time response, 332
  triple, 332

T
Trade-off, 114
Triangle inequality, 324
Twisting algorithm analysis, 193

U
Ultimate boundedness, 342
Uncertainty, 341
Unit-Vector Control design, 68, 82
Unmatched
    disturbance, 95
    uncertainty, 59, 73
Unobservable modes, 123
Utkin, viii, 113

V
Variable structure system, vii
Vector norm, 323

Printed in the United States
By Bookmasters